Bridging Research Disciplines to Advance Animal Welfare Science

A practical guide

Bridging Research Disciplines to Advance Animal Welfare Science

A practical guide

Edited by

Irene Camerlink

Polish Academy of Sciences, Poland

CABI

CABI is a trading name of CAB International

CABI
Nosworthy Way
Wallingford
Oxfordshire OX10 8DE
UK

Tel: +44 (0)1491 832111
Fax: +44 (0)1491 833508
E-mail: info@cabi.org
Website: www.cabi.org

CABI
WeWork
One Lincoln St
24th Floor
Boston, MA 02111
USA
T: +1 (617)682-9015
E-mail: cabi-nao@cabi.org

A catalogue record for this book is available from the British Library, London, UK.

Library of Congress Control Number: 2021939492

ISBN-13: 9781789247886 (hardback)
9781789247893 (paperback)
9781789247909 (ePDF)
9781789247916 (ePub)

DOI: 10.1079/9781789247886.0000

Commissioning Editor: Caroline Makepeace
Editorial Assistant: Lauren Davies
Production Editor: Marta Patiño

Typeset by Exeter Premedia Services Pvt Ltd, Chennai, India
Printed and bound in the USA by Integrated Books International, Dulles, Virginia

Contents

List of contributors

Faical Akaichi, Land Economy, Environment & Society, Scotland's Rural College (SRUC), West Mains Road, EH9 3JG Edinburgh, UK. Faical.Akaichi@sruc.ac.uk

Gareth Arnott, School of Biological Sciences, Queen's University Belfast, University Road, Belfast BT7 1NN, UK. g.arnott@qub.ac.uk

Marianne Benard, Athena Institute, Vrije Universiteit Amsterdam, De Boelelaan 1085, 1081 HV Amsterdam, The Netherlands. m_benard@live.nl

Irene Camerlink, Institute of Genetics and Animal Biotechnology, Polish Academy of Sciences, Ul. Postepu 36a, 05-552 Jastrzebiec, Poland. i.camerlink@igbzpan.pl

Grace A. Carroll, Animal Behaviour Centre, School of Psychology, Queen's University Belfast, 18–30 Malone Road, Belfast BT9 5BN, UK. g.carroll@qub.ac.uk

Tjard de Cock Buning, Athena Institute, Vrije Universiteit Amsterdam, De Boelelaan 1085, 1081 HV Amsterdam, The Netherlands. tjard.de.cockbuning@vu.nl

Jack C.M. Dekkers, Department of Animal Science, Iowa State University, 1221 Kildee Hall, 50011-3150, Ames, Iowa, USA. jdekkers@iastate.edu

Paul Forsythe, Department of Medicine, University of Alberta, 8440 112 St. NW, Edmonton, Alberta T6G 2R7, Canada. forsytp@mcmaster.ca

Robert Frodeman, independent researcher, 1660 J-D Drive, Jackson, Wyoming, USA. robert.frodeman@gmail.com

Dietmar Fuchs, Medical University of Innsbruck, Christoph-Probst-Platz 1, Innrain 52 A, 6020 Innsbruck, Austria. dietmar.fuchs@i-med.ac.at

Luis Gomez-Raya, Departamento de Mejora Genética Animal, Instituto Nacional de Investigación y Tecnología Agraria y Alimentaria (INIA-CSIC), Ctra. de la Coruña km 7.5, 28040 Madrid, Spain. gomez.luis@inia.es

Johanna M. Gostner, Medical University of Innsbruck, Christoph-Probst-Platz 1, Innrain 52 A, 6020 Innsbruck, Austria. johanna.gostner@i-med.ac.at

Lisa Graham-Wisener, Centre for Improving Health-Related Quality of Life (CIHRQoL), School of Psychology, Queen's University Belfast, 18–30 Malone Road, Belfast BT9 5BN, UK. L.Graham-Wisener@qub.ac.uk

Jenny M. Groarke, Centre for Improving Health-Related Quality of Life (CIHRQoL), School of Psychology, Queen's University Belfast, 18–30 Malone Road, Belfast BT9 5BN, UK. j.groarke@qub.ac.uk

Carlos Guerrero-Bosagna, Environmental Toxicology Program, Department of Organismal Biology, Uppsala University, Norbyvägen 18D, 752 36 Uppsala, Sweden; IFM Biology, Linköping University, SE-581 83 Linköping, Sweden. carlos.guerrero.bosagna@ebc.uu.se

Alexandra Harlander, Campbell Centre for the Study of Animal Welfare, Department of Animal Biosciences, University of Guelph, 50 Stone Road East, Guelph, Ontario N1G 2W1, Canada. aharland@uoguelph.ca

Julie Thompson Klein, Department of English, Wayne State University, 5057 Woodward, Detroit, MI 48202, USA; TdLab, Department of Environmental Systems Science, ETH Zurich, Universitätstrasse 16, CH-8006 Zurich, Switzerland. julietklein@comcast.net

Victoria E. Lee, Animal & Veterinary Sciences, Scotland's Rural College (SRUC), West Mains Road, Edinburgh EH9 3JG, UK; Queen's University Belfast, Belfast, UK. Victoria.Lee@sruc.ac.uk

Claire Mindus, Campbell Centre for the Study of Animal Welfare, Department of Animal Biosciences, University of Guelph, 50 Stone Road East, Guelph, Ontario N1G 2W1, Canada. cmindus@uoguelph.ca

Andrea Nolan, Edinburgh Napier University, 9 Sighthill Court, Edinburgh EH11 4BN, UK. A.Nolan@napier.ac.uk

Rebecca E. Nordquist, Division of Animals in Science and Society, Department of Population Health Sciences, Faculty of Veterinary Medicine, Utrecht University, Yalelaan 7, 3584 CL Utrecht, The Netherlands. r.e.nordquist1@uu.nl

Rachel S.E. Peden, School of Natural and Environmental Sciences, Newcastle University, King's Road, Newcastle upon Tyne NE1 7RU, UK. rachpeden91@hotmail.co.uk

Wendy M. Rauw, Departamento de Mejora Genética Animal, Instituto Nacional de Investigación y Tecnología Agraria y Alimentaria (INIA-CSIC), Ctra. de la Coruña km 7.5, 28040 Madrid, Spain. rauw.wendy@inia.es

Jacky Reid, NewMetrica (Research) Ltd, 19 Woodside Place, Glasgow G3 7QL, UK. jacky.reid@newmetrica.com

T. Bas Rodenburg, Animals in Science and Society, Faculty of Veterinary Medicine, Utrecht University, Yalelaan 2, 3584 CM Utrecht, The Netherlands. t.b.rodenburg@uu.nl

Marian Scott, School of Mathematics and Statistics, University of Glasgow, University Place, Glasgow G12 8SQ, UK. Marian.Scott@glasgow.ac.uk

Janice M. Siegford, Department of Animal Science, Michigan State University, 474 S. Shaw Lane, East Lansing, Michigan 48824-1225, USA. siegford@msu.edu

Simon P. Turner, Animal & Veterinary Sciences, Scotland's Rural College (SRUC), West Mains Road, Edinburgh EH9 3JG, UK. Simon.Turner@sruc.ac.uk

Jerine A.J. van der Eijk, Group Animal Health and Welfare, Department of Animal Sciences, Wageningen Livestock Research, Wageningen University and Research, De Elst 1, 6708 WD Wageningen, The Netherlands. jerine.vandereijk@wur.nl

Nienke van Staaveren, Campbell Centre for the Study of Animal Welfare, Department of Animal Biosciences, University of Guelph, 50 Stone Road East, Guelph, Ontario N1G 2W1, Canada. nvanstaa@uoguelph.ca

Kaitlin Wurtz, Department of Animal Science, Michigan State University, 474 S. Shaw Lane, East Lansing, Michigan 48824-1225, USA. wurtzkai@msu.edu

Preface

Animal welfare science is a fusion of many different research fields, including among others ethology, physiology, zoology and veterinary sciences. The different subdisciplines have been applied alongside each other (i.e. in a multidisciplinary manner) or have been integrated with each other to gain new perspectives (i.e. in an interdisciplinary manner). As animal welfare is inherently related to human perception of animals and societal concerns, researchers have engaged non-academic stakeholders in their projects, making it transdisciplinary. The overarching term for these three categories of collaboration across disciplines is interdisciplinary research (formerly also named cross-disciplinary research). Interdisciplinary research is increasingly encouraged, and sometimes made compulsory, by research organizations and funders. This also applies to animal sciences, where over the past years there has been a surge in such projects. Interdisciplinary projects have the potential to gain insights from novel perspectives, and this can accelerate improvements in animal welfare and its related research. Most researchers, however, do not have any training on how to integrate different disciplines, or how to communicate and collaborate effectively across disciplines, and may not have a thorough understanding of other disciplines. In Part 1, this book gives guidance on how interdisciplinary research can advance animal welfare science. Part 2 touches on some of the major links between animal welfare and other disciplines and gives insight into the techniques and potential pitfalls that may occur when applying a certain discipline to animal welfare research. The aim of this book is to provide practical information about interdisciplinary research, particularly to students and researchers of animal behaviour, animal welfare and veterinary sciences, to better understand how the integration of disciplines can advance animal welfare science.

The book first gives an introduction to animal welfare, animal welfare science and interdisciplinarity in Chapter 1. This is followed by a contribution from Professor Julie Thompson Klein, one of the main pioneers in the development of interdisciplinary studies. In Chapter 2, she sheds light on collaboration and communication in interdisciplinary projects. Chapter 3 provides an example of how an animal welfare project can move from mono-disciplinary to transdisciplinary research, and the potential hurdles that can occur in this process. It provides suggestions on how to set up interdisciplinary projects and how communication within such teams can be made more effective. Part 1 concludes with the perspective of Robert Frodeman, founding director of the Center for the Study of Interdisciplinarity, on epistemic sustainability in the dilemma of knowledge generation.

In Part 2, the book gives various examples of how key research fields are integrated in animal welfare research. As there is cross-fertilization with many (sub)disciplines, a selection has been made for some that are currently receiving much research attention. Chapters 5 and 6 cover techniques that originate from human psychology, including survey design, stakeholder involvement and human behaviour change models. These two chapters will be of much value to transdisciplinary projects. Chapter 7 provides an excellent overview of various techniques from economy, and describes with practical examples how these techniques can be used to gain insight as to whether farmers (here the focal stakeholders) are willing to make changes in animal welfare on their farm, and at what financial costs. Animal welfare research is gradually moving towards inclusion of positive welfare, with the consideration of how animals can be provided a good life or a life worth living. The concept of quality of life, which originates from human medical sciences, definitely contributes to this. The developers of the health-related quality of life tool for dogs share their expertise in Chapter 8. Chapter 9 gives an overview of animal cognition research and its techniques. Nordquist also points out that differences between disciplines in the journals, and their impact factor, can complicate collaboration between researchers from human science and animal science. In Chapter 10, a brief overview of the vast field of neuroscience is given in relation to animal welfare science. The outlined techniques provide examples of how neuroscience can help us to gain insight into animals' mental experiences. An example of a fusion between multiple highly interconnected disciplines, including nutrition, physiology, neuroscience and behavioural research, is given in Chapter 11 on the microbiota–gut–brain axis. This chapter introduces the mechanisms through which microorganisms can influence animal welfare and how this knowledge can be used to keep animals physically and psychologically healthy. Genetics and genomics are well-established concepts in animal welfare sciences. Given the importance of genetics to animal behaviour and health, and the rapid advances in technologies, Chapter 12 provides a useful overview of how the disciplines interrelate with welfare. An aspect of genetics that deserves special attention is epigenetics, discussed in Chapter 13, as it has the potential to provide ways to assess chronic stress. Behavioural ecologists and animal welfare scientists both study animal behaviour but from different angles. How these two research fields differ, and how they can support each other, is neatly explained in Chapter 14. Last but not least, precision livestock technology is developing at an enormous pace, with many ongoing research and industry projects to facilitate implementation in practice. Research on this topic is not without challenges, and Siegford and Wurtz provide a comprehensive guide on how to conduct such studies. In the reference list for each of these chapters, recommendations for further reading are marked in bold.

As editor of the book, I have been very fortunate to find excellent authors from many different disciplines who were willing to share their expertise. Many of the multi- and interdisciplinary projects in animal welfare research are conducted without collaboration outside our own field and, as the chapters show, this can be a missed opportunity. I hope that this book will inspire you to have a look into different academic disciplines and to establish new collaborations with experts of other fields who can bring a fresh perspective to the study of animal welfare.

1 Interdisciplinary Research to Advance Animal Welfare Science: An Introduction

Irene Camerlink*

Institute of Genetics and Animal Biotechnology, Polish Academy of Sciences, Poland

Animal welfare science is a blend of so many different subdisciplines (e.g. ethology, physiology, zoology, veterinary science) that animal welfare scientists hardly notice that they are applying different disciplines to address a research question. When animal welfare researchers then speak about multidisciplinary or interdisciplinary research, it usually considers the involvement of stakeholders, which is actually transdisciplinary research. The use of multiple terms, the blending and vague distinction between many subdisciplines, and the lack of information and training on this subject often make the application of interdisciplinary research in animal welfare science undefined and unstructured. Research funders increasingly require a certain level of interdisciplinary research, but in practice, the execution of the envisaged plans mostly do not deliver to their maximum potential. The aim of this book is to provide practical information about interdisciplinary research, particularly to students and researchers of animal behaviour, animal welfare and veterinary sciences. The aim of this chapter is to provide a brief introduction to animal welfare and animal welfare science, and to interdisciplinary research within this context.

1.1 A Definition of Animal Welfare

Animal welfare has many different definitions, and across the decades the discussion about 'What is animal welfare?' and 'What is the most appropriate definition to use?' has varied in intensity (reviewed by Carenzi and Verga, 2009; Ohl and van der Staay, 2012). The need for a unified

*i.camerlink@igbzpan.pl

 Bridging Research Disciplines to Advance Animal Welfare Science: A practical guide (ed. I. Camerlink)
DOI: 10.1079/9781789247893.0001

definition becomes particularly important when it influences international law and regulations. The consequence of a specific definition of animal welfare has widespread implications when it is implemented in large international organizations such as, for example, the World Organization for Animal Health (OIE), the Food and Agriculture Organization (FAO), the Royal Society for the Prevention of Cruelty to Animals (RSPCA) and World Animal Protection. For the operational definition, the specific use of words, the level of detail and the suitability for applying the definition to practical situations are of great importance, especially when the definition is used for welfare assessment in practice and when certification schemes are based on such assessment.

Historically, animal welfare was approached from three angles, whereby researchers remained largely with their own perspective and dismissed the others (Fraser, 2008). One would view animal welfare as (i) the physiological state of the animal; (ii) whether the animal's life resembled the natural situation of the species; or (iii) whether the animal expressed a positive emotional state. Nowadays, animal welfare embraces all three viewpoints and accepts them all as relevant. However, some researchers lean more towards one perspective than the other two whereas others embrace an integration of the three. Definitions of animal welfare reflect these views, with some referring to one of the viewpoints whereas others combine them.

The lack of a single definition does, however, make sense. It is the greatest blessing to animals that animal welfare is not a static concept. If it were, then we would still be stuck with the idea that animals are like machines and do not feel pain. This does not disqualify all the old definitions. Simply, humans cannot express what they do not know, and in the past there was not much knowledge about the physical and especially mental experience of animals. A definition that captures this concept would be that *'animal welfare is a societal concept that evolves with human knowledge and humans' acceptance of animals' capacity to experience and perceive.'*

My proposition is that, for this reason, there cannot be one conceptual definition of animal welfare that can stand across time, as society and human knowledge is ever-evolving. Research is constantly providing new information that feeds into the body of knowledge that eventually leads to human perception of animals (Fig. 1.1). The progress of human knowledge and acceptance about animals' capacity to experience (e.g. emotions) and perceive (e.g. pain) is frustratingly slow (as also expressed and explained by Millman *et al.*, 2004). It has gradually been accepted that (some) animals can experience pain and that they are sentient (Duncan, 2006). For the majority of species, especially invertebrates, this is still under development. The a priori assumption of a lack of sentience in unstudied or understudied species, even when scientific evidence is present (Duncan, 2006), has led to massive intentional and unintentional legally accepted abuse of species. Jeremy Bentham's (1789) famous quote: 'The question is not, Can they reason?, nor Can they talk? but, Can they suffer?' does still today deserve more

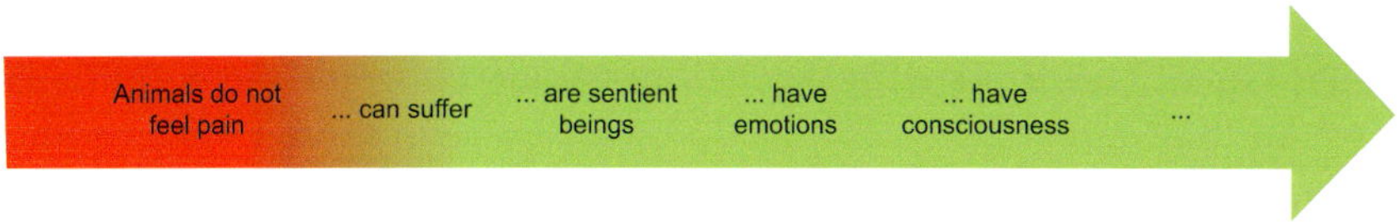

Fig. 1.1. Animal welfare as a societal concept that evolves with human knowledge and humans' acceptance of animals' capacity to experience and perceive. The word 'animals' is in most cases to be replaced by 'at least part of the vertebrate species' as the knowledge on many vertebrates (especially other than mammals) and invertebrates is lacking. Consequently, the majority of species are still treated under the assumption that they do not feel pain.

attention. For a historical overview of the concept of animal sentience, see Duncan (2006). It is only relatively recent that animal emotions have been considered, again mainly for mammals. Initially, this entailed only the basic emotions such as fear and anger, whereas more complex emotions such as jealousy are nowadays studied but looked upon with scepticism (Webb and de Waal, 2018). Likewise, the notion of animal awareness and consciousness has developed slowly. Animal awareness was addressed early on by Griffin in his book *The Question of Animal Awareness* (Griffin, 1976). Griffin presented an overview in which the term 'consciousness' was determined as being the most taboo in animal behaviour. Few continued to address animal consciousness (e.g. Dawkins, 1993; Griffin, 2001), and a recent review on animal consciousness shows major progress partly due to the integration of neuroscience (Paul *et al.*, 2020). The end of Fig. 1.1 is absent for a reason. We are still in the kindergarten of understanding human consciousness and the full human potential, and therefore it would be unjustified to place any limitations on what we believe to be animals' maximum capacity to experience.

The concept that there cannot be one definition of animal welfare across time may be true but is clearly unsuitable in practice. Operational definitions almost all refer fully or in part to the Five Freedoms (described in section 1.2 below), which were formulated in 1965. For example, the OIE defines animal welfare as 'the physical and mental state of an animal in relation to the conditions in which it lives and dies', but uses as guiding principles the Five Freedoms. Although the Five Freedoms are still very applicable, we have to acknowledge that using definitions that were written many decades ago may not capture the full needs of animals as we see them today. The need to update long-standing concepts has indeed been recognized (Ohl and van der Staay, 2012; Mellor, 2016; see also section 1.2 below), and there is a clear call for quality of life and a life worth living, rather than merely keeping animals on the edge of their ability to adapt to a situation. For example, the well-cited definition of Broom (1986, p. 524) where the welfare of an animal is 'its state as regards its attempts to cope with its environment' is still relevant but may have become outdated as the field of science is increasingly emphasizing the need to include positive welfare states (e.g. Rault *et al.*, 2020). The former concepts and definitions were an appropriate reflection of the situation at that time but in some cases may need to be reconsidered regarding the aspects that are relevant in the current time. This is therefore an encouragement for researchers and policy makers not just to rely on long-standing definitions but to reformulate concepts where needed, and to keep doing so, in order to reflect the recent scientific knowledge and human perception of animals and their needs.

1.2 The Five Freedoms

The Five Freedoms penetrate through all levels of animal welfare engagement, from research to international organizations and industry. The framework was formulated by the Farm Animal Welfare Advisory Committee (FAWC) in 1965, which was commissioned by the UK Government and led by Professor Brambell. The Five Freedoms are part of the Brambell Report on the welfare of animals in livestock husbandry (Brambell Committee, 1965). Although the report considered livestock, it is to some extent also applicable to other domesticated animals. The Five Freedoms state that an animal should have the following:

1. Freedom from hunger or thirst by ready access to fresh water and a diet to maintain full health and vigour.
2. Freedom from discomfort by providing an appropriate environment including shelter and a comfortable resting area.
3. Freedom from pain, injury or disease by prevention or rapid diagnosis and treatment.
4. Freedom to express (most) normal behaviour by providing sufficient space, proper facilities and company of the animal's own kind.
5. Freedom from fear and distress by ensuring conditions and treatment that avoid mental suffering.

Although the Five Freedoms seem obvious requirements for basic animal welfare, they are hardly ever satisfied in commercial farming, particularly in intensive farming systems. In many farming practices, at least one, but often multiple, of the Five Freedoms is violated by default. For example, feed restriction is a common management procedure in poultry and pig husbandry to regulate productivity (hindering freedom of hunger); adult cattle, poultry and pigs often suffer from injuries due to inappropriate resting areas (impairing freedom from discomfort and pain); disease and injury are common in all livestock sectors (no freedom from injury and disease); the ability to express natural behaviour is heavily restricted in intensive farming systems, resulting in behavioural problems (limited freedom to express natural behaviour); and fear and distress is common across species due to rough or unpredictable handling by humans (impeding freedom from fear and distress). Judging a situation by the Five Freedoms is therefore not easy and should be considered in relation to the percentage of animals that are affected and the duration of the suffering. Webster (1994) indeed wrote that it is unrealistic that all Five Freedoms are attained, but that it is an attempt to improve the situation. A reason that it has remained in use for so many years may be due to its simple and clear formulation of basic needs of animals, and that it is still relevant. Suggestions have been made for updating the Five Freedoms. Ohl and van der Staay (2012) suggest rephrasing the Five Freedoms: instead of having 'freedom from' hunger, pain, etc. they suggest 'freedom to adequately respond to' hunger, pain, etc., and further add '...and adapt to changing living conditions up to a level that is perceived as positive'. Mellor and Beausoleil (2015) developed the Five Domains model based on the Five Freedoms. This model was extended several times to include relevant changes that occurred within animal welfare science, such as positive welfare states (Mellor and Beausoleil, 2015) and human–animal interactions (Mellor *et al.*, 2020). For non-livestock species, especially companion or wild animals, other approaches, such as the Duty of Care framework (about the need to take proper care of an animal under your responsibility), may be more appropriate. Modification or proposed new frameworks are, however, only marginally taken up in practice. One reason for this might be that students are taught to use original (old) definitions and concepts, and that they maintain these as they advance in their careers.

1.3 Animal Welfare Science

Comprehensive overviews of the history and development of animal welfare science have been given by, among others, Millman *et al.* (2004), Broom (2011) and Keeling *et al.* (2011). In brief, animal welfare was for a long time not regarded as scientific. Whereas some researchers considered animals as response machines and disregarded animal sentience and animal emotions as non-scientific, others commenced working on the topic of behaviour and sentience and established a basis for what later developed into animal welfare science. The interest in animal welfare markedly grew after the publication of the book *Animal Machines* by Ruth Harrison in 1964. She criticized the intensive way of farming and gained widespread social support in her appeal for farm animals. Following this, the Brambell Committee was established, which eventually led to the adoption of the Five Freedoms (Brambell Committee, 1965).

Ethology had a key role in the development of animal welfare science (Millman *et al.*, 2004), especially in Europe. In 1973, Nikolaas Tinbergen, Karl von Frisch and Konrad Lorenz jointly won the Nobel Prize in 'Physiology or Medicine' for their discoveries on social behaviour

Box 1.1. Tinbergen's four questions, with the terminology that has been used since shown in parentheses. Adapted from Bateson and Laland, 2013.

Tinbergen's four questions:

1. Causation (or mechanism): What is it for?
2. Survival value (or adaptive significance): How does it work?
3. Ontogeny (or development): How did it develop during the lifetime of the individual?
4. Evolution (or phylogeny): How did it evolve over the history of the species?

in animals. Tinbergen formulated four questions (Box 1.1) that have profoundly impacted the study of animal behaviour (Tinbergen, 1963). The questions on causation and ontogeny can be considered proximate questions, whereas the questions on survival value and evolution can be considered ultimate questions (see also Chapter 14, this volume), although the distinction of the questions into proximate and ultimate has been debated (Bateson and Laland, 2013). Interestingly, Tinbergen mentions in his famous 1963 publication (p. 416) that the ethological studies 'are moving towards a fusion with the fields conventionally covered by neurophysiology and physiological psychology' ... 'down to molecular biology'. In the conclusion, he emphasized that 'we are witnessing the fusion of many sciences, all concerned with one or another aspect of behaviour, into one coherent science', thereby describing the early interdisciplinary developments in ethology.

The increase in public awareness of the circumstances in which animals were kept, especially on farms and in laboratories, continued to provide fuel for discussions on animal ethics. Gradually, the societal demand for better conditions for animals required science-based evidence on animal sentience and animal welfare (Fraser *et al.*, 1997), prominently in Europe (Veissier and Miele, 2014). With the increasing global awareness of animal welfare (a term that in some countries does or did not exist in the native vocabulary until recently), the need for answers to questions regarding animal welfare also increases.

1.4 Interdisciplinary Research

Interdisciplinary research is an umbrella term used for multidisciplinary, interdisciplinary and transdisciplinary research. The terms have changed over time (e.g. 'cross-disciplinarity' was used earlier instead of interdisciplinarity as the general designation for the types of collaboration), are used in different ways and are notable for conflicting meaning (Wagner *et al.*, 2011; Frodeman, 2017). This is also seen in the animal welfare science literature, where the terms are often used incorrectly (Lund *et al.*, 2006). Therefore, some explanation is warranted.

1.4.1 Disciplines and subdisciplines

An academic discipline is not easy to define, as disciplines are not static entities (Tight, 2020), with the boundaries between disciplines changing over time. Menken and Keestra (2016, p. 27) defined a discipline as 'a field of science with a particular object of research and a corresponding body of accumulated specialist knowledge'. Academic disciplines in the natural sciences are, for example, biology, physics, mathematics and geology; in the social sciences are, for example, economics, politics, sociology and psychology; and in the humanities are, among others, philosophy, history, art and linguistics. The main disciplines consist of multiple subdisciplines. In biology, some of the subdisciplines are ecology, ethology, genetics, developmental biology and zoology. Animal welfare science has been described as a young scientific discipline (e.g. Millman *et al.*, 2004), which was established after the Brambell Report (reviewed by Carenzi and Verga, 2009). To be precise, animal welfare science is a subdiscipline of the discipline *biology*, within *natural sciences* (also termed *life sciences*). For an overview of academic disciplines, see for example Menken and Keestra (2016). Some subdisciplines are in themselves already a fusion between disciplines, such as computational genetics. Animal welfare science could arguably also be considered such a subdiscipline. Researchers in animal welfare science vary in the subdiscipline they affiliate themselves with, as some have, for example, a stronger ethology or a stronger zoology background.

1.4.2 Interdisciplinarity, complexity and wicked problems

A widely used definition of interdisciplinarity is that of the National Academy of Sciences, National Academy of Engineering, and Institute of Medicine (2005, p. 2): 'Interdisciplinary research is a mode of research by teams or individuals that integrates information, data, techniques, tools, perspectives, concepts, and/or theories from two or more disciplines or bodies of specialized knowledge to advance fundamental understanding or to solve problems whose solutions are beyond the scope of a single discipline or area of research practice.'

Other definitions make it more explicit that interdisciplinarity is about 'addressing a topic that is too broad or complex to be dealt with adequately by a single discipline or profession' (Klein and Newell, 1997, p. 393). A middle way is taken in the review of Wagner *et al.* (2011, p. 16): 'Interdisciplinary approaches integrate separate disciplinary data, methods, tools, concepts, and theories in order to create a holistic view or common understanding of a complex issue, question, or problem. The critical indicators of interdisciplinarity in research include evidence that the integrative synthesis is different from, and greater than, the sum of its parts ...'

'Complex' here is not used as a generic term for something difficult. By 'complexity' is meant 'a system that is difficult to understand because it has lots of varied parts whose nonlinear, asymmetrical relationships combine to produce a self-organizing pattern of behaviour that takes on a life of its own' (Newell, 2004; Flood and Carson, 2013) and is 'partly or wholly unobservable' (Flood and Carson, 2013, p. 120). Central to complexity is non-linearity (i.e. the output is not necessarily proportional to the input), which denotes that a complex system cannot be fully predicted from its separate elements, thus producing a dynamic pattern that is only partly ordered and stable (Newell, 2004). Examples of complex systems are ecosystems and societies, as well as the human brain, with the related complex issues of, for example, climate change, stock market crashes and migraine attacks. Over the years, complexity has become tightly connected to interdisciplinarity (Klein, 2004).

Complexity also incorporates the so-called 'wicked problems' (Rittel and Webber, 1973). Wicked problems can be defined, in brief, as 'complex and multifaceted issues that have no single solution, and are perceived by different stakeholders through contrasting views' (Elia and Margherita, 2018, p. 279). They typically have no definite formulation, are ambiguous and are strongly stakeholder dependent, and there is little consensus about what the problem is or how to solve it (Rittel and Webber, 1973; Ritchey, 2013). Not all complex issues are wicked, but wicked problems are complex, and thus need to be addressed in an interdisciplinary manner. Animal welfare is a complex issue and has also been suggested to be a wicked problem (Whyte and Thompson, 2012; Fernandes *et al.*, 2019). Rittel and Webber (1973) listed ten characteristics of wicked problems. These characteristics match the situation of animal welfare: (i) there is no definite description; (ii) there is no definite solution; (iii) solutions are neither true nor false but rather better or worse; (iv) there is no ultimate test for a solution to the problem; (v) every implemented solution is consequential and leaves traces that cannot be undone; (vi) it does not have an enumerable set of potential solutions; (vii) every situation is likely to be one of a kind; (viii) it can be considered a symptom of another problem (e.g. animal welfare issues as a symptom of animal management in livestock industry); (ix) it can be explained in many ways and the choice of explanation determines the nature of the problem's solution (e.g. explaining animal welfare from a health perspective, or from the perspective of animal emotion); and (x) decision makers are liable for the consequences of the actions they generate. From this, it can be concluded that animal welfare is a wicked problem, and thus also a complex issue. Animal welfare as a whole should therefore be approached by using an interdisciplinary perspective (as also advocated by Carenzi and Verga, 2009).

1.4.3 Multidisciplinary, interdisciplinary and transdisciplinary research

There are different levels of integration that are distinguished in interdisciplinarity. The initial description of Rosenfield (1992) into multidisciplinary, interdisciplinary and transdisciplinary has changed over time, and the below descriptions outline how the terms are currently used.

Multidisciplinary research involves more than one (sub)discipline, but the disciplines do not integrate, and the results remain separate. The product is no more and no less than the simple sum of its parts (Wagner *et al.*, 2011). For example, in a research project, animal behaviour and physiology can be assessed alongside each other and presented as separate parts.

Interdisciplinary research integrates the relevant theories, concepts and/or methodologies from different (sub)disciplines to create a holistic view or common understanding of a complex issue, question or problem (Wagner *et al.*, 2011). It brings a larger and more holistic understanding of the problem. An example is the development of algorithms to automatically detect animal behaviour from videos, for which ethologists and engineers/computer scientists need to collaborate closely in order to develop and validate the resulting novel product (see Chapter 15, this volume, for further discussion of this topic).

Transdisciplinary research occurs when interdisciplinary academic projects involve non-academic stakeholders. Academic knowledge and stakeholder perspectives are integrated to determine the problem and to investigate potential solutions. In animal welfare sciences, many of the projects that are labelled as interdisciplinary are actually transdisciplinary. An example in animal welfare science is the development of farming systems whereby alongside evaluation of animal welfare, the environmental impact and economics, farmers and retailers are invited as stakeholders to provide their input (e.g. Schodl *et al.*, 2015).

1.5 Animal Welfare in an Interdisciplinary Landscape

Animal welfare science has from its origin blended aspects of ethology, zoology and veterinary sciences. McGlone (2001) wrote that the assessment of animal welfare has evolved through a multidisciplinary approach, including animal behaviour, physiology, anatomy, health and immunity. Across disciplines in general, the fusion between (sub) disciplines is rapidly increasing, and this is similarly the case for animal welfare science. Increased (international) collaborations, enhanced by funding schemes that encourage researcher mobility, has contributed to a surge in cross-disciplinary projects. Nowadays, there are numerous links between animal welfare science and other (sub)disciplines

Table 1.1. Examples of how different perspectives of animal welfare have influenced the application and integration of various (sub)disciplines.

Context	Perspectives	Evaluation	(Sub)disciplines
Schools of thought			
Animal welfare is…	…when the animal is healthy	Health	Veterinary sciences, immunology, virology, proteomics, etc.
	…when the animal leads a natural life	(Evolution of) behaviour	Ethology, evolutionary biology, genetics, behavioural ecology
	…when the animal has positive emotions	Indirect evaluation of emotional state	Cognition, cognitive psychology, (affective) neuroscience, psychology, ethology, comparative psychology
	…when the animal has a good life	Quality of life	Human medical sciences, positive psychology, neuroscience, cognition
The Five Freedoms			
Animal welfare can be assessed through…	…freedom from hunger and thirst	Health, (abnormal) behaviour	Veterinary sciences, physiology, ethology
	…freedom from discomfort	Preference	Cognition
	…freedom from pain, injury or disease	Health	Veterinary sciences, anatomy, ethology, microbiology
	…freedom to express normal behaviour	Behaviour, stress response	Ethology, physiology, developmental biology
	…freedom from fear and distress	Behaviour, stress response	Ethology, physiology, psychology

Continued

Table 1.1. Continued.

Context	Perspectives	Evaluation	(Sub)disciplines
As an aspect of a larger system			
Animal welfare should be seen within the context of...	...sustainability	Profitability	Economics, agricultural economics, consumer studies
		Societal acceptance	Psychology, ethics, sociology, political science, law, philosophy
		Environmental impact	Ecology, climate science, environmental science
	...the livestock sector	Farming systems	Technology, engineering, infrastructure, sociology, sociocultural studies
		Housing conditions	Zoology, ethology, engineering, technology
		Animal production	Animal genetics, genomics, engineering, technology, proteomics, food science
	...consumer demands	Consumer behaviour	Consumer studies, communication sciences
	...human–animal relationships	Human–animal interactions	Psychology, ethology, physiology
		Human attitudes, human behavioural change	Psychology, communication sciences
		Conservation	Conservation biology, genetics, biotechnology, ecology

(Table 1.1). Lund *et al.* (2006) emphasized that the collaboration is, however, mainly within the natural sciences, and that it does not sufficiently incorporate the social sciences. Moreover, they pointed out that many of the projects that are meant to be interdisciplinary are at best multidisciplinary (i.e. not integrated), especially between natural and social sciences. In projects, the focus has mostly been on the natural sciences, with the social scientists in the project having a 'service' role to only solve specific aspects (Lund *et al.*, 2006; Aagaard-Hansen, 2007). Both Lund *et al.* (2006) and Aagaard-Hansen (2007) warned about this power imbalance. Social science methods, which are largely qualitative, have been undervalued by researchers in natural sciences, who are largely quantitatively oriented. Possibly as a consequence, animal scientists have been attempting to conduct the social sciences parts themselves, especially when it comes to stakeholder involvement and surveys, largely without collaboration or guidance of social scientists. Attracting and engaging stakeholders is often a challenge, and most animal welfare researchers are not trained in communication strategies or techniques to obtain information from stakeholder engagements in a scientific manner. This has regrettably led to some poor methods in studies, as pointed out in Chapters 5 and 6 (this volume). Reports that clearly describe the process that was taken to involve the stakeholders in transdisciplinary projects, as done by Miele *et al.* (2011) in the Welfare Quality® project, provide a learning opportunity for fellow researchers and are therefore of much value. So are reports on aspects of transdisciplinary processes, such as the perspectives of different stakeholders on terminology, as was done for the term 'play behaviour' (Miller, 2017). Involving stakeholders, including society, in the discussions and research on animal welfare is essential as it is a wicked problem in which contrasting stakeholder opinions are central. Therefore, we have to recognize that a transdisciplinary approach is required to come to sustainable solutions (as also advocated by Veissier and Miele, 2014).

The collaboration across disciplines is not without hurdles, regardless of which disciplines are combined, although it may be more difficult for disciplines that are very different from each other (Aagaard-Hansen, 2007). Some of the challenges between animal welfare and ethics are described by Fraser (1999). He wrote that the use of different concepts, assumptions and vocabulary divided researchers into distinct 'cultures'. This is definitely not exclusive to the collaboration between animal welfare scientists and philosophers, or even between academics and non-academics, as also becomes clear from the later chapters in this book. The integration of disciplines requires thoughtful planning and management, especially in cases where input from stakeholders would inform the direction of animal experiments. This was concluded by Benard and de Cock-Buning (2014), who described the transition of an animal welfare project from its monodisciplinary to transdisciplinary stage (described in Chapter 3, this volume). Some guidance on techniques and how to overcome the challenges have been described throughout this book.

The most important discipline that animal welfare science has yet to connect with is, however, the science of interdisciplinarity itself. Interdisciplinary studies is being taught as a degree at various universities. There is a need for the inclusion of training in interdisciplinarity in animal welfare science, starting with undergraduate education (Lund *et al.*, 2006). Klein and Newell (1997) provided suggestions on how to establish (aspects of) an interdisciplinarity curriculum in education programmes, and applying this would be valuable for making progress in animal welfare science.

1.6 Conclusion

Animal welfare science has been from its origin a blend of different subdisciplines, but this has predominantly occurred in an unstructured and trial-and-error manner. When applying the basic concepts used in interdisciplinarity, such as complexity, it becomes clear that animal welfare in itself should be considered as a complex and 'wicked' problem. This implies that there is no single definition or solution, as it is a complex, dynamic and non-linear system. Animal welfare should thus be approached accordingly. Interdisciplinary studies provide methods to address complex problems, and therefore a proper understanding of interdisciplinary research is needed to advance animal welfare science. To this aim, it is important that animal welfare students and researchers are trained in interdisciplinarity so that they can appropriately apply the skills and techniques required for collaboration and integration across disciplines.

References

Aagaard-Hansen, J. (2007) The challenges of cross-disciplinary research. *Social Epistemology* 21, 425–438.

Bateson, P. and Laland, K.N. (2013) Tinbergen's four questions: an appreciation and an update. *Trends in Ecology & Evolution* 28(12), 712–718. DOI: 10.1016/j.tree.2013.09.013.

Benard, M. and de Cock-Buning, T. (2014) Moving from monodisciplinarity towards transdisciplinarity: insights into the barriers and facilitators that scientists faced. *Science and Public Policy* 41, 720–733.

Bentham, J. (1789) *An Introduction to the Principles of Morals and Legislation.* Clarendon Press, London.

Brambell Committee (1965) *Report of the Technical Committee to Enquire into the Welfare of Animals Kept Under Intensive Livestock Husbandry Systems.* Her Majesty's Stationery Office, London, UK.

Broom, D.M. (1986) Indicators of poor welfare. *The British Veterinary Journal* 142(6), 524–526. DOI: 10.1016/0007-1935(86)90109-0.
Broom, D.M. (2011) A history of animal welfare science. *Acta Biotheoretica* 59(2), 121–137. DOI: 10.1007/s10441-011-9123-3.
Carenzi, C. and Verga, M. (2009) Animal welfare: review of the scientific concept and definition. *Italian Journal of Animal Science* 8, 21–30.
Dawkins, M.S. (1993) *Through Our Eyes Only? The Search for Animal Consciousness.* Oxford University Press, Oxford, UK.
Duncan, I.J. (2006) The changing concept of animal sentience. *Applied Animal Behaviour Science* 100, 11–19.
Elia, G. and Margherita, A. (2018) Can we solve wicked problems? A conceptual framework and a collective intelligence system to support problem analysis and solution design for complex social issues. *Technological Forecasting and Social Change* 133, 279–286.
Fernandes, J., Blache, D., Maloney, S.K., Martin, G.B. and Venus, B. (2019) Addressing animal welfare through collaborative stakeholder networks. *Agriculture* 9, 132.
Flood, R.L. and Carson, E.R. (2013) *Dealing with Complexity: An Introduction to the Theory and Application of Systems Science*, 2nd edn. Springer Science and Business Media, New York.
Fraser, D. (1999) Animal ethics and animal welfare science: bridging the two cultures. *Applied Animal Behaviour Science* 65, 171–189.
Fraser, D. (2008) Understanding animal welfare. *Acta Veterinaria Scandinavica* 50, S1.
Fraser, D., Weary, D.M., Pajor, E.A. and Milligan, B.N. (1997) A scientific conception of animal welfare that reflects ethical concerns. *Animal Welfare* 6, 187–205.
Frodeman, R. (2017) The future of interdisciplinarity. In: Frodeman, R., Klein, J.T. and Pacheco, R.C.D.S. (eds) *The Oxford Handbook of Interdisciplinarity.* Oxford University Press, Oxford, UK, pp. 3–8.
Griffin, D.R. (1976) *The Question of Animal Awareness: Evolutionary Continuity of Mental Experience.* Rockefeller University Press, New York.
Griffin, D.R. (2001) *Animal Minds: Beyond Cognition to Consciousness.* University of Chicago Press, Chicago, Illinois.
Keeling, L.J., Rushen, J. and Duncan, I.J. (2011) Understanding animal welfare. In: Appleby, M.C., Mench, J.A., Olsson, I.A.S. and Hughes, B.O. (eds) *Animal Welfare*, 2nd edn. CAB International, Wallingford, UK, pp. 13–26.
Klein, J.T. (2004) Interdisciplinarity and complexity: an evolving relationship. *Emergence: Complexity and Organization Special Double Issue* 6, 2–10.
Klein, J.T. and Newell, W.H. (1997) Advancing interdisciplinary studies. In: Ratcliff, J.L. and Gaff, J.G. (eds) *Handbook of the Undergraduate Curriculum: A Comprehensive Guide to Purposes, Structures, Practices, and Change.* Association of American Colleges and Universities, Washington, DC, pp. 393–415.
Lund, V., Coleman, G., Gunnarsson, S., Appleby, M.C. and Karkinen, K. (2006) Animal welfare science – working at the interface between the natural and social sciences. *Applied Animal Behaviour Science* 97, 37–49.
McGlone, J.J. (2001) Farm animal welfare in the context of other society issues: toward sustainable systems. *Livestock Production Science* 72, 75–81.

Mellor, D.J. (2016) Updating animal welfare thinking: moving beyond the "Five Freedoms" towards "a life worth living". *Animals* 6, 21.

Mellor, D.J. and Beausoleil, N.J. (2015) Extending the 'Five Domains' model for animal welfare assessment to incorporate positive welfare states. *Animal Welfare* 24, 241–253.

Mellor, D.J., Beausoleil, N.J., Littlewood, K.E., McLean, A.N., McGreevy, P.D. *et al.* (2020) The 2020 Five Domains Model: including human–animal interactions in assessments of animal welfare. *Animals* 10, 1870.

Menken, S. and Keestra, M. (2016) *An Introduction to Interdisciplinary Research: Theory and Practice.* Amsterdam University Press, Amsterdam.

Miele, M., Veissier, I., Evans, A. and Botreau, R. (2011) Animal welfare: establishing a dialogue between science and society. *Animal Welfare* 20, 103–107.

Miller, L.J. (2017) Creating a common terminology for play behavior to increase cross-disciplinary research. *Learning and Behavior* 45(4), 330–334. DOI: 10.3758/s13420-017-0286-x.

Millman, S.T., Duncan, I.J., Stauffacher, M. and Stookey, J.M. (2004) The impact of applied ethologists and the International Society for Applied Ethology in improving animal welfare. *Applied Animal Behaviour Science* 86, 299–311.

National Academy of Sciences, National Academy of Engineering, and Institute of Medicine (2005) Executive summary. In: *Facilitating Interdisciplinary Research.* The National Academies Press, Washington, DC, pp. 1–15.

Newell, W.H. (2004) Complexity and interdisciplinarity. In: Kiel, L.D. (ed.) *Knowledge Management, Organizational Intelligence and Learning, and Complexity.* UNESCO Encyclopedia of Life Support Systems, Oxford, UK.

Ohl, F. and van der Staay, F.J. (2012) Animal welfare: at the interface between science and society. *Veterinary Journal* 192, 13–19.

Paul, E.S., Sher, S., Tamietto, M., Winkielman, P. and Mendl, M.T. (2020) Towards a comparative science of emotion: affect and consciousness in humans and animals. *Neuroscience and Biobehavioral Reviews* 108, 749–770. DOI: 10.1016/j.neubiorev.2019.11.014.

Rault, J.-L., Hintze, S., Camerlink, I. and Yee, J.R. (2020) Positive welfare and the like: distinct views and a proposed framework. *Frontiers in Veterinary Science* 7, 370. DOI: 10.3389/fvets.2020.00370.

Ritchey, T. (2013) Wicked problems: modelling social messes with morphological analysis. *Acta Morphologica Generalis* 2, 1–8.

Rittel, H.W.J. and Webber, M.M. (1973) Dilemmas in a general theory of planning. *Policy Sciences* 4, 155–169.

Rosenfield, P.L. (1992) The potential of transdisciplinary research for sustaining and extending linkages between the health and social sciences. *Social Science and Medicine* 35(11), 1343–1357. DOI: 10.1016/0277-9536(92)90038-r.

Schodl, K., Leeb, C. and Winckler, C. (2015) Developing science–industry collaborations into a transdisciplinary process: a case study on improving sustainability of pork production. *Sustainability Science* 10, 639–651.

Tight, M. (2020) Higher education: discipline or field of study? *Tertiary Education and Management* 26, 415–428.

Tinbergen (1963) On aims and methods of ethology. *Zeitschrift für Tierpsychologie* 20, 410–433.

Veissier, I. and Miele, M. (2014) Animal welfare: towards transdisciplinarity – the European experience. *Animal Production Science* 54, 1119–1129.

Wagner, C.S., Roessner, J.D., Bobb, K., Klein, J.T. and Boyack, K.W. (2011) Approaches to understanding and measuring interdisciplinary scientific research (IDR): a review of the literature. *Journal of Informetrics* 5, 14–26.

Webb, C.E. and de Waal, F. (2018) Situating the study of jealousy in the context of social relationships. *Animal Sentience* 3, 22.

Webster, A.J.F. (1994) *Animal Welfare: A Cool Eye Towards Eden.* Wiley-Blackwell, Oxford, UK.

Whyte, K.P. and Thompson, P.B. (2012) Ideas for how to take wicked problems seriously. *Journal of Agricultural and Environmental Ethics* 25, 441–445.

2

Pragmatics for Cross-disciplinary Collaboration

Julie Thompson Klein*

Wayne State University, Detroit, USA, and Transdisciplinarity Lab, ETH Zurich, Zurich, Switzerland

2.1 Introduction

Increases in the number and the size of teams have heightened interest in collaboration across boundaries of expertise. Animal science is only one of many areas where this need is recognized but not always matched by capacity for working across boundaries. The study by Lund *et al.* (2006) on the field cited animal science as a pioneer example of interdisciplinary work. However, they also reported that teamwork occurs primarily among researchers in natural sciences, without engaging social scientists. Subsequently, Marchant-Forde (2015) called animal welfare science a relatively new discipline that emerged from veterinary medicine in the latter half of the 20th century. The initial focus was ethology, but the scope expanded to include behaviour, physiology, pathology, health, immunology, endocrinology, neuroscience and ethics. He was equally mindful of a societal 'big picture', prompting calls for new forms of research in other domains as well. Challenges in animal welfare science, in particular, intersect with concerns about population growth, food security, sustainability and climate change. Looking ahead, Marchant-Forde (2015) advocated drawing from three main overarching approaches in the field: biological function, the naturalness of an animal's life and the animal's feelings. Working across all three approaches, however, would require bridging boundaries of knowledge, methodologies, priorities, forms of data, conceptual frameworks and contexts. Towards that end, he concluded with an exhortation to address 'big issues' that global society is facing today by engaging in interdisciplinary collaboration. There is no single primer for doing so. However, the US-based National Cancer Institute's *Collaboration and Team Science Field Guide* (Bennett *et al.*, 2018) is an excellent place to start.

*julietklein@comcast.net

 Bridging Research Disciplines to Advance Animal Welfare Science: A practical guide (ed. I. Camerlink)
DOI: 10.1079/9781789247893.0002

Bennett *et al.* (2018) steer readers through the *Field Guide*'s major topics, including: preparing for collaboration, fostering trust and a shared vision, communicating, assembling and leading teams, managing difference and conflict, and sharing credit while leveraging networks and systems. Individual sections include definitions, short case studies that aid in knowing whether collaboration is working or not, and key questions for participants. This chapter draws on the *Field Guide*, as well as on a recent study of the boundary work of collaborating in cross-disciplinary and cross-sector work (Klein, 2021), with added insights from literature on team science. Heeding the editors' mandate to focus on pragmatics, it depicts the 'how' of collaboration across boundaries of expertise in short essays that define major dimensions. Each essay is framed at the start by keywords and ends with pertinent resources that individuals and groups may use in whole or in part for training modules and workshops, interventions in the course of actually conducting research, and formal curricula in higher education. The initial section of the chapter describes the overarching topic of collaboration, including the role of a collaboration plan and the centrality of communication. The remaining sections discuss three subtopics that are often linked with collaboration. The first, cross-disciplinary and cross-sector work, reveals distinctions in kinds of teamwork. The second, integration, discloses degrees of interaction and synthesis. The third, leadership, describes typical needs and responsibilities. These shorter accounts of related concepts and approaches begin with definitions of cross-disciplinarity (spanning multi-, inter- and transdisciplinary approaches) and cross-sector work (bridging the academy, government, industry, and communities in the north and global south). After a short summary, the conclusion reflects on the importance of integrative expertise among all team members along with needed requisite competencies.

Box 2.1 outlines the highlights of cross-disciplinary collaboration.

2.2 Collaboration

- *Keywords of collaboration*: interdependence, shared vision, collective identity, infrastructure, readiness, dialogue, mutual learning, communication competence, culture of communication, team functioning, conflict, translation, interlanguage.

Collaborations vary in size, levels and composition. To begin with, the number of participants ranges from two individuals in a small project co-located in the same institution to as many as 2000 in large and dispersed research programmes such as the Large Hadron Collider. A recent state-of-the-art report on team science from the US-based National Research Council defined a team as two or more members who work together interdependently. Small teams typically have ten or fewer individuals and

Box 2.1. Highlights of cross-disciplinary collaboration.

- Collaboration and technological readiness.
- Collaboration plan to define expectations and to monitor team functioning.
- Interaction, cognitive and social integration.
- Common purpose, shared vision and collective identity.
- Active listening and mutual learning.
- Trust, psychological safety, equal voice, and power sharing.
- Negotiation of differences and conflict management.
- Leadership, integrative expertise and communication competence.
- Interlanguage and a culture of communication.
- Constructive feedback and iteration.
- Adaption to alternative viewpoints and methods.
- Iteration and consensus building.
- Joint decision making and problem solving.

large teams more than ten (NASEM (National Academies of Science, Engineering, and Medicine), 2015). The dynamics of collaboration also span levels: from intrapersonal and interpersonal relationships to organizational and institutional parameters to socio-political concerns on local, regional, national and global scales. In addition, the composition of teams varies. Many science-based initiatives are limited to biological and physical disciplines, although they may involve professions such as engineering and medicine. Social sciences are sometimes included, but humanities and arts are often neglected with the exception of the subfield of ethics in philosophy. Formation of a collective identity is crucial to successful collaboration. Young teams, Stone (1969) reported, tend to exhibit secondary-group relationships. They think in terms of 'I' rather than primary-group relationships characterized by thinking in terms of 'we'. Brister's (2016) concept of disciplinary capture, however, impedes moving from secondary to primary relationships. It occurs when one discipline takes precedence over others, dictating the structure and methods of research. It also prioritizes one specialist language and worldview. Effective collaboration avoids capture, but it is not a simple concept. Introducing a book on the subject of collaboration and communication in interdisciplinary research, Crowley *et al.* (2014) identified affective, organizational and instrumental dimensions along with linguistic, relational, cultural and socio-political variables.

The terms coordination, cooperation, and collaboration are sometimes used synonymously, but Gunawardena *et al.* (2010) made an important distinction. Like coordination and cooperation, collaboration involves exchanges of information, mutually beneficial activities, and a common purpose. However, it also entails increased risk, comprehensive planning, communication across levels, a durable and pervasive relationship among team members, commitment to a shared mission, and authority in an infrastructure anchored by a clear work plan. The 'collaborator's pre-nup' (Ledford, 2008) and the template for a collaborative agreement in the *Field Guide* outline typical components, although Hall *et al.*'s (2019) description of a collaboration planning approach paints a fuller picture of what is entailed. It is a deliberative approach to assessing a team's environment for collaborative success. Hence, a collaboration planning approach can not only be used for establishing responsibilities and relationships at the outset but can also be revisited periodically to evaluate how well a team is functioning, in the process identifying required improvements in structure, coordination, communication, leadership, and resources.

Given that teams vary in goals, size, diversity of members, and their expertise, as well as the degree of knowledge integration, not all collaboration plans will be identical. Yet, Hall *et al.* (2019) identified ten generic influences on likelihood and degree of success:

1. Rationale for team approach and configuration.
2. Collaboration readiness.
3. Technological readiness.
4. Team functioning.
5. Communication and coordination.
6. Leadership, management and administration.
7. Conflict prevention and management.
8. Training.
9. Quality improvement activities.
10. Budget and resource allocation.

Hall *et al.* (2019) likened collaboration plans to roadmaps but cautioned that they should not be static or prescriptive. They should be 'living documents' because conditions change and the process of teamwork needs monitoring. With particular regard to the purpose of this book, defining the pragmatics of collaboration, several of their generic influences stand out:

- *Collaboration readiness* assesses existing capabilities, including components of team functioning such as assembled expertise and individuals' prior experience with teamwork, as well as institutional parameters that span resources, infrastructure, policies, and protocols. Determining existing capacity is crucial in identifying both strengths and gaps that need filling through formal training, interventions in the course of research, and/or mentoring by a leader or consultant.
- *Technological readiness* applies to both local and dispersed collaboration. Berente and Howison's (2019) chapter in the National Research Council's report on team science and Olson and Olson's (2014) book on the topic describe tools for data sharing, teamwork, project management, and monitoring progress. In animal welfare science, they may also include devices for automatic behaviour recording. In all cases,

determining appropriate hardware and software, individuals' ability to use them and the feasibility of local platforms is crucial.

- *Team functioning* encompasses conditions for success, including 'generating a shared vision, mission, and goals; creating shared mental models of the team structure and the collaborative scientific project; and externalizing group cognition throughout the collaboration' (Hall *et al.*, 2019, p. 591). Visualization can also aid in depicting a problem space, research questions, and collaborative workflow, and in this field depicting levels of improvement for animal welfare.

2.2.1 Communication

The topic of communication looms large in literature on collaboration, and indeed was a major factor in Lund *et al.*'s (2006) recommendations for animal welfare science. The National Research Council's state-of-the-art report on *Facilitating Interdisciplinary Research* asserted that communication lies at the heart of interdisciplinarity (NASEM (National Academies of Science, Engineering, and Medicine), 2005). Communication is not just a linguistic phenomenon, however. Joann Keyton (1999) called it a social process and, while introducing a book on communication in teams, Klein *et al.* (2020) reported that the topic entails both pragmatic and conceptual dimensions while spanning intrapersonal, interpersonal, organizational, institutional, and socio-political levels. Describing the influence of 'communication and coordination', Hall *et al.* (2019) urged specifying the frequency and modalities of meetings in collaboration plans. Thompson (2009) went a step further by placing collective communication competence at the heart of interdisciplinary teamwork. It is nurtured by spending time together, developing trust, focusing on designated tasks, discussing differences in language and engaging in reflexive talk. Laursen's (2018) concept of collaborative interdisciplinary reasoning, in turn, aligned productive dialogue with co-applying, co-revising and co-creating intersubjective standards to generate shared assumptions about what counts as 'good' reasons and inferences. Freeth *et al.* (2019) also emphasized the importance of moving from serial monologue to engaged dialogue and, comparably, the *Field Guide* distinguished dialogue from debate. Dialogue is interactive and reciprocal, whereas debate is competitive with the aim of winning an argument.

2.2.2 Conflict and translation

Two additional aspects of communication are also prominent in discussions of collaboration: conflict and translation. Conflict arises for multiple reasons, although the *Field Guide* distinguished scientific disagreements from interpersonal tensions that can scuttle even the most promising projects. Scientific disagreement does not necessarily do so if the focus remains on common goals. Even so, the authors added, teams evolve over time and go through periods of 'storming', following Tuckman's classic 1965 model of group development in four stages of forming, storming, norming and performing, along with Tuckman and Jensen's (1977) subsequent addition of a fifth stage for adjourning or transforming. Likewise, in describing the influence of 'conflict prevention and management', Hall *et al.* (2019) advised that conflict is inevitable to some degree due to differences in ownership of data and patents, intellectual property rights, order of authorship, potential fault lines on a team, and scientific challenges. Disciplinary and demographic diversity are added factors, in the former instance including preference for quantitative or qualitative approaches and in the latter gender, age and cultural background.

Conflict also arises in animal welfare science. As described in Chapter 3 (this volume), despite efforts to overcome inequality in the field, bias exists in disciplinary aspects, such as valuing quantitative data over qualitative data. A variety of strategies lessen the likelihood of conflict derailing a project, including establishing ground rules at the outset for how to communicate during meetings and how to handle differences. The *Field Guide* cites several models including one in particular that can scaffold building trust. A 'Welcome to my team' letter can outline expectations and procedures for a particular group, while regular meetings and iterative reflection on progress of a project are valuable for monitoring progress and handling problems that arise early, rather than letting them haunt later stages. Documenting decisions also renders agreements transparent to everyone. The process of communication, however, is not a simple set of tasks on a checklist. Effective communication requires creating an environment conducive to constructive dialogue linked with the notion of a safe place for talking across boundaries and ranks in academic hierarchy. It empowers rather than inhibits individuals' willingness to speak in the midst of dominant voices. Mindful of this need, Vosskamp (1994) contended that the quality of outcomes is directly related to establishing a shared language culture. Schmithals and Berkenhagen (2004) also deemed cooperation and communication culture essential to transdisciplinary research, and Carpenter (2009) emphasized the importance of creating a communication boundary space.

In addition, translation and interlanguage are major topics in discussions of collaboration. Translation is necessary because even the same word, including one as basic as the 'welfare' of animals, may have different meanings in different domains. Understanding them is crucial to a consensus on meaning for a particular initiative. Individuals often retain separate connotation within specialist domains. The interlanguage that develops through deliberations may also contain specialist terms adjudicated for relevance to a particular initiative. However, participants must agree on shared meaning. Frey (1973) reported that the language of collaboration usually begins at a level similar to a popular scientific presentation. It becomes more precise as individuals combine everyday and specialist languages. The process was subsequently likened to contact languages known as pidgins and creoles. In linguistics, a pidgin is a technical term for an interim tongue enabling different groups to communicate. A creole constitutes a new first language. Galison (1997), Holbrook (2012) and Klein (2013) all linked pidgins and creoles to cross-boundary work. In research, a pidgin is typically a mode of communication that may be temporary and is linked to the companion notion of a trading zone. In contrast, a creole is a new language of a community of practice. Individuals learn its connotations in the course of being socialized into a domain where underlying meanings are taken for granted.

Further Reading

Bennett, L.M., Gadlin, H. and Marchand, C. (2018) *Collaboration and Team Science: A Field Guide*, 2nd edn. NIH publication no. 18-7660. National Cancer Institute, National Institutes of Health, Bethesda, Maryland.

Hall, K.L., Vogel, A.L. and Crowston, K. (2019) Comprehensive collaboration plans: practical considerations spanning across individual collaborators to institutional supports. In: Hall, K.L., Vogel, A.L. and Croyle, R.T. (eds) *Strategies for Team Science Success: Handbook of Evidence-based Principles for Cross-disciplinary Science and Practical Lessons Learned from Health Researchers.* Springer, Cham, Switzerland, pp. 587–611.

Klein, J.T. (2017) Typologies of interdisciplinarity: the boundary work of definition. In: Frodeman, R., Klein, J.T. and Pacheco, R. (eds) *The Oxford Handbook of Interdisciplinarity.* Oxford University Press, Oxford, UK, pp. 21–35.

Klein, J.T. (2021) Communicating and collaborating. In: *Beyond Interdisciplinarity: Boundary Work, Communication, and Collaboration.* Oxford University Press, New York (in press).

Klein, J.T. (2021) Learning. In: *Beyond Interdisciplinarity: Boundary Work, Communication, and Collaboration.* Oxford University Press, New York.

NASEM (National Academies of Science, Engineering, and Medicine) (2015) *Enhancing the Effectiveness of Team Science.* National Academies Press, Washington, DC.

Swiss Academies of Arts and Sciences *Methods for Coproducing Knowledge.* Network for Transdisciplinary Research (td-net). Swiss Academies of Arts and Sciences, Bern, Switzerland. Available at: https://naturalsciences.ch/co-producing-knowledge-explained/about (accessed 16 June 2021).

2.3 Cross-disciplinary and Cross-sector Work

- *Keywords of cross-disciplinarity*: multidisciplinary, interdisciplinary, transdisciplinary.
- *Keywords of cross-sector work*: stakeholders, participation, co-production of knowledge, community engagement, participatory research.

Three widely recognized terms mark differences in practice:

1. *Multidisciplinary* approaches present separate forms of expertise without integrating them.
2. *Interdisciplinary* approaches typically integrate existing methods and tools, concepts and theories. Sometimes, they result in a new community of practice, although more often a solution to a particular complex problem or an answer to a complex question.
3. *Transdisciplinary* was traditionally associated with the quest for unity of knowledge, but in the latter half of the 20th century became aligned with new synthetic paradigms and co-production of knowledge with stakeholders in society.

Implications follow for collaboration at each point of the spectrum. *Multidisciplinary* contributions are typically separate, submitted and arranged in serial or encyclopaedic order. A project leader or an editor might synthesize them, but the effort is not collaborative, the status quo of disciplines remains intact, and individuals may never meet. Both individuals and groups perform *interdisciplinary* research, but teamwork requires proactive interaction. Even then, the degree of integration varies and bodies of expertise are not necessarily interrogated or transformed. Although the term *transdisciplinary* was traditionally associated with a philosophical quest for unity of knowledge, it also became a label for new overarching frameworks: notable among them were general systems theory, feminist theory, cultural critique, postmodernism and sustainability. In addition, it became linked with a new connotation grounded in the worldview of complexity in science, advanced by the International Center for Transdisciplinary Research. During the closing decades of the century, the term became associated with two additional connotations in which collaboration is a norm. The first, co-production of knowledge with stakeholders in society, arose in the late 1980s and early 1990s in environmental research and is linked today with sustainability. The second, new conceptual and methodological frameworks, is linked with the growing field of team science. It emerged in health research, although it is being promoted in all areas where complex problems are a primary focus. The emergence of these two connotations has resulted in increased attention to the process of collaboration, expanding beyond but still encompassing earlier interests in epistemology and organizational structure.

Researchers in animal welfare science are not strangers to cross-sector work, either. The field encompasses not only disciplines of laboratory science but also occupational professions such as farming, veterinary care, zoo keeping and aquarium management, in addition to the public's roles in domestication of animals, food consumption, health consequences of interactions between humans and animals, and ecological sustainability of natural habitats. The degree of stakeholder participation beyond the academic sphere hinges on whether individuals and groups are limited to a secondary role or are active agents in co-producing knowledge. Patient engagement illustrates the range of involvement in the arena of health. Arnstein's (1969) ladder of participation extends from tokenism or illusory inclusion to a meaningful role in decision making accompanied by reciprocal learning among all parties to an initiative. Implications also follow for the communication practice of translation. The National Research Council's report on team science aligned the concept with application and transfer (NASEM (National Academies of Science, Engineering, and Medicine), 2015). This narrow equation, however, ignores a large body of scholarship in humanities and translation studies on material, textual and cultural factors that belie a universal meaning. Moreover, Yanow (2004) explained, translators are not passive conduits of academic knowledge. Even so, numerous case studies document a one-directional flow of knowledge, delivered from academic research to external stakeholders with the expectation that they will adopt it (see also Chapter 3, this volume). As a result, many projects stop short of making them full partners, circumscribing their contributions while minimizing or outright bypassing knowledge they bring from their cultural backgrounds and lived experience of a particular problem.

Further Reading

Frodeman, R., Klein, J.T. and Pacheco, R. (eds) (2017) *The Oxford Handbook of Interdisciplinarity*. Oxford University Press, Oxford, UK, and New York. In particular, see the following chapters: Klein, J.T., Typologies of interdisciplinarity, pp. 21–35 (for definitions); Burggreen, W., Chapman, K., Keller, B.B., Monticino M. and Torday, J.S., Interdisciplinarity in biological sciences, pp. 101–113; Jasanoff, S., A field of its own: the emergence of science and technology studies, pp. 173–187; Pohl, C., Truffer, B. and Hirsch Hadorn, G., Addressing wicked problems through transdisciplinary research, pp. 319–331; and Hall, K.L., Stipelman, B.A., Vogel, A.L. and Stokols, D., Understanding cross-disciplinary team-based research: concepts and conceptual models from the science of team science, pp. 338–356.

Hirsch Hadorn, G., Hoffmann-Riem, H., Biber-Klemm, S., Grossenbacher-Mansuy, W., Joye, D. *et al.* (eds) (2008) *Handbook of Transdisciplinary Research*. Springer, Cham, Switzerland, and New York. In particular, see the following chapters: Elzinga, A., Participation, pp. 345–359; Hollaender, K., Loibl, M.C. and Wilts, A., Management, pp. 385–397; Pohl, C., van Kerkhoff, L., Hirsch Hadorn, G. and Bammer, G., Integration, pp. 411–424; and Pohl, C. and Hirsch Hadorn, G., Core terms in transdisciplinary research, pp. 427–432.

2.4 Integration

- *Keywords of integration*: types, stages, iteration, integrative process, cognitive and social integration, synthesis, positioning, bounding, purposing.

Integration is widely regarded as central to both cross-disciplinary and cross-sector work. Klein and Newell (1997, p. 404) called it the 'acid test' of interdisciplinarity in higher education, while Pohl *et al.* (2008, p. 42) deemed integration 'the core methodology underpinning the transdisciplinary research process'. However, there is no universal model. Zierhofer and Burger (2007) differentiated thematic, product or problem-oriented, and social types, while Jahn *et al.* (2012) distinguished epistemic, cognitive, social-organizational, and communicative approaches to integration. O'Rourke *et al.* (2016) identified multiple means: including unification by reduction, a global theory or an overarching model, interconnections between fields, local theories, and micro-level integrations. In addition, they identified four fault lines of definition. The first fault line differentiates linear, algorithmic step models from heuristic and constructivist frameworks. The second hinges on whether integration is cognitive in nature or factors in social and communicative aspects. The third is the conception of integration as an individual or a collaborative phenomenon that, in turn, requires commitment to joint research, decision making and determination of outcomes. Finally, the fourth is emphasis on disciplines versus societal perspectives.

The first and third fault lines are both evident in the difference between stage models, such as Repko's (2008) linear scheme of ten sequential steps for individual student research projects and my adoption of a socio-linguistic approach that Laursen and O'Rourke (2019) likened to bootstrapping up in a group process of iterating, patterning and back-and-forth testing. Even the latter process, however, can vary. Hall and O'Rourke (2014) reported that participants might use a variety of tools when launching a project in order to externalize individuals' values, assumptions and knowledge. When moving to integration, participatory modelling may range from simple cognitive mapping to complex models of system dynamics. Mapping also helps visualize spatial aspects of a project, including stakeholder knowledge and project design. When generating alternatives, scenarios can focus on solution-oriented knowledge and operate in conjunction with co-constructing a systems model and visualizing tools. At the final stage, the Delphi method can help a group weigh alternatives (also applicable to determining animal welfare priorities, e.g. Rioja-Lang *et al.*, 2020). No one best approach, then, dominates. Consequently, participants need to be familiar with a repertoire of strategies and mechanisms. Even an iterative process, however, involves stages. Piso *et al.*'s (2016) account of a workshop on the ontology of fog systems on the California coast and methodologies for studying them illustrates interaction across stages also identified in accounts of other projects.

Piso *et al.* (2016) identified five capabilities as individuals in the workshop moved through stages of negotiating an integrated conception of fog systems and means of implementing it. At the first stage of *positioning*, individuals located themselves within group conversations while recognizing the limits of their individual expertise. In doing so, they also began generating a transactive memory system, which consists of shared awareness of who knows what. During *bounding*, they refined the scope of research while suggesting interfaces for integration, informed by mutual learning that begins at the first stage. During *layering*, they continued to sharpen relationships among entities and processes initially located in separate domains but channelled for relevance to the task at hand. Selection continued during *focusing*, however, through complexity reduction, and then in *purposing* produced projects that could connect variables within a system. The five stages did not constitute an absolute linear order. *Positioning* occurred throughout other stages as individuals continued to situate themselves, making adjustments as needed. Dialogue facilitated *bounding* around a construct that did not erase differences but further contextualized them during *layering*. During *focusing*, intersections were also refined, and *purposing* produced possible (sub)projects. The underlying process, to echo Piso *et al.*'s (2016) subtitle, is 'catalysing integrative capacity', enabling a group to synthesize contributions from different domains. Identifying relationships and interactions is crucial, leveraging the expertise of other participants while negotiating the scope of a project. Therefore, the authors linked integrative and collaborative capacity.

Further Reading

Bergmann, M., Jahn, T., Knobloch, T., Krohn, W., Pohl, C. and Schramm, E. (2010) *Methoden Transdisziplinärer Forschung: Ein Uberblick Mit Anwendungsbeispielen.* Campus Verlag, Frankfurt, Germany; English translation (2012) *Transdisciplinary Research Methods.* Campus Verlag, Frankfurt, Germany, and University of Chicago Press, Chicago, Illinois.

McDonald, D., Bammer, G. and Deane, P. (2009) *Research Integration Using Dialogue Methods.* Australian National University, Canberra.

O'Rourke, M. (2017) Comparing methods for cross-disciplinary research. In: Frodeman, R., Klein, J.T. and Pacheco, R. (eds) *The Oxford Handbook of Interdisciplinarity.* Oxford University Press, Oxford, UK, and New York, pp. 276–290.

O'Rourke, M., Crowley, S. and Gonnerman, C. (2016) On the nature of cross-disciplinary integration: a philosophical framework. *Studies in History and Philosophy of Biological and Biomedical Sciences* 56, 62–70.

Pohl, C., van Kerkhoff, L., Hadorn, G.H. and Bammer, G. (2008) Integration. In: Hirsch Hadorn, G., Hoffmann-Riem, H., Biber-Klemm, S., Grossenbacher-Mansuy, W., Joye, D. *et al.* (eds) *Handbook of Transdisciplinary Research*. Springer, Cham, Switzerland, pp. 411–444.

2.5 Leadership

- *Keywords of leadership*: styles, unilateral and multilateral dependence, cognitive-structural-processual tasks, brokering, mediating conflict, psychological safety.

Although comparable to integration, there is no universal model of leadership. Salazar *et al.* (2012) asserted that integrative capacity and collaborative capacity are analogous. Leaders are widely regarded as essential to achieving both. In defining 'leadership, management and administration' in their generic influences, Hall *et al.* (2019) emphasized that a leader's approach varies depending on the immediate project, team, institutional factors, and styles of leadership including hierarchical, heterarchical, transformational and transactional approaches. Integration by a leader or a subgroup was also the first of four types in Rossini and Porter's (1979) study of 24 technology assessments. The other three were negotiation among experts, common-group learning, and joint integration. Reflecting on their typology, Andersen and Wagenknecht (2013) invoked Hardwig's (1985) distinction between unilateral and multilateral epistemic dependence in interdisciplinary collaborations. Integration by a leader is unilateral, while common-group learning is multilateral. It derives from mutual learning and shared mental models. Leaders of cross-disciplinary and cross-sector teams are often chosen on the basis of reputation in an academic speciality. However, they need to be familiar with integrative and collaborative processes as well.

When delineating integrative capabilities for leaders of cross-disciplinary teams, Salazar *et al.* (2019) highlighted in particular visioning, reflexivity, perspective seeking, conflict management, and coordination. Leaders of cross-disciplinary and cross-sector teams are often chosen on the basis of reputation in an academic speciality. However, their litany of capabilities and Gray's (2008) set of related tasks go beyond intellectual stature alone to include fostering integrative and collaborative processes. Gray identified three sets of leadership tasks in transdisciplinary research:

- *Cognitive tasks* manage meaning and sense making by creating a mental model or map of goals and methods in a project. Two activities stand out. Visioning occurs on two levels: in content and in the process of collaboration. Reframing, in turn, requires suspending individual assumptions in favour of building a shared conceptual framework.

- *Structural tasks* foster coordination and exchange through the actions of boundary spanning, conveying information across (sub)groups, brokering the flow of information, and facilitating transactions. In addition, leaders need to tend to transferring and translating individual approaches while building linkages across all components and subprojects.
- *Processual tasks* develop team dynamics that move from Stone's (1969) notion of the singularity of 'I' to the collective identity of 'we'. Doing so entails deploying skills for joint decision making and problem solving, as well as exchanging information, coordinating efforts, managing boundaries, and negotiating differences throughout all stages of a project.

The three sets of leadership tasks are not isolated. Indicative of their interrelationship, Mäkinen (2018, 2020) identified a shift from leader-centric models to systems thinking. Comparably, Salazar *et al.* (2012) contended that knowledge and social integration coevolve through emergent states in an interactive system that links social, psychological and cognitive processes. Leaders can foster social integration, in particular by nurturing trust, knowledge sharing, and a common vision and identity. They also need to manage conflict while ensuring that communication practices facilitate an equal voice. In addition, the *Field Guide* emphasized that leaders need to model how to give complementary and constructive feedback. Gray (2008) described leaders as brokers between groups by helping to clear up misunderstandings, translate jargon, create psychological safety for all team members, and mediate conflict when it arises. Moreover, they can aid in resolving questions of legitimacy and ameliorating power differences. In the aggregate, these leadership actions can overcome or at least minimize failures and shortfalls in decision making, planning and cognition. Although much of the literature focuses on individuals, Gray (2008) and others also acknowledge the role of multiple leaders, especially in large and dispersed initiatives. A collaboration plan should also designate how to coordinate across sites, local leaders and their networks.

Further Reading

Eigenbrode, S. *et al.* (2017) *Leading Large Transdisciplinary Projects Addressing Social–Ecological Systems: A Primer for Project Directors.* National Institute of Food and Agriculture, Washington, DC. Available at: https://nifa.usda.gov/leading-transdisciplinary-projects (accessed 2 June 2021).

Gray, B. (2008) Enhancing transdisciplinary research through collaborative leadership. *American Journal of Preventive Medicine* 35, S124–S132.

Part VII: Leadership and management of teams. In: Hall, K.L., Vogel, A.L. and Croyle, R.T. (eds) (2019) *Strategies for Team Science Success: Handbook of*

Evidence-based Principles for Cross-disciplinary Science and Practical Lessons Learned from Health Researchers. Springer, Cham, Switzerland, and New York, pp. 313–388. Includes chapters on integrative capabilities, an organizational perspective, attaining goals, an interdisciplinary executive scientist, and research development professionals.

2.6 Conclusion: Towards Integrative Expertise

Summing up, collaboration is a recursive process that is vital to constructing a shared outcome in both cross-disciplinary and cross-sector work. It entails negotiating across boundaries of expertise and the stakes they represent in order to communicate, to engage in mutual learning, to build trust, and to craft a common vision while also managing conflict. Doing so also requires moving from the self-positioning singularity of 'I' to the interdependent collective identity of 'we'. A collaboration plan contributes to effective teamwork by providing a template for assessing levels of readiness, expectations at both individual and team levels, designated roles and responsibilities, and ongoing monitoring of how a team is functioning while revealing any areas where improvement is needed. Communication looms especially large because successful collaboration requires an environment and a culture in which team members translate domain-specific meanings, enabling participants to move from serial monologue to non-competitive dialogue and to develop an interlanguage with shared terminology. Degrees of integration will still vary, however, driven by the particular mix of expertise in any one initiative and the level of engagement among all stakeholders. In addition, all collaborations require informed leadership, raising the question of whether an integration expert should be a member of a team. Leading examples include a specialist in integration and implementation sciences, an interdisciplinary executive scientist, and a research development professional. Leaders, however, are not the only ones responsible for successful collaboration. All participants need integrative expertise.

Integrative expertise (Klein, 2021) includes the ability to:

- contribute a pertinent form of expertise;
- articulate one's own approaches and meanings in an accessible manner;
- listen actively to others and ask for translation if needed;
- engage in mutual learning;
- provide and receive constructive feedback;
- adapt flexibly to alternative viewpoints and methods;
- negotiate differences without personalizing them;
- build trust and rapport;
- view problems and questions holistically;
- surrender the exclusivity of 'I' for the collective identity of 'we'; and
- participate in building consensus and integrating separate inputs.

Collins and Evans' (2007) distinction between interactional and contributory expertise also marks an important difference in individuals' ability to participate. The first level is sufficient acquaintance with another discipline to intermingle during teamwork, while the second depends on fuller knowledge of its language, concepts and theories, methods and tools, and data with a potential to influence transactions. In both cases, requisite skills should be made explicit from the outset, informed by pertinent literature. Moreover, they need to become part of the traditional curriculum in order to make good on the rhetoric of preparation of the 'next generation' of researchers. Fiore *et al.* (2019) reported that the majority of interdisciplinary experiential learning for team science comes through coursework, with supplementary mentoring. Team-based projects are often also cited, and Bosque-Pérez *et al.* (2016) highlighted student collaboration on manuscripts and grant proposals, fieldwork and stakeholder workshops. Workshops and seminars, internships and summer research experiences are additional learning opportunities, although in an overview of transdisciplinary training, Nash (2008) concluded that knowledge and skills are best developed by combining formal didactics, mentoring and research experiences. Misra *et al.*'s (2015) concept of transdisciplinary orientation also underscores the complexity of what integrative expertise entails.

Transdisciplinary orientation is a composite of values, attitudes, beliefs, skills, knowledge, and behavioural repertoires that predispose individuals to collaborate. Nurius and Kemp (2019) built on Misra *et al.*'s (2015) model in defining individual-level competencies for team collaboration. Predisposition to integrating knowledge from different disciplines and stakeholders is linked with a cluster of values, attitudes and beliefs. They added willingness to balance assertiveness with patience in working with others. 'Habits of mind' include broad intellectual curiosity coupled with recognizing the limits of one's own expertise while making modifications based on the influence of others. 'Knowledge-based competencies' include deep familiarity with a particular body of expertise while also making connections across other forms, including stakeholder knowledge with its attendant concepts, theories and methods. Echoing the roster of abilities above, as well, 'interpersonal competencies' include learning the language and methods of other disciplines, communicating one's perspectives in a manner understandable to others, navigating tensions and conflicts, and building effective relationships with sensitivity to differences of culture and power. Many studies separate individual and group learning. However, Decuyper *et al.* (2010) maintained that their relationship is dialectical, while Volet *et al.* (2009) cautioned against reducing analysis to either level. Fam *et al.*'s (2017) findings from interviews with transdisciplinary leaders provide a closing gloss on qualities conducive to collaboration.

Fam *et al.*'s (2017) informants were involved in practice-based research and theoretical development of transdisciplinarity in Australia, the USA, Europe and South America. They also worked in both cross-disciplinary and cross-sector contexts. The six categories of qualities that emerged from the interviews span curiosity, creativity, commitment, critical awareness, communication, and connectedness. Hence, they include many qualities needed for collaboration cited in this chapter. The authors acknowledged that some can be taught, such as communication skills and pattern recognition. Others, however, might require experiential learning. Some individuals may already possess qualities that heighten willingness to engage in collaboration. However, as we have seen, levels of readiness still vary. The capacity needed for successful collaboration, then, is not a narrow set of skills, although they are needed. It also includes attitude, temperament, orientation, ability, dispositions, and predispositions. Commitment, for example, requires investment in building an effective team, valuing related skills, and being patient and willing to engage in negotiation. Critical awareness requires humility and empathy while suspending a priori judgements. Furthermore, communication practices, as we have also seen, are the responsibility of all participants, not just leaders. Moreover, connectedness depends on both readiness to contribute and thinking collectively across components of a system. On a final note, many of Fam *et al.*'s (2017) informants felt that resources exist, including online materials by communities dedicated to advancing collaboration across boundaries. However, they are often unfamiliar to individuals, teams and leaders. This chapter has aimed to close the gap between the abundance of resources that are available and the need for them on the ground of research practice. Otherwise, shortfalls in collaboration and integration will occur in cross-disciplinary and cross-sector contexts.

References

Andersen, H. and Wagenknecht, S. (2013) Epistemic dependence in interdisciplinary groups. *Synthese* 190, 1881–1898.

Arnstein, S.R. (1969) A ladder of citizen participation. *Journal of the American Planning Association* 35, 216–224.

Bennett, L.M., Gadlin, H. and Marchand, C. (2018) *Collaboration and Team Science: A Field Guide*, 2nd edn. NIH publication no. 18-7660. National Cancer Institute, National Institutes of Health, Bethesda, Maryland.

Berente, N. and Howison, J. (2019) Strategies for success in virtual collaboration: structures and norms for meetings, workflow, and technological platforms. In: Hall, K.L., Vogel, A.L. and Croyle, R.T. (eds) *Strategies for Team Science Success: Handbook of Evidence-based Principles for Cross-disciplinary Science and Practical Lessons Learned from Health Researchers*. Springer, Cham, Switzerland, pp. 563–574.

Bosque-Pérez, N., Klos, P.Z., Force, J.E., Waits, L.P. and Cleary, K. (2016) A pedagogical model for team-based, problem-focused interdisciplinary doctoral education. *BioScience* 66, 477–488.

Brister, E. (2016) Disciplinary capture and epistemological obstacles to interdisciplinary research: lessons from Central African conservation disputes. *Studies in History and Philosophy of Biological and Biomedical Sciences* 56, 82–91. DOI: 10.1016/j.shpsc.2015.11.001.

Carpenter, R. (2009) Boundary negotiations: electronic environments as interface. *Computers and Composition* 26, 138–148.

Collins, H. and Evans, R. (2007) *Rethinking Expertise.* University of Chicago Press, Chicago, Illinois.

Crowley, S., Eigenbrode, S., O'Rourke, M. and Wulfhorst, J.D. (2014) Introduction. In: O'Rourke, M., Crowley, S., Eigenbrode, S. and Wulfhorst, J.D. (eds) *Enhancing Communication and Collaboration in Interdisciplinary Research.* SAGE Publications, Thousand Oaks, California, pp. 1–10.

Decuyper, S., Dochy, F. and van den Bossche, P. (2010) Grasping the dynamic complexity of team learning: an integrative model for effective team learning in organizations. *Educational Research Review* 5, 111–133.

Fam, D., Smith, T. and Cordell, D. (2017) Being a transdisciplinary researcher: skills and dispositions fostering competence in transdisciplinary research and practice. In: Fam, D., Palmer, J., Riedy, C. and Mitchell, C. (eds) *Transdisciplinary Research and Practice for Sustainability Outcomes.* Routledge, London and New York, pp. 62–77.

Fiore, S.M., Gabelica, C., Wiltshire, T.J. and Sokols, D. (2019) Training to be a (team) scientist. In: Hall, K.L., Vogel, A. and Croyle, R. (eds) *Strategies for Team Science Success: Handbook of Evidence-based Principles for Cross-disciplinary Science and Practical Lessons Learned from Health Researchers.* Springer, Cham, Switzerland, pp. 421–444.

Freeth, R., Clarke, E. and Fam, D. (2019) Engaging creatively with tension in collaborative research: harnessing the "I" and "we" through dialogue. In: Brown, V., Harris, J. and Waltner-Toews, D. (eds) *Independent Thinking in an Uncertain World: A Mind of One's Own.* Routledge, London, pp. 262–283.

Frey, G. (1973) Methodological problems of interdisciplinary discussions. *RATIO* 1, 161–182.

Galison, P. (1997) *Image and Logic: A Material Culture of Microphysics.* University of Chicago Press, Chicago, Illinois.

Gray, B. (2008) Enhancing transdisciplinary research through collaborative leadership. *American Journal of Preventive Medicine,* 35(2 Suppl), S124–S132.

Gunawardena, S., Weber, R. and Agosto, D. (2010) Finding that special someone: interdisciplinary collaboration in an academic context. *Journal of Education for Library and Information Science* 51, 4210–4221.

Hall, K.L., Vogel, A.L. and Crowston, K. (2019) Comprehensive collaboration plans: practical considerations spanning across individual collaborators to institutional supports. In: Hall, K.L., Vogel, A.L. and Croyle, R.T. (eds) *Strategies for Team Science Success: Handbook of Evidence-based Principles for Cross-disciplinary Science and Practical Lessons Learned from Health Researchers.* Springer, Cham, Switzerland, pp. 587–611.

Hall, T.E. and O'Rourke, M. (2014) Responding to communication challenges in transdisciplinary sustainability science. In: Huutoniemi, K. and Tapio, P. (eds) *Heuristics for Transdisciplinary Sustainability Studies: Solution-oriented Approaches to Complex Problems.* Routledge, London and New York, pp. 119–139.

Hardwig, J. (1985) Epistemic dependence. *Journal of Philosophy* 82, 335–349.

Holbrook, J.B. (2012) What is interdisciplinary communication? Reflections on the very idea of disciplinary integration. *Synthese* 190, 1865–1879.

Jahn, T., Bergmann, M. and Keil, F. (2012) Transdisciplinarity: between mainstreaming and marginalization. *Ecological Economics* 79, 1–10.

Keyton, J. (1999) Relational communication in groups. In: Frey, L., Gouran, D. and Poole, M.S. (eds) *The Handbook of Group Communication Theory and Research*. SAGE Publications, Thousand Oaks, California, pp. 192–222.

Klein, J.T. (2013) Communication and collaboration in interdisciplinary research. In: O'Rourke, M., Crowley, S., Eigenbrode, S. and Wulfhorst, J.D. (eds) *Enhancing Communication and Collaboration in Interdisciplinary Research*. SAGE Publications, Thousand Oaks, California, pp. 11–30.

Klein, J.T. (2021) *Beyond Interdisciplinarity: Boundary Work, Communication, and Collaboration*. Oxford University Press, Oxford, UK, and New York.

Klein, J.T. and Newell, W.H. (1997) Advancing interdisciplinary studies. In: Gaff, J.G. and Ratcliff, J.L. (eds) *Handbook of the Undergraduate Curriculum: A Comprehensive Guide to Purposes, Structures, Practices, and Change*. Jossey-Bass and Association of American Colleges and Universities, San Francisco, California, pp. 393–415.

Klein, J.T., Misra, S. and Lotrecchiano, G. (2020) Introduction: communication in inter- and trans-disciplinary teams. In: Lotrecchiano, G. and Misra, S. (eds) *Communication in Transdisciplinary Teams*. Informing Science Press, Santa Rosa, California, pp. 2–17.

Laursen, B.K. (2018) What is collaborative interdisciplinary reasoning: the heart of interdisciplinary team research. In: Lotrecchiano, G. and Misra, S. (eds) *Communication in Transdisciplinary Teams*. Informing Science Press, Santa Rosa, California, pp. 55–80.

Laursen, B.K. and O'Rourke, M. (2019) Thinking with Klein about integration. *Issues in Interdisciplinary Studies* 37, 33–61.

Ledford, H. (2008) Collaborations: with all good intentions. *Nature* 452(7188), 682–684. DOI: 10.1038/452682a.

Lund, V., Coleman, G., Gunnarssond, S., Appleby, M.C. and Karkinen, K. (2006) Animal welfare science – working at the interface between the natural and social sciences. *Applied Animal Behavior* 97, 37–49.

Mäkinen, E. (2018) Complexity leadership theory and the leaders of transdisciplinary science. *Informing Science: The International Journal of an Emerging Transdiscipline* 21, 133–155.

Mäkinen, E. (2020) A new paradigm for research organization: academic leadership in transdisciplinary science teams. In: Lotrecchiano, G.R. and Misra, S. (eds) *Communication in Transdisciplinary Teams*. Informing Science Press, Santa Rosa, California, pp. 123–147.

Marchant-Forde, J.N. (2015) The science of animal behavior and welfare: challenges, opportunities, and global perspective. *Frontiers in Veterinary Science* 2, 16. DOI: 10.3389/fvets.2015.00016.

Misra, S., Stokols, D. and Cheng, L. (2015) The transdisciplinary orientation scale: factor structure and relation to the integrative quality and scope of scientific publications. *Journal of Translational Medical Epidemiology* 3, 1042.

Nash, J.M. (2008) Transdisciplinary training: key components and prerequisites for success. *American Journal of Preventive Medicine* 35(2 Suppl), S133–S140. DOI: 10.1016/j.amepre.2008.05.004.

NASEM (National Academies of Science, Engineering, and Medicine) (2005) *Facilitating Interdisciplinary Research*. National Academies Press, Washington, DC.

NASEM (National Academies of Science, Engineering, and Medicine) (2015) *Enhancing the Effectiveness of Team Science.* National Academies Press, Washington, DC.

Nurius, P.S. and Kemp, S.P. (2019) Individual-level competencies for team collaboration with cross-disciplinary researchers and stakeholders. In: Hall, K., Vogel, A. and Croyle, R.T. (eds) *Strategies for Team Science Success: Handbook of Evidence-based Principles for Cross-disciplinary Science and Practice Lessons Learned from Health Researchers.* Springer, Cham, Switzerland, pp. 171–187.

Olson, G.M. and Olson, J.S. (2014) *Working Together Apart.* Morgan Claypool, San Rafael, California.

O'Rourke, M., Crowley, S. and Gonnerman, C. (2016) On the nature of cross-disciplinary integration: a philosophical framework. *Studies in History and Philosophy of Biological and Biomedical Sciences* 56, 62–70. DOI: 10.1016/j.shpsc.2015.10.003.

Piso, Z., O'Rourke, M. and Weathers, K.C. (2016) Out of the fog: catalyzing integrative capacity in interdisciplinary research. *Studies in History and Philosophy of Science* 56, 84–94. DOI: 10.1016/j.shpsa.2016.01.002.

Pohl, C., van Kerkhoff, L., Hadorn, G.H. and Bammer, G. (2008) Integration. In: Hirsch Hadorn, G., Hoffmann-Riem, H., Biber-Klemm, S., Grossenbacher-Mansuy, W., Joye, D. *et al.* (eds) *Handbook of Transdisciplinary Research.* Springer, Cham, Switzerland, and New York, pp. 411–424.

Repko, A. (2008) *Interdisciplinary Research: Process and Theory.* SAGE Publications, Thousand Oaks, California.

Rioja-Lang, F.C., Connor, M., Bacon, H. and Dwyer, C.M. (2020) Determining a welfare prioritization for horses using a Delphi method. *Animals* 10(4), 647. DOI: 10.3390/ani10040647.

Rossini, F.A. and Porter, A.L. (1979) Frameworks for integrating disciplinary research. *Research Policy* 8, 70–79.

Salazar, M.R., Lant, T.K., Fiore, S.M. and Salas, E. (2012) Integrative capacity facilitating innovation in diverse science teams through integrative capacity. *Small Group Research* 43, 527–558.

Salazar, M.R., Widmer, K., Doiron, K. and Lant, T. (2019) Leader integrative capabilities: a catalyst for effective interdisciplinary teams. In: Hall, K., Vogel, A. and Croyle, R.T. (eds) *Strategies for Team Science Success: Handbook of Evidence-based Principles for Cross-disciplinary Science and Practice Lessons Learned from Health Researchers.* Springer, Cham, Switzerland, pp. 313–328.

Schmithals, J. and Berkenhagen, J. (2004) Beziehungen und motivation. In: Schophaus, M., Schön, S. and Dienel, H.-L. (eds) *Transdisziplinäres Kooperaationsmanagement.* Ökom Verlag, Munich, Germany, pp. 69–102.

Stone, A.R. (1969) The interdisciplinary research team. *The Journal of Applied Behavioral Science* 5, 351–365.

Thompson, J.L. (2009) Building collective communication competence in interdisciplinary research teams. *Journal of Applied Communication Research* 37, 278–297.

Tuckman, B.W. (1965) Developmental sequence in small groups. *Psychological Bulletin* 63, 384–399. DOI: 10.1037/h0022100.

Tuckman, B.W. and Jensen, M.A. (1977) Stages of small-group development revisited. *Group and Organization Studies* 2, 419–427.

Volet, S., Vaurus, M. and Salonen, P. (2009) Self- and social regulation in learning contexts: an integrative perspective. *Educational Psychologist* 44, 215–226.

Vosskamp, W. (1994) Crossing of boundaries: interdisciplinarity as an opportunity for universities in the 1980s? *Issues in Integrative Studies* 12, 43–54.

Yanow, D. (2004) Translating local knowledge at organizational peripheries. *British Journal of Management* 15, S9–S25.

Zierhofer, W. and Burger, P. (2007) Disentangling transdisciplinarity: an analysis of knowledge integration in problem-oriented research. *Science Studies* 20, 51–74.

3

Doing Animal Welfare Transdisciplinary Research with Stakeholders: Advantages, Pitfalls and Hands-on Solutions

Tjard de Cock Buning* and Marianne Benard

Athena Institute, Vrije Universiteit Amsterdam, Amsterdam, The Netherlands

3.1 Introduction

This chapter describes the case of a successful transdisciplinary academic research project in animal welfare, resulting in four PhD graduations. It is written from the perspective of the transdisciplinary PhD student (M.B.) and the supervisor (T.C.B.). It contains an introduction to the context, sections on barriers in mutual learning, lessons learned, and common pitfalls and solutions for those who want to design transdisciplinary learning circles.

3.1.1 Historical context

A previous large Dutch national programme on animal welfare, called 'Limits to welfare and animal production', was successful in knowledge production (20 PhD dissertations each containing four or more international publications) but was criticized because animal welfare problems cannot be understood by more biological knowledge. What was missing was the relationship between biology, societal production forces and consumers' demands. Therefore, the Dutch Research Council (NWO) and the Ministry of Agriculture started a wide consultation, which was a unique combination and an unusual time investment around 2008 in The Netherlands. In order to define the new national research programme, called 'The value

*Corresponding author: tjard.de.cockbuning@vu.nl

 Bridging Research Disciplines to Advance Animal Welfare Science: A practical guide (ed. I. Camerlink)
DOI: 10.1079/9781789247893.0003

of animal welfare', as well as ethologists and genetic breeding companies, farmers and animal welfare organizations also joined in co-design work-groups. This programme aimed to organize inter- and transdisciplinary research teams to explore the routes to animal husbandry that is accepted by society at large while remaining economically vital. 'Value' is defined in the socio- and agroeconomic sense. Only five projects received grants, each having four students or postdocs from different disciplines, with the explicit mission to let the synergy work towards added value. It had to be a joint learning process. Our 'Sociable swine' project was one of the five, and with a €765,000 grant, we started in 2009 with three life-science students (quantitative genetics, behavioural biology and physiology) and a biology and society student educated in transdisciplinary methods (M.B.). All students and supervisors were trained as biologists, but were at the same time experts in diverse subdisciplines. M.B. was trained as a biomedical scientist but specialized in a 2-year gamma-beta Masters in Biology and Society (Vrije Universiteit Amsterdam). Her supervisor (T.C.B.) is a biophysicist and philosopher who has specialized in transdisciplinary action research to bridge science-based conflicts with society (Athena Institute, Vrije Universiteit Amsterdam). The four students were part- or full-time and were located in the Animal Sciences building at Wageningen University.

3.1.2 Science–society tension in the 'Sociable swine' project

Without going in detail into the research challenges of the 'Sociable swine' project, some explanation is necessary to understand the potential science–society conflict. The project was based on the theory of indirect genetic effects where individuals in a group affect each other's production performance. The genetic selection method was further developed at Wageningen University (Bijma *et al.*, 2007), following on from studies on group selection (Griffing, 1967; Muir, 2005). The genetics department collaborated closely with a pig-breeding company to investigate the hypothesis that pigs are able to reduce each other's growth performance through an indirect genetic effect on each other, presumably through behaviour such as aggression and tail biting. The new selection method for social behaviour might offer opportunities to remove the standing practice of mutilations such as tail docking, teeth clipping and castration, which has been heavily criticized by animal protection and nature conservation organizations. Their solution is to transition to more extensive husbandry systems instead. This logical conclusion was echoed by the urban citizens in The Netherlands.

At the start of the project, it was unclear how the public would respond to the 'creation' of a sociable swine. Would they regard it as a better alternative than tail docking or as an unnatural modification to adapt animals to the bare cages of industrial farming? Therefore, the genetics student investigated the genetic potential of the sociable swine, the behavioural student focused on the behavioural patterns of several traits in relation to barren and enriched pens, and the physiologist looked for subtle changes in the pigs' stress-coping system. We, as the transdisciplinary experts, faced the challenge of engaging our colleagues in a learning circle (interdisciplinary), while at the same time bringing us all together in a learning circle with the societal stakeholders, such as farmers, citizens, animal protection organizations, industries and policy makers (transdisciplinary) (Benard and de Cock-Buning, 2014).

3.1.3 Support by the national programme committee

The national programme committee of 'The value of animals' was highly committed to integration of the life sciences and social sciences, and was well aware that interdisciplinary teamwork and transdisciplinary joint research do not spontaneously emerge among engaged young students and postdocs who have been educated in traditional disciplinary institutes. Therefore, all 20 of the students/postdocs in the five projects jointly followed five intensive training courses of 2 days each, on the topics of sustainability, product marketing, qualitative behavioural assessment, agroethics and political sciences. The aim of each crash course was for the participants to learn from each other (situated in a remote hotel) and to provide insight into the leading models and concepts used in the various projects and social sciences. The agenda of the courses included, among others, short presentations, group work, excursions, stakeholder guest speakers and discussions.

The programme committee was composed of high-level policy makers from the Ministry of Agriculture and the NWO, the Dutch Society for the Protection of Animals (Nederlandse Dierenbescherming) and the Governmental Directorate for Agricultural Research. The programme committee committed themselves to monitoring the five projects closely, through annual site visits, and by acting as observers and advisers in 'speed-dating' sessions in the courses.

3.2 Setting the Stage for Inter- and Transdisciplinary Learning

In our research proposal, we used an image symbolizing our approach: 'all relevant stakeholders at the research table'. By this, we meant that, in this project, all our joint progress meetings would be held together with each other and the relevant stakeholders, i.e. the societal actors. We would not be talking *about* farmers or citizens, but always *with them* about choices in research questions. In practice, this meant that, in addition to the supervisors and PhD students, the relevant experts from the Dutch Society for the Protection of Animals, a meat company and the Farmers' Union (LTO)

Phases	Purpose	Methods and skills
Background research	Identifying relevant actors	Social sciences methods: actor–network analysis, qualitative data handling, grey literature analysis, policy documents, interviews, surveys, focus groups, group interviews
Transdisciplinary group learning	Facilitating learning circle and frame reflection	Tailor-made: design focus groups, design dialogue meetings
Internalization and supporting further implementation	Joint planning to anchor institutional and governmental support	System analysis Tailor-made: focus groups, dialogue meetings, round table, interactive root cause analysis

Reflective monitoring in action

Eye-opening designs

Dynamic research agenda

Fig. 3.1. Phases and skills needed for a transdisciplinary (co)researcher.

were invited. However, we realized that none of the participants in this transdisciplinary project, except the authors, had first-hand experience in interdisciplinary or transdisciplinary research. The life-science students in particular were expected to dive into the latest literature of their subproject, plan experiments, analyse them, write as many publications as possible and at the same time also coordinate with their life-science colleagues. This is a typical multidisciplinary practice.

Although the supervisors reached a shared understanding and were motivated to experiment with transdisciplinary methods, the invited stakeholders had a more passive attitude. We anticipated that everyone would be insecure about their roles, the efforts needed and the risk of failure. In ethological terms, unproductive fight/flight behaviour would be boiling under the surface. This implied that our transdisciplinary learning activities had to be in a safe and relaxed setting through inviting them in conceptual and methodological questions without addressing these personally. M.B. took on the role of explorer in the non-academic society. Figure 3.1 outlines the methods and skills of the transdisciplinary expert role. As a team, but not individually, the members were confronted with this information and had to decide jointly what this meant for the project and their own subprojects.

3.2.1 Our approach

A challenge was how to get everybody on board with team learning when each student and expert had their own targets to reach. As facilitators of transdisciplinary research, our job was to build the interface with the stakeholders and society at large. We did not have to plan 6 months ahead or wait for 6 months in the stables to observe the animals. Our data collection could start any moment with open in-depth interviews, semi-structured interviews, theory-laden structured interviews, and tailor-made, designed, interactive meetings to explore real or imagined conflicts between the scientists and other stakeholders. Our flexibility enabled us to work from an emergent design. Our cycle of problem definition, data collection, analysis, feedback and action depended on the intervals between the joint transdisciplinary dialogues at the research table.

Figure 3.1 shows the three successive phases: (i) background research (in society); (ii) designing and executing joint learning sessions with all four students, supervisors and stakeholders; and (iii) support follow-up meetings. Presentations from 'society background investigations' were presented to the joint research table to discuss what these findings meant for the research questions of the other student projects. This activity was meant to open the eyes of the discipline-focused student to the wider context. An example of such an 'eye-opener session' is discussed in section 3.3.1 below. Such meetings ended with a decision on further steps. The introduction of a 'dynamic research agenda' was a jointly agreed action list. This might be steps taken by the life-science students to present their adapted research questions next time, or stakeholders to interview, or some literature search that might be initiated to look for publications of societal-driven research questions in animal behaviour. When a transition to innovative farming is aimed for, the transdisciplinary researcher will have to manage long-term networks to find a niche and support from government, industry and academia. The related tools have been well described in *Reflexive Monitoring in Action* (van Mierlo *et al.*, 2010). During the 4 years, we organized five successive cycles. We will briefly summarize the intentions of each cycle below, along with the lessons learned for us and for the project team.

3.2.2 Focus groups and dialogue tools

There are many definitions and related applications of 'focus groups' (Kitzinger, 1994). We followed the recommendation of Rowe and Frewer (2004) in their critical article: instead of lumping all kinds of methods under 'focus group', one should describe in detail the design and analysis steps of the focus group collection of data. In our case, a focus group was an interactive group learning activity between unrelated people but with a similar background (same discipline, sector, ideology). Usually, there were five to 12 participants and it took 4–6 hours. The facilitator would design a tailor-made sequence of analytical steps (script) to explore the shared or not-shared perspective of an issue. In Fig. 3.2(a), the approach for the focus groups is visualized in seven steps with a round table and five random

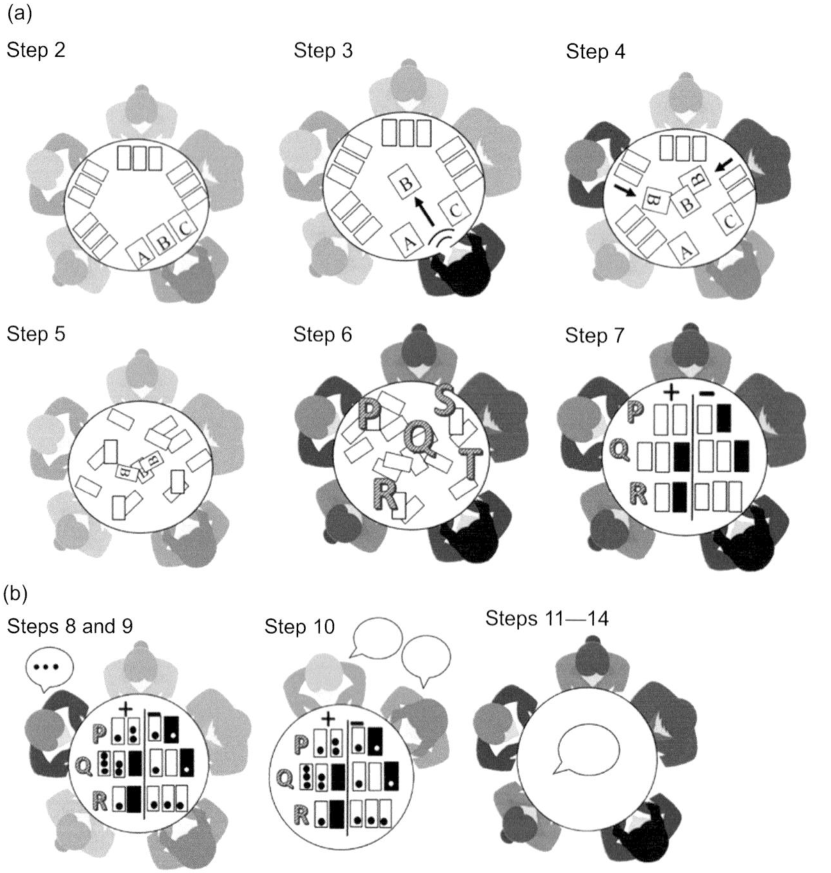

Fig. 3.2. Stepwise focus group meeting. Steps 2–14 describe a generalized script during a focus group. © T. de Cock Buning.

unrelated citizens who were invited 'to talk about new technologies and what you might think about them'.

- **Step 1:** At the start of the meeting, a short presentation of 15 minutes was given about genetic selection in animal breeding and the profits for the farmers.
- **Step 2:** Everyone received three Post-it notes and a marker (large tip) with which they were asked to write only one keyword per Post-it note that reflected their perspective about this technology.
- **Step 3:** The facilitator asked a participant to place one Post-it note in the centre of the table and to explain the association. This was an exercise to formulate and rationalize neutral/positive/negative associations with

the other participants. Based on the explanation given, the facilitator might add other relevant keywords to the note in a different colour.

- **Step 4:** If another participant had the same association, they placed their Post-it note next to the first one.
- **Step 5:** After 30–45 minutes, a lot of information had been exchanged and the table looked as depicted in Fig. 3.2.
- **Step 6:** In the next reflection exercise, all participants jointly regrouped the Post-it notes in clusters and labelled them. This was an exercise in naming the shared characteristic of each cluster and brought the participants to the conceptual level of their discussion. The facilitator observed the discussion between the participants and intervened only with 'why' questions if a label appeared not to be self-explanatory. This took approximately 20 minutes. A voice recorder was run for detailed analysis afterwards (e.g. transcription of what was said), and a picture was taken at the end.
- **Step 7:** Next, participants reflected on the internal variation of the conceptual clusters by splitting the table into two fields. The four Post-it notes in cluster P were divided into positive and negative associations. During this discussion, the participants spontaneously formulated other related associations, which they wrote on Post-it notes with a different colour. This was repeated for the remaining clusters. The positive/negative associations could also be expensive/cheap or sustainable/not sustainable, etc., depending on the topic of the focus group. This step was finalized by a joint reflection. Here, the first participant was invited to summarize the 'two faces' of the P cluster. The moderator asked 'why' questions, and invited comments from the others (these were also added as Post-it notes to the cluster). Then, the second participant summarized the Q cluster, etc. This took approximately 45 minutes. The discussion was recorded and a picture of the Post-it notes was taken at the end.
- **Step 8:** Every participant received three dots to stick on the three Post-it notes that they regarded as most relevant. This made clear which aspects cluster (P, Q or R) was most relevant, but also whether this was because of negative aspects or positive aspects. The focus group could continue in a follow-up session with steps 9–14 (Fig. 3.2b).
- **Step 9:** After prioritizing, the group reflected on the results. Why does Q have high results? Why does R have low results? Where is most trouble (negative results) in R?
- **Step 10:** The group was broken up into couples to think up solutions to mitigate the negative aspects of P, Q and R. They reported their discussion back to the group.
- **Step 11:** Continuation of the plenary session. Regarding these solutions, who obstructs the positive aspects of P and Q, and why?
- **Step 12:** Who will obstruct the solutions? Which interests are at stake? Which values?

- **Step 13:** Which actors have shared values?
- **Step 14:** Where are potential constructive coalitions?

'Dialogue sessions' differ from focus groups as the participants are known to have different, or even opposite, interests. Conflict reduction and stimulation of joint reflection with emotional distance is the challenge for an effective design, i.e. lifting the discussion to the abstract system level (second-order learning, which is about how to solve a system problem related to a project; Grin and van de Graaf, 1996). Our dialogue meetings were based on the 'power-free' dialogue rules of Habermas (1997). For this, we designed a setting in which opponents with an open mind and subscribing a respectful and curious attitude to understand the joint problem are invited. In contrast to the focus group, the participants in the dialogue meetings were carefully selected. The invitation called the *wicked problem* (a complex problem that seems to be impossible to solve by the usual monodisciplinary solutions; see Chapter 1, this volume) by its name. The designs were based on the assumption that: (i) the wicked problem was a shared problem; (ii) there were deeper shared causes and relationships to this problem; and (iii) there were potential solutions and expected barriers. In order to avoid the habit of stakeholders pointing to the other as the cause and/or solution, such exercises/workshops should reduce dominance (i.e. be well balanced based on careful selection). Joint learning from each other's interpretations is the main focus. In order to avoid the pitfall of 'jumping to the solution', typical stakeholder statements about solutions, barriers and actions are postponed to the end of the session, preferably in a follow-up session. Box 3.1 outlines some useful instruments to enable this.

Box 3.1. Useful instruments to realize transdisciplinary problem definitions, solutions and actions.

- 'Write & Shift' exercise based on *root cause analysis* to mix framings for out-of-the-box creative joint problem analysis and options for solutions. This can be conducted over two sessions.
- *White papers* to redefine or translate discipline specific jargon in one shared common terminology (shared working model). This can take months to complete.
- Investigate the problem jointly in the local practice and context (e.g. farmers or patients reflect on why the observations are different between experts and professionals).

3.3 Barriers in Mutual Learning, and How We Dealt With Them

In order to function successfully as a transdisciplinary team, three interdependent barriers needed to be resolved: two were related to interdisciplinary collaboration and one to transdisciplinary research specifically:

1. The first barrier was the ability to relate to two different research fields, i.e. social sciences and life sciences, with their different approaches, methods, models, framings and valuation of types of knowledge.
2. The second barrier was to find possible shared activities and to determine the roles within the team of the differently trained researchers.
3. The third barrier was to acknowledge the relevance of experiential knowledge of non-academics and to incorporate this into the research design of the animal science projects.

During the project, four successive cycles emerged that built up over time to conquer these three transdisciplinary barriers.

3.3.1 Cycle 1: The ability to relate to two different research fields (months 1–13)

The motivation of learning from the outside world and from each other's disciplines should be part of the benefits brought to the team and their own project. As there are hardly any benefits at the start, the motivation is then usually quite low, as it was in our case. M.B. designed a solution for this barrier. To illustrate this point, we describe the transdisciplinary learning activity of the third transdisciplinary dialogues at the research table.

We changed the format of the usual progress meeting (which was five PowerPoint presentations and an open discussion) into one suitable for joint learning. In the first months, during which the students were reviewing the literature and arranging their animals for the not yet fully defined experiments, M.B. collected data from the wider circle of stakeholders, her colleagues and all supervisors. She used in-depth and semi-structured interviews to get a picture of the background and the perspectives of everyone in and around the project. She discovered that some of the central concepts were interpreted quite differently, e.g. social skills (of the sociable swine), sustainable husbandry, productive practice and unwanted behaviour. Her 'progress' presentation for the 'Dialogue at the research table' meeting was about these differences, and she took the lead in a

Table 3.1. Example of an eye-opening exercise on different interpretations of one of the key concepts of the project.

	Associations and interpretations about the central concept *Social skills*				
	'Dealing with conflicts'	'Buzz word'	'Amount of social interactions'	'Showing affection'	'Don't know'
Supervisor 1	✓				
Supervisor 2	✓				
Supervisor 3				✓	
Student 1	✓				
Student 2			✓		
Student 3			✓		
Industry	✓				
Farmer					✓
Animal welfare organization			✓		

round-table discussion. M.B. used a quiz exercise by showing a PowerPoint slide depicting a central concept and some associations and interpretations, selected as representative for various stakeholders (including the team members). Everyone (including the non-academic members of the meeting) decided on their own which of the five options corresponded with their own definition. Table 3.1 shows the voting results. The diversity among the supervisors and students and stakeholders was unexpected. The discussion that followed formed the start of the learning exercise to build an interdisciplinary shared model and a transdisciplinary synchronized process that was understood and valued.

In interdisciplinary engineering projects, this is worked out in so-called *white papers*, defining the concepts and the precise translation of disciplinary jargons into joint interdisciplinary jargon for efficient communication. The pitfall of this process is that it takes more time than five short presentations. In general, this time aspect is one of the main barriers to success in transdisciplinary research. This led to a negotiation about the desired frequency of these meetings, which became two to three meetings a year instead of one every 2 months.

3.3.2 Cycle 2: Possible shared activities and shared challenges (months 16–24)

Shared activities intend to result in new insights but are not always a holiday outside one's comfort zone. This was the case in most of the organized shared activities, such as the joint intensive courses and writing joint scientific or lay articles.

This was also reflected in the joint symposium organization. Tail biting in pigs is a behaviour with a multi-factorial origin, about which current scientific research states that the absence of suitable enrichment materials for rooting and chewing is a major risk factor (reviewed by Nannoni *et al.*, 2014). Around halfway through the project, the life-science students had analysed their different datasets and aimed to exchange their acquired knowledge on the causes of tail biting and the relevance of a sociable swine with farmers in an interactive science–society symposium designed together with M.B. entitled 'Peaceful pigs'. It embodied a double goal: interdisciplinary learning and transdisciplinary learning with non-academics. For details, see Benard *et al.* (2014).

During the symposium, farmers and researchers disagreed on the causes of tail biting, and some farmers appeared only moderately open to the presented scientific insights. The initial assumption by the researchers that the scientific insights speak for themselves did not work out that way. Another barrier to reaching the farmers was the relatively low degree of applicability of the scientific insights. They were frequently not concrete enough (e.g. no cost–benefit data), too uncertain (e.g. no guarantee that environmental enrichment will certainly stop tail biting) or not related to the complex real world of the farmers (e.g. focusing on animal welfare as an isolated aspect, and not taking into account the effects of the environment, the farm context and costs).

Another barrier was that some of the involved farmers and researchers had different framings of animal welfare and the human–animal relationship, leading to polarization during the symposium. The farmers emphasized the biological functioning of the animal when defining good welfare. In contrast, the researchers emphasized the natural behaviour pattern of animals. They framed animals primarily as sentient beings, acknowledging pigs' ability to experience emotions such as pleasure. Some of the farmers resisted the use of such anthropomorphic projections on pigs and framed pigs primarily as production animals.

Obviously, the frames of the researchers and these farmers did not have enough in common to start a dialogue. This was not different from the value frames and conflicts that were reported by M.B. on the basis of her investigation among stakeholders in several dialogue meetings at the research table. This message appeared to be true as the confrontation did not leave any room for another interpretation. Although not all of the participants were that critical, the straightforward denial of the presented cause of tail biting was a shock for all researchers present. We handled it by making a follow-up arrangement immediately with those critical farmers to turn the conflict into a learning process, aiming for mutual understanding.

3.3.3 Cycle 3A: The relevance of experiential knowledge (months 28–48)

We changed the group approach and organized two-to-one dialogues between two scientists and one farmer with M.B. as the mediator, asking the following questions:

1. What is the most important reason for pigs to start tail biting?
2. What is a potential solution for tail biting?
3. What is good animal welfare (definition)?
4. How do you assess whether the welfare is good?
5. What is natural behaviour (definition)?

It appeared that some scientific relationships between tail biting and farm practices were not recognized by farmers in their husbandry, and, vice versa, researchers did not agree with some of the causalities mentioned by the farmers. The farmers and researchers had their own observations, resulting in different convictions and beliefs, and thereby strengthening and even 'validating' their own framing. Different observations were the result of different framings and practices, which is understood here as the way of handling and understanding the farm versus laboratory environment, or the 'usual practices' coherent with two professionalisms. The practice appeared to function as an a priori filter, influencing what was actually observed in their environment, and inducing 'blind spots' (Table 3.2). For example, the farmers noticed abnormalities when scanning their pigs, e.g. tail biting when the weather changed. The researchers instead modelled and moulded reality by eliminating 'disturbing' factors, e.g. fluctuations in light, temperature and humidity. These results imply that observational differences may lead to confusion when discussing welfare and are a serious barrier when both parties stay at the first-order learning ('my observations/methods are better than yours'). Mutual learning at a second-order level can be reached by accepting, from a system perspective, that both truisms belong to the same pig.

3.3.4 Cycle 3B: Taking the vocabulary of citizens and farmers seriously using qualitative behaviour assessment (months 18–30)

Observed differences among stakeholder groups were studied jointly in more depth. The participating stakeholders were pig farmers, animal scientists and urban citizens. They were asked to observe the behaviour of a pig on nine videos and to score 21 given terms on moods for each one, such as 'happy' or 'irritated', i.e. the qualitative behaviour assessment (QBA) approach (Rutherford *et al.*, 2012). This method was introduced

Table 3.2. Examples of differences in observation between farmers and scientists.

	Researcher	Farmer
Observation		
Indicator	Primarily behaviour/physiology	Primarily production numbers/health status
Timespan	Several hours per day	One or two short moments per day
Focus	Both individual animals and groups	The group as a whole
Number of animals	Relatively small	Relatively large
Animal		
Breed	One breed	Different breeds
Interventions	Intact tails	Docked tails
Husbandry		
Space	Animals/m^2	Animals/0.8 m^2
Enrichment material	Pens enriched with straw and barren pens	Barren pens
Contact with human	Frequent	Minimal
Environment	Controlled for light, temperature and humidity	Not controlled

in one of the central courses organized by the programme committee (month 18) as an alternative systematic, qualitative, common language scoring system alongside quantitative measures in behavioural scoring systems used by ethologists (weight and frequencies of countable variables). The results of the QBA analysis showed that the pig farmers systematically observed the behaviour of pigs more positively than the urban citizens and animal researchers (Duijvesteijn *et al.*, 2014). Urban citizens and animal researchers observed the behaviour of pigs similarly. Intra-observer analysis indicated that the differences were probably not due to interpretation of the terms. From the additional questionnaire, it appeared that the farmers framed pigs primarily as production animals (similar to the farmers who participated in the studies described by Benard *et al.*, 2014), while the citizens and researchers primarily considered them sentient beings. This study underlined in quantitative terms that differences in observation form a root cause that hampers the development of a shared understanding of pigs and their welfare, and that there is a need to address observed differences during multi-stakeholder learning processes. For details, see Duijvesteijn *et al.* (2014).

3.3.5 Cycle 4: Workshop on 'View on tail biting' (month 42)

Based on the striking differences in the way pig farmers and animal scientists observed pig behaviour, the second symposium (a workshop entitled 'View on tail biting') had a completely different design (Box 3.2). Hardly any presentations by the students were given, and key was the direct focus on the common problem – tail biting (avoiding any scientific or farming bias). Pig behaviour in video fragments on laptops was observed and discussed together in small groups. In groups where the participants were optimally balanced by different backgrounds and professions, their observations differed the most, resulting in a constructive in-depth discussion. This way of organizing a learning circle was generally appreciated by the participants.

3.4 Lessons Learned

Tip 1: Structural and recurrent embedding of learning interventions

The first conclusion is that structural and frequent embedding of learning interventions is of primary importance for realizing successful learning during a research programme lasting 4 years. Active sessions are needed: (i) to become aware of and reflect on one's own framing; (ii) to open up to other's knowledge and underlying framing; (iii) to synchronize expectations, language and terminologies; and (iv) to search for the shared activities needed to realize win–wins.

Tip 2: Organizing mutual learning at the level of the professional practice

During daily practices, learning tends to be limited to a first-order level, the skills and capabilities within the profession, with underlying framings being taken for granted (Schön, 1983). To increase the chance of creating solutions that match the diversity of societal perspectives, learning should take place at a second-order level, by problematizing these frames of reference that steer daily activities and influence perspectives (Schön and Rein, 1995).

The second conclusion of our study was that the motivation of the life-science students for second-order learning might be realized better or faster by eye-opening exercises and by stepping jointly into the practice at stake, i.e. tail biting at the farm. Frequent mutual visits to each other's local environment to become aware of each other's practices, which is understood as the way of living, handling and understanding the environment, helps to make differences explicit and at the same time difficult to ignore and escape from. A successful example in this multi-stakeholder learning process was the use of video fragments showing pigs as a boundary

Box 3.2. Programme of the symposium 'Peaceful pigs' and masterclass 'View on tail biting'.

	Symposium 'Peaceful pigs'		Masterclass 'View on tail biting'
Opening (chairman)	Linking pig behaviour to future policy vision; introducing the research team		*A workshop in four active rounds on 'Do we observe the same?'*
Introduction	Relevance of pig behaviour for farmers and their perspectives on it	**Introduction**	Participants distributed over four groups; one per group presents the summary of the discussions
Talk 1	Consequences of pig behaviour and tail biting for productivity, welfare and health	**Round 1**	Short film shots of pig behaviour: tail biting shots were scored and discussed (15 minutes)
Talk 2	Influence of breeding on social behaviour and tail biting	**Round 2**	Another set of film shots were scored and discussed (15 minutes)
Talk 3	Influence of environment on pig behaviour and tail biting	**Round 3**	Another set of film shots were scored and photos of tails were discussed (15 minutes)
Talk 4	Influence of rearing conditions on social pig behaviour and tail biting	**Round 4**	Presentations workgroups: what was learned? Scoring rounds presented and discussed together
Workshop	1. Inventorying ideas for solutions 2. Narrowing down to undocked pigs 3. Prioritizing ideas based on impact 4. Inventorying bottlenecks that prevent implementation and filter questions	**Presentation**	Presentation on tools to prevent tail biting
Plenary	Presentations by the workgroups	**Video**	Farmer who was able to keep tails undocked
Round-up	Project leader sums up implications of outcomes of workshops for future research	**Panel discussion**	Panel with members of multi-stakeholder working group 'Curly tail' (Werkgroep Krulstaart)

object (i.e. as a physical object that connects two worlds of interpretation because it exists and functions in both worlds), and to discuss around the table how to assess and interpret their welfare.

Tip 3: Designing eye-opening sessions

Participants often had assumptions that they took for granted that supported their staying in their comfort zone where there is no need to undertake action (e.g. 'We have a shared perspective on animal welfare, and therefore we do not need to discuss it.' 'I think it is the responsibility of the other person and not mine.' 'I think it is the other person who has a wrong understanding and not me.')

The third conclusion is therefore that learning interventions are most successful when different perspectives are used to problematize the issue at stake and to create an eye-opening effect. It is necessary to show the 'unusual' in what is taken for granted or to show the wide disparity between what is expected and what the real situation turns out to be (a 'shock/confusion'). This was most effective when a real-life, inescapable setting was created (a direct experience at the farm or at the experimental barns). This did not necessarily mean a face-to-face meeting; other tools appeared to be sufficient too, when they represented the other perspective realistically and explicitly. Examples of tools that appeared to be successful were film fragments and documents based on quotes of other stakeholders. Often, due to the inherent explorative nature, the participants found it difficult to operationalize the derived insights of the learning interventions, e.g. because the issue at stake was too big or not concrete enough. Therefore, from a facilitator's perspective, it is key to discuss an issue that is manageable and can be translated into direct actions (the experience of 'self-efficacy'), or to plan a follow-up session.

3.5 Common Pitfalls in Design

De Cock Buning and others have described several pitfalls in the design and execution of interactive methods (van Mierlo *et al.*, 2010). These pitfalls are used to anticipate the interactive learning designs (e.g. focus groups, dialogue sessions, workshops, symposia) that were the tools in the 4-year, multi-stakeholder learning processes. Each meeting was composed of sets of tools to make the learning experience as rich as possible in the time available. What are the risks of these design pitfalls and how can we avoid them? These puzzles were the main part of the supervisor task by T.C.B. in support of M.B. We will illustrate this by six examples.

3.5.1 Pitfall 1: Dominance by status or expertise

If there is too large a difference between the participants' knowledge related to the topic of the intervention, or in intellectual and verbal competences between the participants, this might lead to dominance, which could negatively affect the atmosphere of equal exchange. This pitfall was expected to be influential in the learning process between farmers and urban citizens. Although urban citizens generally have no or little direct experience with animal husbandry, they are well able to deliberate on how to keep animals in animal production. In a dialogue session, however, farmers might label them automatically as uninformed and might use this as an excuse not to listen to them, as we indeed observed in a dialogue between researchers and farmers. To ensure a safe environment, it was decided to design a dialogue at a distance. Homogeneous focus groups were conducted, consisting of either urban citizens or farmers (Fig. 3.3). Although the indirect communication (e.g. the use of role play and film fragments showing the perspectives of the other party) might have drawbacks for the learning potential, the reflection and learning process of both groups could be observed in explicit argumentations and not hampered by group conflicts over the table (Benard and de Cock Buning, 2013).

	Design for urban citizens	**Design for pig farmers**
Introduction	Assessing familiarity with pigs and husbandry	Assessing familiarity with animal welfare and husbandry
Exercise 1	First associations regarding pig husbandry (writing on Post-it notes)	First associations regarding urban citizens (writing on Post-it notes)
Exercise 2	Sketch the living conditions of pigs and discuss them in terms of natural living, mental and physical welfare Presentation will be **filmed**	1. Sketch and discuss the living conditions of pigs from the perspective of the urban citizen 2. Reflect on **video fragments** of urban citizens presenting their drawing
Exercise 3	List farmers' motives	1. Forecast how urban citizens see farmers' motives 2. Reflect on video fragments of urban citizens about farmers' objectives
Exercise 4	1. Role-play exercise on the farmers' frame 2. Reflect on role play and define solutions for conflicting interests	1. Reflect on **video fragments** of urban citizens about their learning progress 2. Define solutions for closing the gap between urban citizens and farmers
Evaluation	Learning progress (will be **filmed**)	Learning progress

Fig. 3.3. Schedules of two successive eye-opener meetings with urban citizens and pig farmers, without being in the same room.

3.5.2 Pitfall 2: A hidden agenda

People with strategic interests are attracted to learning interventions, potentially resulting in convincing strategies and power play, instead of explorative deliberations. In the case of dialogue sessions with the researchers, which was a structural element of the multi-stakeholder learning process, this was prevented by inviting the same panel of four stakeholder representatives each time. They had committed themselves to the learning process at the start of the research programme. However, we avoided placing these four societal actors, the supervisors and the students in a political or strategic situation in which they felt it necessary to defend their sector or colleagues around the discussion table. Instead, challenging anonymous contrasts were used as input for these sessions to stimulate second-order (reflective) learning, resulting from other but related stakeholder interviews or focus groups. As part of the learning process, citizens were consulted using focus groups. They were invited by a recruitment company without mentioning the topic of the session. The invitation only mentioned that the session concerned a societal issue. In none of our focus groups did we encounter strategic behaviour.

3.5.3 Pitfall 3: Selection bias of participants

An invitation may not be motivating enough for potential participants to join. This pitfall was relevant to the first farmer symposium, which was organized around the main theme of the research programme, 'social pig behaviour'. It appeared to be difficult to motivate busy conventional farmers for an animal welfare-related theme. This might be due to the fact that the majority of conventional farmers tend to have an agro-industrial paradigm, which is associated with intensifying in order to reduce input costs (Horlings and Hinssen, 2014). In describing good animal welfare, conventional farmers tend to have a *biological functioning approach* (as described in Chapter 1, this volume), and emphasize the health, fertility and productivity of the animals (Te Velde *et al.*, 2002). Therefore, it was decided to link the theme of social pig behaviour at the second symposium to production efficiency, health and the issue of tail biting, which was a high priority on the agenda of the governmental and sector agenda at that time. Nevertheless, their attendance was still relatively low. This did not affect the liveliness of the discussion but made it difficult to scale-up the learning processes and disseminate our project results. As a second-best solution, branch media and farm consultants were additionally invited. Their input for the open deliberation worked constructively, and the potential polarity between researchers and farmers was absent.

3.5.4 Pitfall 4: The facilitator is skilful in terms of process but not knowledgeable in terms of content

This concerns a skilful facilitator of learning interventions who is not familiar with the content. This pitfall was not applicable because the life-science and transdisciplinary students formed one research team and learned from each other. The transdisciplinary student gained the knowledge needed from the life-science students, while the life-science students were trained in process skills in order to facilitate workshops for the second symposium (in the last research year). Farmers who participated in the multi-stakeholder process (e.g. in focus groups, at symposia and in interviews) often assumed that the transdisciplinary student (M.B.) was unfamiliar with the content because she was living in an urban city. However, they appreciated that she was interested in animal production. The open attitude of M.B. had an advantage over T.C.B., as a professor, in making it easier to facilitate such sessions. For M.B., it was legitimate to ask 'why' questions, which are very important to problematize the participants' easy rhetoric (e.g. 'this can't be true', 'every farmer knows') in order to derive deeper insights and validation of undirected input by the participants. It is important to realize that the self-introduction of the facilitator at the start of the focus group is crucial for the participants to be open, and thus for the success of the data collection. Often, it is better to introduce oneself by means of a personal anecdote related to the topic, avoiding any suggestion about expertise in the topic, as the participants should be central.

3.5.5 Pitfall 5: Absence of oversight

Too often the facilitator of learning interventions misses the moments of knowledge co-creation, formulates summaries that do not cover the essence of the participants or uses keywords that are not in line with their frame of reference. This pitfall was well anticipated in the chosen design of the learning process ensuring that needs, perspectives and underlying framings were inventoried prior to multi-stakeholder dialogue sessions. The transdisciplinary student was responsible for each preparatory step, through which she became familiar with the perspectives, language and underlying framings in question. This helped to anticipate value conflict or moments of knowledge co-creation during the facilitation of learning interventions. Just like all teachers and trainers, they need to know how to guide and praise the pupils when they achieve new insights. In our approach, this meant that every learning session was written out in detail in a script divided into 5-minute sections regarding the precise text of the explanation, the possible types of answers, the ordering of the answers (in a table, in mind maps or in causal trees (root cause analysis; for applications see van Mierlo *et al.*, 2010)), follow-up steps to a new exercise, how to prioritize barriers, how to come to solutions (or which are dead ends), etc.

3.5.6 Pitfall 6: Stakeholder representativeness

To deliberate with the researchers on future research directions, we invited a panel on a regular basis, consisting of a senior researcher of the breeding organization and three spokespersons at a national level, representing pig farmers, animal protection and the food industry (inner-circle stakeholders). They were not invited to defend the stakes they represent, as that increases the risk of strategic power play. Instead, during a carefully designed dialogue session, they were asked to act as an individual, with an open deliberative attitude, while bringing their knowledge of the practice of the stakeholder groups in which they are embedded. They were expected to have a broad understanding of animal production and to function outside the bias related to individual or institutional interests. In addition, a broad range of stakeholders was consulted by M.B., and these results were used as individual and institutional testimonial input from the sectors during these sessions (outer-circle stakeholders), thus reducing the drive of our inner-circle stakeholders to defend their stake personally.

How representative were the inner-circle stakeholders during these sessions? During the dialogues at the research table, the animal and food industry representatives appeared to show a scientific approach and to be very familiar with scientific insights, methodologies and language. Therefore, discussions on work progress and future research directions often tended to have a scientific methodological character. Nevertheless, they still represented the practice for which they were invited. The animal protection representative argued from an animal welfare perspective but was open to deliberating on the other perspectives too. This was in line with the vision of this organization that animal welfare can best be realized by searching for shared solutions instead of rebelling against other stakeholders. The food industry representative was critical of the experiments and the social pigs hypothesis, because he had high-quality requirements regarding the health status of the meat. This is in line with the general culture in the food sector, where health claims and the safety of the consumer product must be guaranteed at all times. The farmer representative did not have a scientific approach but was clearly able to bridge science and practice. She was aware of the knowledge needs of farmers and argued which strategies to improve welfare were feasible and which were not. The representatives had a full agenda and therefore were not always able to participate in the dialogue sessions. Because the input for the sessions was based on societal input derived from interviews and focus group sessions with diverse stakeholders (outer-circle stakeholders), the reflection and learning of the researchers were only impaired to a limited extent.

3.6 Managing a Transdisciplinary Project

3.6.1 Planning a high frequency of shared activities

As stated by Davis (1995) on interdisciplinary research, 'The greater the level of integration desired, the higher the level of collaboration required.' This is not about teambuilding but about scientific action to connect the scientific and non-scientific frames and research theories and models. For this reason, M.B. was also physically embedded in the research department of Wageningen University. She and the life-science students formed a team with no hierarchical difference between them. Based on the input provided by M.B. during learning interventions, transdisciplinary activities were decided upon and undertaken together (this typical interdisciplinary collaboration to reach out for data mining with stakeholders is defined here as *transdisciplinary team research*).

3.6.2 Managing the right moment

Halfway through the programme, conducive circumstances were provided to fully embrace transdisciplinary inclusion activities. The first long-term experiments were finished, which was beneficial in two ways. First, there were results ready to share with society. Second, new experiments had to be decided upon, creating room for reflection and societal input. The international mid-term evaluation of the programme functioned as a catalyst for undertaking transdisciplinary activities together, as it called for a critical reflection on the societal and other ambitions of the NWO research programme 'The value of animal welfare'. As a matter of fact, it took us 2 years to implement a fruitful interdisciplinary collaboration. Interventions in the first 2 years functioned primarily to:

- stimulate the development of a shared understanding of the research programme and how it relates to its wider context; and
- align expectations on the interdisciplinary collaboration.

For the first intervention, it was a challenge that the students were relatively inexperienced at the start with the research and unfamiliar with its context. This meant that the first year was primarily taken up with setting up their own project and developing an understanding of it. In addition, this made it complex to respond adequately to societal input and to each other. The five courses (each 18–29 hours) that were designed over the 4 years for the students, made a positive contribution to the development of a shared understanding of their projects. Each course related the research programme to its wider socio-economic context (Box 3.3).

For the second intervention, it became apparent that it is difficult to imagine in advance what transdisciplinary teamwork really involves, resulting in different expectations and confusion about the collaboration. As confirmed in the literature, there must be practical aspects of role clarification, task distribution and negotiation in teamwork and in integration (DuRussel

Box 3.3. Interventions that we used in order to align expectations and the development of a shared understanding of the research programme and its context.

Development of shared understanding

- Lectures from university staff and sector to broaden vision.
- Excursions to broaden vision (e.g. husbandry systems, production processes, organizations related to animal production).
- Interactive sessions with stakeholders to broaden visions or to practise with shared vision (e.g. dialogue sessions or speed-date sessions).
- Shared assignments to practise with a shared vision (e.g. writing shared texts such as a folder or paper, shared presentations, the building of a shared research design and problem definition).

Aligning of expectations

- The use of management tools to identify team skills (e.g. Belbin team roles test and core quality quadrants).
- Discussing each other's key literature to get familiar with each other's disciplines, theories and models.
- Stepping into each other's practice (e.g. assisting in focus group sessions or observing an animal experiment).
- Negotiating the intensity of collaboration and task division.

and Derry, 2005; Klein, 2005). Therefore, much time can be gained when the expectations and attitudes of the team members are aligned in the early phase (Box 3.3) (Klein, 1996; Bruun *et al.*, 2005; Stokols *et al.*, 2008).

In hindsight, it seems beneficial to take into consideration the following questions during joint progress meetings:

- What skills can we expect from each other?
- How much time do we each have available?

The first question requires insights into competences such as personal (team) skills and disciplinary skills. To raise (self-)awareness about each other's team skills and qualities, team management tools might be used, such as the Belbin team roles test and core quality quadrants for individual qualities and second-order conflict reflection. The first tool helps to identify which team role one could take up easily and which one is more difficult, while the second tool helps to identify and reflect on one's own team qualities, characteristics and their related pitfalls.

3.6.3 What was the influence of skills and roles in this team?

A way to reflect on the team skills in the research programme is by using the Belbin roles. Belbin identified nine team roles that clarify the effectiveness of a team or potential trouble based on team composition (Isaac and Carson, 2016). Most people have two or three dominant roles, and one is pre-eminent. The more widely the team roles are distributed over the team, the better they can work together in demanding contexts. Two *shapers* (visionary leaders) come with a high risk of conflict. The role that was taken up by some of the students from the beginning was the *completer*, who is accurate and has an eye for detail. Three roles were initially taken up by seniors but were slowly taken over by the students as a result of their shared development in the interdisciplinary team. Two of these roles have in common that they keep a certain distance and overview: the *coordinator*, who sees the big picture and consults each team member before decisions are made, and the *monitor evaluator*, who has a broad view on problems and detaches themself from bias. The third role was the *team worker*, who is a good listener and keeps the team running. Additionally, due to their individual development, they also became a *specialist*, who is dedicated to a particular topic and is more single-minded.

Three roles were not typically taken up by one of the students, but their absence was not problematic. They were the *plant*, who is the generator of ideas, and the *resource investigator*, who explores opportunities and is a good networker. For example, the set-up of the research programme quickly brought us into contact with several stakeholders, and the centrally organized courses were designed to stimulate creativity. The role that should be of primary importance, but was originally not dominant in our team, was the *implementer*, who turns ideas (of others) into action and gets work done. M.B. took up this role during the project.

An interesting aspect was the absence of a *shaper* role, which made this team quite democratic, but the downside is that nobody will take decisions. There was a frequent struggle with the operationalization of what was decided during dialogue sessions and deciding how to deal with the different framings of the research programme that we learned at the courses. It was challenging to translate moral deliberation into practical research design strategies, or to combine different visions into a joint scientific paper. In hindsight, it can be concluded that this team composition of roles enabled the team members to work with each other and reflect in a sustainable and robust way, which turned out to be very beneficial.

3.6.4 Time management

How much time has to be reserved requires agreements on the intensity of the collaboration, because inter- and transdisciplinary research is typically more time-consuming than mono- and multidisciplinary research (Epstein, 2005). Issues to discuss include, for example, how much time to spend on society-related activities or the number of shared papers that need to be written. It can be expected that some negotiation and compromises come along with these agreements. For example, the life-science students were expected to join in the organization of a science–society symposium, but could have received in return the assistance of M.B. in observing during animal experiments. Another option is that the writing of shared papers or organizing of science–society activities might in turn lead to a contract extension or release from other duties.

The major lesson that can be distilled from these insights is that transdisciplinary teamwork demands a different organization and management from mono- or multidisciplinary research. The aligning of expectations and development of a shared understanding needed for optimal transdisciplinary teamwork demand intensive facilitation during the first 2 years. Although it is tempting to start directly with the planning of experiments, it seems a better strategy to start with teambuilding instead, followed by structural and frequent embedding of learning interventions. Teambuilding facilitates the alignment of expectations and the formation of an open mind for shared research in a very early phase. Follow-up learning interventions are needed because, as the results presented in this study show, mutual learning is a skill that increases the quality of the design and outcomes. This will drive a continuous process consisting of different phases and different learning needs.

3.7 Follow-up of the Joint Transdisciplinary Project

As a result of the data from student projects, the breeding industry became more interested in pig behaviour and the related animal welfare issue of tail biting. This was relatively unique because they were usually more interested in traits directly related to production data (e.g. weight gain or food conversion). Science and industry then started an experiment that included observational studies, preparatory for a possible selection against tail biting. Additionally, scientists involved in this research programme and one representative of the breeding industry were asked to participate in the 'Curly tail working group' (Dutch: Werkgroep Krulstaart). This Dutch initiative united more than eight representatives of a broad range of organizations from the farming branch, animal protection, research, government, feed industry and meat industry, and started on its own to search for integral strategies to stop tail docking. Currently (2021), this autonomous, multi-stakeholder initiative launched by the Ministry of Agriculture has created an animal welfare benchmark based on 50 criteria, which will become compulsory for all Dutch pig farmers. The idea is that the farmers can use this benchmark annually to monitor for themselves what needs to improve. The three criteria that were the most debated during our symposia, but are now linked in the benchmark format, were: draught (improper climate/ventilation), low body condition score and impoverished housing conditions.

Another development that showed the added value of transdisciplinary learning was that two of the senior scientists became involved in research programme initiatives in which science–farmer interactions had a major role, and which was experienced by them as inspiring. In addition, the three life-science students now occupy research positions in which they support animal welfare in discussion with the relevant stakeholders.

References

Benard, M. and de Cock Buning, T. (2013) Exploring the potential of Dutch pig farmers and urban-citizens to learn through frame reflection. *Journal of Agricultural and Environmental Ethics* 26(5), 1015–1036. DOI: 10.1007/s10806-013-9438-y.

Benard, M. and de Cock-Buning, T. (2014) Moving from monodisciplinarity towards transdisciplinarity: insights into the barriers and facilitators that scientists faced. *Science and Public Policy* 41(6), 720–733. DOI: 10.1093/scipol/sct099.

Benard, M., Schuitmaker, T.J. and de Cock Buning, T. (2014) Scientists and Dutch pig farmers in dialogue about tail biting: unravelling the mechanism of multi-stakeholder learning. *Journal of Agricultural and Environmental Ethics* 27, 431–452.

Bijma, P., Muir, W.M. and Van Arendonk, J.A.M. (2007) Multilevel selection 1: quantitative genetics of inheritance and response to selection. *Genetics* 175(1), 277–288. DOI: 10.1534/genetics.106.062711.

Bruun, H., Hukkinen, J., Huutoniemi, K. and Klein, J.T. (2005) *Promoting Interdisciplinary Research: The Case of the Academy of Finland.* Academy of Finland, Helsinki.

Davis, J.R. (1995) *Interdisciplinary Courses and Team Teaching.* Amsterdam University Press, Amsterdam.

Duijvesteijn, N., Benard, M., Reimert, I. and Camerlink, I. (2014) Same pig, different conclusions: stakeholders differ in qualitative behaviour assessment. *Journal of Agricultural and Environmental Ethics* 27(6), 1019–1047. DOI: 10.1007/s10806-014-9513-z.

DuRussel, L.A. and Derry, S.J. (2005) *Schema (Mis)alignment in Interdisciplinary Teamwork.* Lawrence Erlbaum Associates, Hillsdale, New Jersey.

Epstein, S.L. (2005) Making interdisciplinary collaboration work. In: Derry, S. and Schunn, C.D. (eds) *Interdisciplinary Collaboration. An Emerging Cognitive Science.* Lawrence Erlbaum Associates, Hillsdale, New Jersey, pp. 245–264.

Griffing, B. (1967) Selection in reference to biological groups. I. Individual and group selection applied to populations of unordered groups. *Australian Journal of Biological Sciences* 20(1), 127–139. DOI: 10.1071/BI9670127.

Grin, J. and van de Graaf, H. (1996) Technology assessment as learning. *Science, Technology, & Human Values* 21(1), 72–99. DOI: 10.1177/016224399602100104

Habermas, J. (1997) *Theorie des Kommunikativen Handelns (2 Bände).* Suhrkamp, Berlin, Germany.

Horlings, L.G. and Hinssen, J.P.P. (2014) Sustainable innovation in intensive animal husbandry; policy and public protests towards a mega-farm in the Netherlands. *Journal for Communication Studies* 7, 125–145.

Isaac, M. and Carson, K. (2016) *A Guide to Belbin Team Roles.* Van Haren Publishing, 's-Hertogenbosch, Germany.

Kitzinger, J. (1994) The methodology of focus groups: the importance of interaction between research participants. *Sociology of Health and Illness* 16(1), 103–121. DOI: 10.1111/1467-9566.ep11347023.

Klein, J.T. (1996) *Crossing Boundaries: Knowledge, Disciplinarities, and Interdisciplinarities.* University of Virginia Press, Charlottesville, Virginia.

Klein, J.T. (2005) Interdisciplinary teamwork: the dynamics of collaboration and integration. In: Derry, S.J., Schunn, C.D. and Gernsbacher, M. (eds) *Interdisciplinary Collaboration: An Emerging Cognitive Science.* Lawrence Erlbaum Associates, Hillsdale, New Jersey, pp. 23–50.

Muir, W.M. (2005) Incorporation of competitive effects in forest tree or animal breeding programs. *Genetics* 170(3), 1247–1259. DOI: 10.1534/genetics.104.035956.

Nannoni, E., Valsami, T., Sardi, L. and Martelli, G. (2014) Tail docking in pigs: a review on its short- and long-term consequences and effectiveness in preventing tail biting. *Italian Journal of Animal Science* 13(1), 3095. DOI: 10.4081/ijas.2014.3095.

Rowe, G. and Frewer, L.J. (2004) Evaluating public-participation exercises: a research agenda. *Science, Technology, & Human Values* 29(4), 512–556. DOI: 10.1177/0162243903259197.

Rutherford, K.M.D., Donald, R.D., Lawrence, A.B. and Wemelsfelder, F. (2012) Qualitative behavioural assessment of emotionality in pigs. *Applied Animal Behaviour Science* 139(3–4), 218–224. DOI: 10.1016/j.applanim.2012.04.004.

Schön, D.A. (1983) *Reflective Practitioner: How Professionals Think in Action.* Basic Books, New York.

Schön, D.A. and Rein, M. (1995) *Frame Reflection.* Ingram Publisher Services, Jackson, Tennesee.

Stokols, D., Hall, K.L., Taylor, B.K. and Moser, R.P. (2008) The science of team science: overview of the field and introduction to the supplement. *American Journal of Preventive Medicine* 35(2 Suppl), S77–S89. DOI: 10.1016/j.amepre.2008.05.002.

te Velde, H., Aarts, N. and Van Woerkum, C. (2002) Dealing with ambivalence: farmers' and consumers' perceptions of animal welfare in livestock breeding. *Journal of Agricultural and Environmental Ethics* 15(2), 203–219. DOI: 10.1023/A:1015012403331.

van Mierlo, B.C., Regeer, B., van Amstel, M., Arkesteijn, M.C.M. and Beekman, V. (2010) *Reflexive Monitoring in Action. A Guide for Monitoring System Innovation Projects.* Communication and Innovation Studies, Wageningen University and Research, Athena Institute, Vrije Universiteit, The Netherlands.

4

Interdisciplinarity and the Dilemmas of Knowledge

Robert Frodeman*

Jackson, Wyoming, USA

4.1 Disciplinarity, Interdisciplinarity and Transdisciplinarity

> The most thought-provoking thing in our thought-provoking time is that we are still not thinking.
>
> Heidegger, *What is Called Thinking?* (1976, p. 6).

This chapter explores the underlying issues that affect attempts to make one's research relevant to wider-than-disciplinary audiences. It does so via reflections on the nature of interdisciplinarity, the growing contradictions of the modern university and the future of knowledge production generally.

Overall, the chapter offers an *environmental* critique of knowledge production within the contemporary university. This critique is environmental in the sense that knowledge production in the modern university is unsustainable. Sustainability is commonly defined in terms of the three parameters of the economic, sociological and ecological limits to growth. Today, a fourth parameter is needed: epistemic sustainability. Epistemic sustainability recognizes the problems of endless disciplinary knowledge production, and entails redirecting our focus towards the blending, contexualizing and translation of already existing knowledge for societal needs. This is the definition of interdisciplinarity that is worth defending.

4.1.1 Interdisciplinarity

Discipline is something that parents impart to their children for an ordered life. Discipline is also needed by anyone who wants to master an art or craft.

*robert.frodeman@gmail.com

DOI: 10.1079/9781789247893.0004

Within an academic setting, discipline means the standardized method of a particular line of inquiry, as well as the administrative structure of the modern university. In recent decades, there has been a growing sense that disciplinarity is in need of revision. Calls for interdisciplinary approaches to knowledge have become obligatory. In some cases, however, they consist of little more than a proclamation of the relevance of one's research. And even when their claims go beyond mere posturing, advocates of interdisciplinarity can mistake what is at stake.

The most common error is to confuse interdisciplinarity for transdisciplinarity. Quite often, the former term is substituted for the latter. Interdisciplinarity identifies the blending of different types of academic knowledge. Transdisciplinarity points to something quite different – a move beyond solely academic actors, whereby one's epistemic efforts are made in concert with one or another part of the public. Transdisciplinarity thus marks a political as well as an epistemic change, for it implies that academics are giving up some of both their authority and autonomy.

If one is cynically minded, the slippage between the two terms looks like no accident, for it allows what sounds like a commitment to greater relevance to become just another occasion where academics produce knowledge of no particular use to the wider world. Interdisciplinarity – the blending of different disciplinary perspectives – offers no guarantee of wider societal relevance. Nor would such fraudulence be surprising: few people willingly give up either their authority or their autonomy.

The switch in terms also points towards another issue: the discordance of ends between academics and non-academics. Teaching aside, academics understand their job as consisting of the production of new knowledge. Non-academics have a different priority: getting the job done. They have limited time or interest in acquiring additional knowledge unless it directly helps them with their task.

The muddle surrounding interdisciplinarity starts with the root term, disciplinarity. To a remarkable degree, given the fact that academics seek to interrogate everything, the concept of an academic discipline goes unanalysed. Inhabiting a discipline is treated as simply a background condition; real work goes on at the subdisciplinary level, as academics specialize and specialize again. Raise wider questions about the future of the university, or the future of knowledge generally, and people beg off: it is not their research focus. And given the structure of the modern university, it is not anyone's research focus. After all, such research might demonstrate the pointlessness of further research. (Granted, a few fields raise such questions: social epistemology sometimes does, and so too interdisciplinary studies. It is also occasionally discussed in education departments, and there is even a new field called critical university studies. But, as with more traditional disciplines, the dynamic of all these fields drives them towards more detailed research questions, a process I have called 'disciplinary capture'.)

As a result, larger questions of the overall purpose and direction of disciplinary knowledge, and of knowledge in general, are neglected – or at least, not asked in a serious way, treated as a matter of sustained study (cf. Frodeman and Briggle, 2016). This is the point of the opening quote from Heidegger (1976): for all our research, in what is undoubtedly the most knowledge-intensive culture in history, we are still not asking the most basic questions concerning thinking. These might be boiled down to two. First: in a given situation, what is pertinent knowledge – knowledge that could make a difference to the larger world, rather than merely satisfying one's professional curiosity? Second: is knowledge even pertinent to the problem at hand? Or is the further pursuit of knowledge an excuse for not facing up to the need for making a decision or changing our behavior?

Another reason we do not interrogate our disciplinary assumptions: the tacit belief that there is some type of epistemic core that unites the disparate parts of a discipline. Here we fall prey to the fallacy of composition, which assumes that when the members of a collection all share a property, the collection as a whole must possess that property as well. Academics are housed within a given discipline and all pursue knowledge of one type or another. But this does not mean that there is an overarching epistemic unity to the discipline. What core epistemic qualities do a mineralogist and a palaeontologist, both in geology, share? Or a physical anthropologist and a linguist, although both study anthropology, or a symbolic logician and an ethicist, both housed in philosophy? I have argued elsewhere (e.g. Frodeman, 2014) that disciplines are defined by social, economic and administrative factors as much as by epistemic ones. Certainly, there are barriers to working across disciplines, but these are as much institutional and linguistic as epistemic in nature. Disciplines are historical accretions that shift over time depending on a variety of social and economic influences. But if disciplines already consist of distinct epistemic elements, it is not clear what interdisciplinarians bring to the table that is unique.

Of course, this point is a matter of degree. There are close connections between stratigraphy and sedimentology, just as there are between Renaissance and Enlightenment historiography. And on the other side, we can find clear cases where a physicist and a philosopher are engaged in interdisciplinary (although not necessarily transdisciplinary) research. But by what measure could we determine that the theoretical distance between two researchers in different disciplines is greater than the distance between parts of a single discipline?

4.1.2 Narrow and wide interdisciplinarity

Given the structure of the university, it is important to distinguish between what can be called narrow and wide interdisciplinarity (Frodeman *et al.*,

2001). The former occurs so regularly that it attracts little notice: think of a chemist and a physicist working together. The latter involves crossing the divide that exists between the natural and social sciences on the one hand and the humanities on the other – between a discussion that is primarily concerned with the way things are versus the way things should be.

This point was once commonly described in terms of the fact–value distinction. But this way of describing the divide is passé. We have learned that every fact-based inquiry presupposes value judgements concerning what issues are worth examining, as well as the manner in which they are examined. Conversely, every value discussion relies on one or another set of facts. Nonetheless, the question remains whether these points are operationalized: whether the ethical, social, political and aesthetic dimensions of an issue are married to the facts that both constrain and open up opportunities for the resolution of a problem.

In closing this section, allow me to emphasize my main point: interdisciplinarity largely functions as a slightly more adventurous form of disciplinarity, when what is needed is a wider questioning of the premises underlying both. Disciplinarity, interdisciplinarity and transdisciplinarity are species of a larger genus. This genus is our knowledge culture; this is what needs a thorough rethinking. When this finally occurs, the attention now given to interdisciplinarity may be seen as a passing phase in the evolution of our knowledge culture.

4.2 The Assumptions of Our Culture of Knowledge

Questioning our knowledge culture has not occurred because academics are committed to the disciplinary status quo. They are good at what they do – narrowly focused research designed to be of interest to their colleagues – and fear that any change in their circumstances will be a change for the worse. In this, they are probably correct. Despite the growing problems within the contemporary university (declining public funding, exorbitant tuition fees, the move to online education, research that goes unread, increasing administrative burdens), the professor's life remains a privileged one. There is teaching to be taken care of, and an irritating amount of administrivia, but then you are free to pursue whatever research you find interesting – with little regard for whether there is any particular demand for your research.

While these assumptions are being challenged in some quarters, there is a substantial amount of inertia to the system. A precipitating event is probably necessary before the assumptions of the modern research university will seriously be called into question. The COVID-19 pandemic may be that event. I am writing in the midst of the pandemic and claim no particular skill at soothsaying. In fact, it could be the better part of a decade before it becomes clear whether the COVID-19-driven changes in academic life will be trivial or profound. But there are already signs that the status quo ante of university life – in-person classes, tenure, the never-ending circuit of academic conferences and the endless production of new knowledge – may be over at many institutions. Indeed, some of these institutions themselves may come to an end. It is worth speculating on the direction things could take.

4.2.1 Knowledge generation

Let us begin with a point that is too often passed over: in terms of formal structures – setting aside craft know-how and folk wisdom – disciplinarity has defined our thinking about knowledge only fairly recently. The creation of the modern research university, which can be dated from the founding of Johns Hopkins University in 1876, made disciplinarity inevitable. The central mission of the university shifted from the preservation of our cultural inheritance to the discovery and invention of new knowledge. Producing new knowledge required disciplinary specialization, for only by focusing on greater and greater detail would continued progress in knowledge be possible.

The modern research university, then, is built on two interdependent assumptions:

1. Knowledge production is an infinite project.
2. Knowledge is flat: no discipline is more fundamental than or superior to another.

These assumptions are so deeply embedded in academic culture and society at large that they are not even subject to debate. When I raise them in talks, I am greeted with disbelief. Take the first point: why would anyone want to stop the pursuit of knowledge? Surely, we want to grow the economy, conquer disease, address environmental problems and generally improve the human condition? To state such goals in a piecemeal fashion, as disciplinary researchers and the public do, makes the point obvious. Of course we want a vaccine for COVID-19 and a cure for cancer, more efficient solar power and better battery storage, more powerful computers and a faster Internet. The list is endless, as are our desires. But there is something missing from this account. Return to our earlier discussion of the fallacy of composition: where does this piecemeal process take us when considered as a whole?

4.2.2 Transhumanism

The movement known as transhumanism (which advocates the transformation of the human condition through technology) provides an answer to this question. Having thought through the knowledge enterprise as a whole, transhumanists offer an account of where advances in science and technology will eventually take us: to a condition of nearly infinite human power. Transhumanists differ on the particularities of this process – it may come through the physical and cognitive augmentation of our simian bodies, or through a silicon future, as artificial intelligence comes to either serve or absorb us. But by whichever means, the end result is the same: something close to deification.

Transhumanism is dismissed as the passion of a few oddballs. This is a mistake: rather than an aberrant interpretation of science and technology, the transhumanists have spotted something important. They have correctly diagnosed the tacit agenda of modern culture. Whether judged in terms of capitalism, or the ideology of continual scientific and technological progress, or simply the nature of human desire, our culture's love of infinity is tacitly transhumanist in orientation. Transhumanism makes explicit the logical end point of the Enlightenment project of *sapere aude* (Frodeman, 2019). How else should we describe the pursuit of scientific and technological progress that has no end? The US National Science Foundation places no limit on its programme of scientific and technological advance, just as the US National Institutes of Health seek to overcome every human infirmity. The same is true for every other nation's research portfolio. The only difference between the transhumanists and the rest of us is their self-awareness. Our trajectory points towards deification on the instalment plan; transhumanists have simply made the point explicit.

Transhumanists deserve praise for achieving a global view of our situation. But this clarity raises a new set of questions, the most basic of which is whether we are taking the dangers of continued knowledge production seriously enough. The perpetual pursuit of technoscientific knowledge will lead to any number of improvements. But as our knowledge increases so does our power, which, like the Sorcerer's Apprentice, is liable to spin out of control. This raises terrible possibilities – of political instability, as society fails to properly integrate new technologies; of totalitarian government, as advances increase the means for surveillance, manipulation and control of the population; and of social or environmental disruption, via catastrophic accidents or the deeds of rogue actors.

Most basically, transhumanism highlights the fact that the overall results of knowledge production may be quite different from the piecemeal outcomes of these efforts. The dismissive response transhumanists have received by sober-minded science and technology policy analysts reflects the latter's focus on the piecemeal aspects of our knowledge culture. This is what Heidegger meant by the 'forgetfulness of being' – the loss of a view of the whole, and a sense of our overall trajectory, as we focus on smaller matters. Our university system, which has the same structure across the globe, possesses no unit whose task is to consider the whole of knowledge – except in terms of administrative functions, where functionaries attend to the practical necessities for keeping the universities running.

When Heidegger claims that we live in the age of technology, he is not primarily interested in the machines and instruments that surround us. Rather, he is concerned with a way of life that focuses on means to the exclusion of ends. The nihilism of our culture, which Postman (1985) described in terms of 'amusing ourselves to death', is visible in the fact that talk about the ends of life has become irrational, reduced to questions of personal preference. Ours is a technological age not because of cars and cell phones but rather because rationality has become instrumental in nature with no larger goal in mind.

4.3 The End(s) of Knowledge

> In the multiversity there is an irreducible plurality of communities, functions, disciplines, and interests, external constituencies, agendas, and beliefs.
>
> Marginson, *The Dream is Over* (Marginson, 2016).

This brings us to the second assumption of contemporary knowledge culture: knowledge is flat. No discipline is subordinate to or preparatory for any other; no discipline represents the end point of knowledge. The university is a self-serve buffet where everyone selects what they want. This is why Clark Kerr, president of the University of California system in the 1950s and 1960s, described the modern research university as a 'multiversity' serving a vast number of interests and constituencies. In other words, the university is a space for knowledge but not for wisdom – if wisdom is understood as a non-relativized claim about the nature of human flourishing and the ends of life.

4.3.1 The goal of the medieval university

Compare the multiversity with what came before. The crucial difference between the modern research university and the medieval university concerns the relationship between means and ends. The medieval university understood that knowledge must have an end in the sense of a goal, and thus would have an end in the sense of a terminus or completion. Rather than being flat, in the medieval university knowledge was hierarchical in nature. Individual knowledge projects were not independent but rather were seen as contributing to an overall goal.

This was reflected in the structure of the medieval university: the division of professors into higher and lower faculties reflected the fact that some types of knowledge were subordinate to others. The three higher faculties of medicine, law and theology trained people for professional roles in society. These faculties were arranged in ascending order: the health of the body was important, the laws governing the polity still more crucial, and the destiny of our immortal souls of supreme consequence. There was a lower faculty as well, consisting of the arts or philosophy, which had their own worth as well as offered preparatory training for the higher degrees.

The meaning of these terms shifted over time. In the medieval university, 'philosophy' was not disciplinary in nature; it included what we now call the natural and social sciences as well as the arts and humanities. 'Natural philosophy' was not merely the name for a more primitive natural science; the natural world was studied for many reasons, but chief among them was the theological goal of understanding God's work and overall plan and our place in the universe. Similarly, the social sciences evolved from moral philosophy, but differing from today's social sciences moral philosophy was normative, and had a clear pastoral element, with an explicit focus on training young men to be virtuous, motivated by a sense of *noblesse oblige*.

The medieval university possessed administrative units, and some of the designations remain the same as ours. But knowledge in the medieval university was not disciplinary in nature. In a lengthy career, a professor could begin in law, move to the faculty of medicine and finish his career in the faculty of theology (Clark, 2006). New discoveries were welcome, but knowledge was essentially conservative in nature: the bulk of efforts were directed towards preserving and passing down the hard-won cultural inheritance that had been achieved across the generations.

In contrast, in the modern university, the acquisition of knowledge is an infinite process because the human desire for power is infinite. The use of this power is essentially libertarian in nature: individuals are free to turn the knowledge gained in whatever direction they desire. What explains the shift to a libertarian approach to knowledge? One account sees it as tied to Enlightenment values and the march of human freedom, where dogmatic Christian claims concerning the *summum bonum* were thrown off and questions concerning the nature of the good life were individualized. People should be free to do what they want with their lives, subject to the minimal condition described by John Stuart Mill in *On Liberty* (Mill, 1859): his harm principle claims that people should be free to act however they wish unless their actions cause harm to others.

4.3.2 The assumption of abundance

'Unless their actions cause harm to others': this brings us back to environmental sustainability. There is a hidden environmental premise to Mill's argument: it assumes the existence of ecological abundance, a 'new world' (i.e. the Americas) to explore and exploit, where there is enough room and resources that people can agree to disagree about fundamental issues. Under conditions of scarcity – of land and resources, which can be polluted without consequence – there are very few actions we can take that do not significantly affect others. These conditions will only become more stringent in an increasingly populated and developed world.

The vaulted pluralism of contemporary culture, where we assume the existence of irreconcilable differences in life plans, presumes that we do not find ourselves in the situation of the inhabitants of a lifeboat. If we are to survive, there needs to be a fairly exacting set of values applicable to us all. The modern university, then, has been built on the ecological assumption of abundance. The libertarian approach to knowledge, where all knowledge can be treated as a means to whatever ends an individual wishes to pursue, assumes an infinite world where we do not need to consider how individual decisions contribute to the whole. But where the whole was once Christian in nature, today it is ecological in nature, as resource shortage, pollution, climate change and the rights of beings other than humans need to be balanced with our desires. Knowledge production does not exist in a vacuum; it opens up opportunities that impinge on the lives of others, both human and non-human.

Which brings us to a final word about the field of animal welfare science. In this discipline, as in others, there will be a significant amount of research at a small and discrete scale that should go forward, where concerns with the larger issues outlined here will not intrude. But at points – impossible to identify beforehand, and thus the possibility must always be kept in view – researchers in animal welfare science will need to consider how their work contributes to the overall goals of an environmental civilization. We can no longer rely on the *laissez faire* assumption that knowledge produced in one or another area of research will naturally combine with other types of knowledge in a benign fashion. Nor can we assume that research is itself an infinite process – and will be infinitely paid for by society. As we pursue our individual projects, we need to keep an eye out for the good of the whole, and to recognize that our work will increasingly consist of translating already existing knowledge for different audiences rather than the single-minded pursuit of new knowledge.

References

Clark, W. (2006) *Academic Charisma and the Origins of the Research University*. University of Chicago Press, Chicago, Illinois.

Frodeman, R. (2014) *Sustainable Knowledge: A Theory of Interdisciplinarity*. MacMillan/Palgrave Press, London.

Frodeman, R. (2019) *Transhumanism, Nature, and the Ends of Science*. Routledge, London.

Frodeman, R. and Briggle, A. (2016) *Socrates Tenured: The Institutions of 21st Century Philosophy*. Rowman and Littlefield, Lanham, Maryland.

Frodeman, R., Mitcham, C. and Sacks, A.B. (2001) Questioning interdisciplinarity. *Science, Technology, and Society Newsletter* 126/127, 1–5.

Heidegger, M. (1976) *What is Called Thinking?* Harper & Row, New York.

Marginson, S. (2016) *The Dream Is Over: The Crisis of Clark Kerr's California Idea of Higher Education*. University of California Press, Berkeley, California.

Mill, J.S. (1859) *On Liberty*. John W. Parker and Son, London.

Postman, N. (1985) *Amusing Ourselves to Death: Public Discourse in the Age of Show Business*. Penguin Press, New York.

5 Human-centred Research Approaches to Improve Animal Welfare: Survey Design and Stakeholder Involvement

Grace A. Carroll[1]*, Jenny M. Groarke[2] and Lisa Graham-Wisener[2]

[1]Animal Behaviour Centre, School of Psychology, Queen's University Belfast, Belfast, UK; [2]Centre for Improving Health-Related Quality of Life (CIHRQoL), School of Psychology, Queen's University Belfast, Belfast, UK

5.1 Introduction

The majority of animals studied by animal welfare scientists are managed by humans; we control what they eat, where they are housed and the conditions in which they live. In addition, animal welfare is largely dependent on the humans who care for the animals, whether it be a laboratory technician, zookeeper, farmer or pet owner. Animal welfare science encompasses several disciplines including ethology, nutrition and genetics, to name but a few (see Chapter 1, this volume). While traditionally the domain of natural scientists, the role of social scientists in animal welfare research is becoming increasingly recognized. Indeed, Lund *et al.* (2006, p. 41) argued that 'if the study of animal welfare includes husbandry and human–animal interactions as part of its subject matter, then the social sciences become relevant'. Humans are an essential consideration if animal welfare is to be improved, and therefore human-centred research methods are of importance to animal welfare scientists. Within animal welfare science, there is a diverse range of stakeholders who may be users of your research, including knowledge producers, knowledge users and knowledge beneficiaries. If we look at farm animal welfare as an example, stakeholders may include those working in industry (e.g. producers, transporters, meat processors, retailers), research and development (e.g. other

*Corresponding author: G.Carroll@qub.ac.uk

 Bridging Research Disciplines to Advance Animal Welfare Science: A practical guide (ed. I. Camerlink)
DOI: 10.1079/9781789247893.0005

researchers), the community sector (e.g. animal rights organizations) and the government sector (e.g. policy makers). There are therefore several different perspectives to be considered and factored into any decision-making process.

Survey designs are the predominant research methodology when working with human stakeholders, with questionnaires used to measure a range of important constructs. For example, questionnaires can be used to measure aspects of animal welfare, such as quality of life (see Chapter 8, this volume), that cannot be assessed directly in the animal. The familiarity that pet owners have with their animals allows them to detect and report subtle changes in their companion animal's disposition. Similarly, veterinarians may be best suited to detecting and reporting clinical signs of abnormalities (Mullan, 2015). Questionnaires can also assess human factors such as attitudes, perceptions and intentions that may have implications for animal welfare (e.g. Pulido *et al.*, 2018; Cornish *et al.*, 2020). Additionally, respondents' views can inform animal welfare policies and practices (e.g. Pejman *et al.*, 2019; Aluwé *et al.*, 2020). However, many animal welfare researchers lack the knowledge, skills and confidence needed to appropriately design and validate questionnaires. For example, Belshaw *et al.* (2015) carried out a rapid review of instruments used to assess quality of life in dogs and found that only 11 out of 52 publications reported the validation process associated with the instrument. In addition, other researchers did not subsequently adopt many of these, raising questions about the need to develop these particular instruments in the first place. Moreover, many authors failed to consult with those who would be completing the questionnaire, such as pet owners and veterinarians (Belshaw *et al.*, 2015). It is, however, very important to actively engage with stakeholders throughout the development of animal welfare instruments.

Stakeholders can make valuable contributions to the development not only of questionnaires and related instruments, but also of animal welfare solutions, practices and policies. However, animal welfare research is often conducted without consulting the very people who will be using the new method or technology. As a result, a substantial amount of money has been spent on concepts developed by animal welfare scientists that have yet to be adopted in practice (Weary *et al.*, 2016). This traditional top-down approach is problematic, as it fails to consider practical implementation knowledge (Fernandes *et al.*, 2019). Rather than assessing acceptability post-development, stakeholders should have an active role in the research process. Animal welfare scientists must now shift towards a bottom-up approach where relevant stakeholders collaborate at each stage of solution development (van Dijk *et al.*, 2019). For example, animal welfare should be considered in light of other stakeholder priorities such as food safety or productivity (Christensen *et al.*, 2012; van Asselt *et al.*, 2016), as these other priorities will impact on the uptake of novel animal welfare solutions. Within health psychology and health-related disciplines, there is a longer history of stakeholder involvement in research. This is referred to as 'patient and public involvement' (PPI). Research involvement can help with establishing priorities for research and agreeing which outcomes to use in a study. Within health psychology and related disciplines, patients are involved in establishing research priorities because it helps prioritize under-researched areas that are valued by patients. This ensures that limited resources are directed towards answering important questions. When it comes to animal welfare, research priorities are likely to vary according to the stakeholder. For example, veterinarians may prioritize farm animal health, while consumers often focus on the ability of animals to perform natural behaviour. In contrast, farmers will prioritize solutions that improve or at least maintain current productivity levels while also being practical to implement. Different stakeholders will also have various levels of influence on the successful implementation of research findings. For example, farmers are a key stakeholder; if they cannot or will not implement a solution on farm, it leads to waste in research (Chalmers and Glasziou, 2009). In addition, welfare outcomes that researchers choose to measure may not be deemed relevant by all stakeholders. A pig farmer will probably be interested in a welfare solution if it has demonstrable beneficial effects on growth rate or carcass quality parameters. The same farmer could be less interested in seeing a reduction in tail or skin lesion prevalence. This is not to suggest that such outcomes should be ignored. Rather, consideration should be given to a variety of welfare outcomes, maximizing the benefit of the solution to a range of stakeholders. While sometimes being a difficult task, bringing diverse stakeholders together allows for a novel mix of perspectives on a given issue (Klerkx and Nettle, 2013). Research involvement can also help in problem solving and improving the study design during the research process. For example, farmers, zookeepers and laboratory technicians will be aware of practical constraints that would make certain ideas unfeasible. In addition, stakeholders can help with interpreting the findings, and can support knowledge transfer and exchange through their existing networks.

Poorly designed research can be the result of inadequate training. Animal welfare science benefits from being a multidisciplinary field, with no one pathway. This does, however, mean that there will be variation in the level of training received in research methods and reporting. Given the diverse nature of animal welfare science and the emergent challenges, there is recognition of the need to collaborate and adopt approaches from outside disciplines. The aim of this chapter is to provide a starting point for conducting high-quality, rigorous research in animal welfare science by focusing on human-centred research approaches that can be used to improve animal welfare. For example, how can we design questionnaires that are valid, reliable and useful in gaining insights into how humans influence animal welfare? How can animal welfare solutions be developed in collaboration with stakeholders to ensure uptake in practice? And how can research findings be successfully communicated to members of the public and industry stakeholders for maximum impact?

5.2 Survey Design

To design a high-quality survey, there is a specific set of tasks that one must perform in sequence. There are also several important psychometric concepts to consider prior to beginning the process of designing a survey.

5.2.1 Concepts

Psychometrics

Psychometrics involves the study of theory and techniques associated with psychological measurement. Psychometrics enables researchers to quantify the accuracy of measurement of qualitative concepts such as attitudes, perceptions and beliefs.

Defining measurement

The terms 'measure', 'questionnaire', 'survey' and 'scale' are often used interchangeably. However, there are subtle differences between them. A measure is any procedure aiming to collect data. Measures can be objective (e.g. salivary cortisol or temperature monitoring) or subjective (e.g. using behavioural observations to infer emotional state). A questionnaire is a set of questions, which can be closed or open ended. A survey is a research method; the survey always incorporates a questionnaire, in addition to broader considerations of distribution, logic and branching (i.e. presentation of questions is dependent on the respondent's previous answers), and analysis. Surveys can include qualitative data collection but primarily use closed questions. Responses can be continuous (e.g. 'not at all satisfied' to 'completely satisfied') or dichotomous (i.e. yes/no). Similar to surveys, scales typically focus on behaviours, attitudes, perceptions and feelings, i.e. constructs for which there is no correct answer. They are called scales because responses vary along a continuous scale (e.g. 'strongly disagree' to 'strongly agree'; 'never' to 'always'). A set of questions is not necessarily a scale; all the items on a scale should be tapping into the same latent variable.

Latent variable

Many, if not most, of the constructs and variables that we wish to measure cannot be measured directly. A latent variable is just that – a variable that cannot be directly observed or measured. For example, we can *infer* emotional state from objective indicators such as heart rate or with subjective indicators such as attitudes from responses on a scale, but we cannot measure 'emotional state' or 'attitudes' directly. The questions on our scale are intended to reflect the underlying construct or phenomenon of interest, i.e. the latent variable. We presume that the latent variable is causing the respondents' scores on the items.

Reliability

A measure that is reliable is one that performs in consistent and predictable ways. For example, a respondent's score on a measure of attitudes towards animal testing should not change unless there has been an actual change in that respondent's attitude. To assess the reliability of a measure, respondents are given the same measure at two separate points in time, and the correlation between their scores is the measure's test–retest reliability or temporal stability, with stronger correlations indicating greater reliability. However, this approach is not suitable for every measure. There are some latent variables that we might expect to fluctuate (e.g. feelings), while others should be more stable (e.g. dispositions). Another aspect of reliability is internal consistency. This concerns the relationships among items intended to reflect the same latent variable. A scale that is internally consistent will have strong intercorrelations among its items, the implication being that the items share a common cause (i.e. the latent variable). There are also specific measures of internal consistency that are frequently used (e.g. Cronbach's α coefficient).

Validity

The validity of a measure is dependent on whether it measures what it is intended to measure, i.e. the latent variable. There are different types of validity. Content validity refers to the adequacy of the scale items to reflect the content domain (i.e. the construct being measured). Criterion-related validity is demonstrated when a newly developed scale has a close association with an existing 'gold-standard' measure. For instance, if the best available measure of stress reactivity in dogs is observation by a behavioural expert, scores on a new scale to measure reactivity reported by owners should have a strong positive correlation with the expert ratings. Construct validity is concerned with the theoretical relationships of the latent variable with other variables. Construct validity can be assessed by examining how well scores on the measure converge with scores on other measures of the same or similar constructs (i.e. convergent validity) and the lack of convergence with scores on measures of different or unrelated constructs (i.e. discriminant validity). For example, scores on a measure of attitudes towards animal testing might be hypothesized to correlate positively with scores on a measure of empathy for animals, and to be unrelated to scores on a measure of knowledge of animal stockmanship practices.

5.2.2 Techniques

The field of psychometrics is one of rapid developments and innovations. As such, any aspiring scale developer should consult the evidence base to determine current best practices in scale design and psychometric evaluation. A set of guidelines for scale development was developed by DeVellis (2016) that involves following a series of steps in the scale development process. A simplified version of these guidelines is described below with examples relevant to animal welfare science.

Conducting a thorough review of the literature surrounding the constructs to be measured

Before setting out on the scale development journey, you must understand precisely what it is you want to measure and why. Perhaps you have a research question that cannot be answered by existing measures. For example, there may be a valid and reliable measure of attitudes towards animal welfare, but you are interested specifically in attitudes towards farm animal welfare. Scales can vary in terms of their content (e.g. attitudes towards animal welfare versus empathy for animals), the settings they were designed to be used in (e.g. farms or abattoirs) or the population they are designed for (e.g. farmers or children). You must therefore have a clear picture of the context in which your scale will be used and who will be using it. Qualitative research with stakeholders can be very informative at this stage of the process, contributing to a deeper understanding of the construct of interest. Qualitative work can also increase the validity of the constructs developed later as they are grounded in real-life experiences and observations.

Generating an exhaustive item pool

No matter how well you understand your construct, a scale is only as good as the items it is made up of. Short, declarative statements that are easy to understand are best (e.g. 'Farm animals should be treated with respect'). Beware of double-barrelled items, such as 'Farm animals should be treated with respect and allowed to freely roam outside.' These items are problematic because the respondent might agree with one half of the statement but not the other. Each and every item needs to reflect the construct underlying it, i.e. the particular behaviour, attitude or feeling the scale is intended to measure. The scale needs to be composed of multiple items tapping into the latent variable in different ways. Aim for at least three times as many items as you think you will need. The idea is to be exhaustive at this stage because only the highest-quality items will be retained later following psychometric assessment.

Selecting an appropriate response format

The most common format by far is the Likert scale. The number of response options typically ranges from five to seven (e.g. 'strongly disagree' to 'strongly agree'), with the option to include a neutral midpoint (e.g. 'neither agree nor disagree'). Another option is to use a visual analogue scale with two bipolar endpoints (e.g. 'no distress' to 'unbearable distress'). A well-known example of the use of a visual analogue scale comes from Wemelsfelder *et al.*'s (2001) qualitative behavioural assessment where the observer rates animals on a number of behavioural descriptors. Respondents mark their response anywhere along a 'ruler', usually ranging from 0 to 10 or from 0 to 100. A binary scale, in contrast, has two response options (e.g. yes/no, agree/disagree). The limited variability in binary responses presents some challenges to scale developers. If the scale is intended to be brief (i.e. fewer than five items), then a larger number of response options should be used (e.g. 0–100) to increase variance in the responses. If the scale will be long (i.e. 50 items), then a smaller number of response options may be sufficient (e.g. yes/no, agree/disagree). There will be sufficient variance due to the number of items. Choosing between fewer response options will also prevent respondents from becoming bored or fatigued, leading to higher-quality data.

Item evaluation by experts

The next step involves having the items you developed reviewed by people who are experts in the content domain. This constitutes a test of the measure's content validity. The reviewers may be experts by research and practice or by experience (e.g. academics, farmers or animal rights activists). It can be a good idea to include a reviewer with expertise in scale development. Experts should be provided with your working definition of the construct you want to measure. They are then asked to rate each item in respect of its relevance to the construct as defined, as well as the item's clarity. The reviewers may also identify aspects of the content domain that have been overlooked. Three to five experts would be sufficient for this procedure.

Psychometric evaluation

Now that you have developed your items, you will want to evaluate them by administering them to a large sample of participants. There are various guidelines to determine how large your sample should be, with estimates ranging anywhere from three to 20 participants per item. It is imperative that the exploratory and confirmatory factor analyses be carried out with separate samples. The original sample can be split in half, or two groups of respondents can be recruited separately. First, carry out exploratory factor analysis (EFA). EFA will determine the dimensionality of the measure, i.e. whether your scale is composed of one or more factors/dimensions. For example, using EFA, Connor *et al.* (2018) found that animal cruelty during adolescence had three underlying dimensions: accidental, neglectful and deliberate animal cruelty. The related but distinct processes of factor analysis and item deletion should be carried out iteratively. This involves removing items from the analysis, one at a time, repeating the EFA and comparing the factor solutions using multiple criterion methods (e.g. the scree plot, proportion of variance explained). Items should be deleted if they do not meet standard psychometric criteria, specifically high factor loadings and no cross-loading. If factor analysis and its associated terminology is new to you, it is recommended that you consult Field's (2017) *Discovering Statistics* (see Further Reading). At this stage, you would also examine the factors and ensure that the inter-item correlations are high. To ensure reliability of the scale, the correlation between each item and the total scale score (i.e. item-total correlations) should also be high. Cronbach's α coefficient can be calculated to assess the reliability of the scale and, in the case of a multidimensional factor structure, its subscales.

Decisions around which factors and items to retain in EFA should not be driven only by statistical output. Ultimately, the developer needs to consider the conceptual and theoretical implications of each statistical decision. Next, carry out confirmatory factor analysis (CFA) to confirm the factor structure identified using EFA, with a separate sample of respondents. CFA is frequently carried out using structural equation modelling software. A full description of the CFA procedure is beyond the scope of this chapter, but detailed explanations can be found in the listed recommended references (in bold in the References).

Determine the validity of the measure

Once you have arrived at the final version of the scale, you will want to examine its validity, i.e. test the direction and strength of hypothesized relationships with other measures of similar and different constructs. You should use correlation analyses for this as described earlier in section 5.2.1.

Use open science principles by sharing your scale and validation process with others

Open science involves reforming traditional science with a number of key principals, which ensure that research is undertaken in a more transparent and collaborative way. There are multiple schools of thought within open science (Fecher and Friesike, 2014). However, a collaboratively developed definition of open science is 'transparent and accessible knowledge that is shared and developed through collaborative networks' (Vicente-Sáez and Martínez-Fuentes, 2018, p. 428). There are several methods that can be used to ensure that data is accessible. These include publishing the data as supplementary material with a journal article or on a public-facing website, or depositing your data in a general, subject-specific or institutional data repository. This is a requirement that is increasingly endorsed by funders, research institutes and journals. Publishing details of the validation process, in addition to the scale itself, will help other researchers replicate aspects of the research, assess the reproducibility of the findings or adopt the scale for their own research purposes.

5.2.3 Pitfalls

Not adhering to the principles of scale development

Too often, researchers will develop a set of questions to measure a construct central to their research question without undergoing the rigorous procedure outlined by DeVellis (2016). As a result, the set of questions cannot be considered a true psychometric instrument and the validity of the data collected is called into question.

Low-quality item development

Following the appropriate steps in the scale development process does not guarantee that a high-quality measure will be produced. A poorly designed questionnaire is one that includes questions that are ambiguous, vague, confusing, complex, double-barrelled or leading. These types of poorly written items can introduce bias into the research (Choi and Pak, 2005). Poorly formed questions can result from failing to develop a clear and coherent working definition of the construct being measured.

Failing to involve stakeholders in the design and pilot stages of the questionnaire design

Piloting the items with prospective participants can help identify flaws in the wording of questions. Consultation with stakeholders helps to ensure that the items of the scale are relevant and representative of the construct being measured.

Failing to carry out robust psychometric assessments

A common limitation of many scales is the use of principal components analysis (PCA) for EFA. The default setting in SPSS Statistics is PCA.

However, PCA should be avoided as it is a method of item reduction rather than a true method of factor analysis, such as principal axis factoring. It is also common for scale developers to skip the important next step of CFA, i.e. confirming the factor structure identified using EFA in another sample or a split sample of participants.

Failure to validate the questionnaire

Validity refers to the appropriateness of the measure to tap into the underlying construct/latent variable. Without validity assessments, we cannot be sure that the scale measures what it intends to measure. The process of validation is relatively straightforward and is essential if the scale is to be useful to the research community or in applied settings.

Creating a new instrument when validated ones already exist

A literature review that is sufficiently deep and broad should identify all existing or related instruments for a given construct. Given the resource-intensive nature of scale development, creating a new measure when one exists is a fine example of waste in research. Nevertheless, there may be circumstances where an existing measure is flawed or unsuitable such that developing a new or updated version of the measure is justified.

5.3 Stakeholder Involvement

The focus here is on conducting research *with* stakeholders, who provide input into decisions made about the research and in doing so ensure that the methods and outcomes are more appropriate to end users (INVOLVE, 2020). This is aligned to PPI principles in health-related research.

5.3.1 Concepts

Stakeholder involvement strategy

A stakeholder involvement strategy is a plan that details how stakeholders will be involved in the research and development process. A stakeholder involvement strategy can be for a research group, a programme of research or a discrete research study, and comprises a number of key components. Example components include specifying stakeholder involvement objectives, deciding when involvement will happen, finding people to involve, and laying out the training and support that stakeholder contributors will receive. You can provide details of your stakeholder involvement strategy in grant applications or as part of a research ethics application.

Stakeholder representatives

Stakeholder representatives can be referred to by many different names in research including 'stakeholder contributor', 'research partner', 'public partner' or 'user research consultant'. Different stakeholders may be involved to a lesser or greater extent throughout the research cycle. Examples of more intensive involvement include stakeholders involved as co-investigators in a research project or in advisory groups involving regular meetings. Less intensive involvement may include stakeholders attending a one-off workshop.

5.3.2 Techniques

Stakeholder involvement

An initial first step in developing a stakeholder involvement strategy is to consider why you want stakeholders involved in your research, what you feel their perspective will bring and what influence they can have on decision making. It is also important to identify whether there has been previous stakeholder involvement in other research that is applicable to your study.

Activities and when involvement will happen

Ideally, stakeholder involvement will take place at an early stage in the research cycle (see Chapter 3, this volume) and should be a sustained process. Traditionally, within agricultural research at least, collaboration in the development of new solutions has focused more on the collaboration process itself and less on how collaboration can be used to overcome encountered challenges (de Olde *et al.*, 2017). However, stakeholders should be involved throughout the research process as much as possible, from agreeing priorities for research to identifying and solving problems, and assisting with dissemination of the study findings. It is worth bearing in mind, however, that too many meetings or other interactions can lead to participation fatigue (Neef and Neubert, 2011), so it is important that a balance is struck. Table 5.1 gives an example of a research study and the stakeholder involvement activities that could be implemented at each stage.

How many people will be involved, and finding people to involve

The reason for the inclusion of certain individuals or groups should always be clear and transparent. For example, you will want to consider which individuals are most likely to bring the experience and knowledge you need to the project. Would you like to involve individuals who can

Table 5.1. Stakeholder involvement at various stages of the research cycle. Adapted from National Institute for Health Research (2020).

Stage in research cycle	Example of stakeholder involvement
Agreeing priorities for research	A round-table discussion with representatives from the farming community, academia and government bodies to discuss welfare priorities for indoor-reared pigs.
Agreeing which outcomes to use in a study	A steering group is formed and is made up of agricultural advisers and representatives from farming, meat processing and consumer groups. In collaboration with the researchers, a set of key outcomes to assess is established.
Agreeing which outcome measures to use in a study	The steering group considers the practical constraints of collecting data and makes suggestions on data collection methods such as the simplification of lesion scoring systems.
Stakeholder perspective on study design	The steering group assesses the feasibility and practicability of the research design to determine whether farmers and meat processors would agree to take part in the proposed study, and suggest changes.
Stakeholder involvement in the design of study documentation	Agriculture and meat technology students at an agricultural college help the researchers to write a plain English summary of the documentation. This documentation will be provided to participants prior to data collection. With many of these students coming from a family farming background, they are well placed to advise the researchers on accessible and engaging language.
Problem identification and solving during the project	The steering group uses their experience and network of contacts to provide solutions to the researchers throughout the data collection period as problems are identified.
Interpreting data	Results from the study are presented at an animal science conference attended by individuals from academic and industry backgrounds. Comments from the audience help in interpreting unexpected results.
Commenting on end-of-study information	The steering group provides guidance on identifying and communicating the findings most important to each stakeholder.
Assistance with disseminating study findings	The steering group uses their existing network of contacts to disseminate the findings to those who will be implementing them in practice.

speak of their own professional or lived experience (e.g. a farmer or consumer)? Would you like an individual who can advocate for a group (e.g. a meat processor's union representative)? Does the individual stakeholder need to provide a local perspective? The importance of a diversity of views and perspectives is increasingly acknowledged, so it is important to consider multiple perspectives, and you should determine whether any pertinent individuals or groups may be hard to reach. For example, you may want to work with individuals who have a criminal record or a history of animal abuse. In order to access this demographic, you could consider working with key organizations that are trusted by this group, such as an ex-offenders support organization. It may take a long time to engage people in your research and to build working relationships, so the process of finding stakeholders needs to start early. Potential candidates may be identified through personal connections, advertisements in relevant media outlets or through professional organizations. You could also approach stakeholders through social media or by asking other stakeholder partners to identify relevant persons or groups (i.e. snowball sampling). Using a variety of approaches to identify stakeholder partners can help to ensure a diversity of views and perspectives.

Formalizing involvement

Getting a close match between the requirements of the task and the individual performing the task is essential. It is best practice to formalize the stakeholder contributor role with a detailed person specification and role description (which you use to advertise the opportunity). Key information to incorporate in these documents includes a description of the project and the role, whether the role is open ended or fixed term, and the time commitment. Time costs should not be underestimated (Neef and Neubert, 2011). You should also detail whether there is travel involved and outline the training or support that would be provided to the individual. You should outline any payment for the role, expenses covered and any other incentives. The main contacts for the role should be provided. The NIHR INVOLVE website (see Further Reading) has templates of these documents available pertaining to PPI, which would also be relevant to stakeholder involvement in animal welfare projects. An expression-of-interest form should be developed to collect any information on stakeholder representatives who declare an interest in the role. It is then good practice on receipt of the expression of interest to contact the applicant and discuss any questions they may have.

Impact of stakeholder involvement

When developing the stakeholder involvement strategy, it is important to consider how you will evaluate the impact of the stakeholder involvement.

This is because stakeholder involvement may have impacts on the research from an early stage. As stakeholder involvement is a relatively new focus within animal welfare science, this will contribute to the discipline's understanding of what impacts are achieved and how involvement works best in different contexts. For example, stakeholder involvement could increase knowledge, skills and confidence in farmers to tackle animal welfare issues. Stakeholders may also blend their own knowledge with scientific knowledge (Neef and Neubert, 2011), and this has the potential to have beneficial impacts in terms of animal welfare. Similarly, stakeholder involvement could also increase the researcher's knowledge of the practical constraints that affect the success of animal welfare solutions. Finally, stakeholder involvement may impact on how the findings are disseminated. The Public Involvement Impact Assessment Framework (PiiAF; see Further Reading) includes a guide for exploring impacts and developing an impact assessment plan.

Provide stakeholders with accessible feedback

Stakeholders should be kept informed of the research findings and any developments that happen along the way. Importantly, research findings should be easy for stakeholders to understand (Neef and Neubert, 2011). Therefore, technical jargon should be avoided, and in-depth details of statistical analyses should be reserved for academic conferences and journal articles. In addition to furnishing stakeholder contributors with accessible information, stakeholders more generally should be able to access the resultant publications associated with any research. This is being facilitated by an increase in open access publishing. This is where reports of research are freely and openly available to all, without a paywall limiting access to a selected few. This increases the accessibility of research, supporting increased scrutiny, reproducibility and replicability. In addition, an increasing number of journals are requiring authors to provide a lay summary of the research in addition to the more traditional abstract. This allows individuals from a variety of backgrounds to read about and understand the key aims, methods and findings associated with a piece of research.

5.3.3 Pitfalls

Using a top-down approach

A top-down approach involves a small number of individuals making decisions on behalf of those who will be directly affected by those decisions. A bottom-up approach is preferable, where stakeholders are consulted throughout the development of animal welfare solutions. This gives the researchers insight into practical constraints, practical implementation knowledge and stakeholder priorities. Each of these is likely to affect the success of any given solution.

Failure to involve all relevant stakeholders

Within animal welfare science, there is a range of stakeholders to consider, and this will vary according to the species and context in which the animal is kept. It is important to give careful consideration to the identification of all individuals or organizations that may be affected by your research. For example, when working on farm animal welfare solutions, you may engage with producers, meat processors and retailers but fail to include consumer perspectives.

Tokenistic stakeholder involvement

It is important to develop a stakeholder involvement strategy before commencing stakeholder involvement activities. This strategy needs to be developed collaboratively and should be agreed between all research team members (including your stakeholder partners). The development of a stakeholder strategy helps to avoid tokenistic stakeholder involvement, which is unlikely to have the desired impacts. Stakeholder involvement should not be viewed as a 'tick box' exercise and should be given adequate consideration.

For further pitfalls, see Chapter 3 (this volume).

5.4 Conclusions and Tips

Humans are an essential consideration when seeking to make improvements to the welfare of animals that are under human control. Human-centred techniques are commonplace within health psychology and other social sciences, yet have been implemented within animal welfare science with limited success. Often, questionnaires and other instruments are not properly designed or validated. They may also be difficult for other researchers to access, and therefore difficult to evaluate or use. Similarly, stakeholder involvement in research projects can be tokenistic or absent altogether. Issues such as these are often due to unfamiliarity with, or lack of confidence in, carrying out research with humans. This chapter provides a starting point for those interested in adopting and integrating psychological research methods into their research. By highlighting key criteria and common pitfalls associated with human-centred research, the aim is to increase awareness in animal welfare scientists of best practice. There are a number of resources available to guide questionnaire design and the process of stakeholder involvement. Existing guidance from health psychology and related disciplines can be applied to an animal welfare context. However, it is always advisable to consult or collaborate with psychologists who have been trained in questionnaire development and stakeholder involvement. Interdisciplinary research is increasingly being valued by funders and employers, and allows animal welfare scientists to combine their expertise with that of researchers from other disciplines. This will give animal welfare scientists confidence in their methods and subsequent results. Finally, open science practices should be integrated into your research approach. For example, questionnaires and other instruments should be made available as supplementary material alongside publications, and research results should be shared in various formats that are accessible to a variety of animal welfare stakeholders.

Further Reading

For further literature on survey design, a definitive guide to scale design is offered by DeVellis (2016) and this is the best place to start for anyone interested in developing a new measure. Other useful guidance (given in bold in the reference list) on best practices in scale development are provided by Costello and Osborne (2005) and Worthington and Whittaker (2006). The seminal paper by Campbell and Fiske (1959) outlines the multi-trait/multi-method matrix, a rigorous methodological and statistical approach to establishing construct validity. Field (2017) provides a detailed explanation of EFA in SPSS, while Byrne (2013) offers a step-by-step guide for conducting confirmatory factor analysis using structural equation modelling in a variety of different statistical packages. Finally, some subject-specific detail is available in the review by Belshaw *et al.* (2015).

For further literature on stakeholder involvement, there are many well-recognized websites and resources relating to PPI in health-related research, which can be applied to an animal welfare research context. The list below refers to resources from well-recognized PPI organizations. Two examples of stakeholder engagement in an animal welfare context are also provided on the topics of managing cat predation of wildlife (Crowley *et al.*, 2020) and the welfare of laying hens (van Dijk *et al.*, 2019).

- *BMC Research Involvement and Engagement.* An interdisciplinary, health and social care journal focusing on patient and wider involvement and engagement in research, at all stages. The journal is co-produced by all key stakeholders, including patients, academics, policy makers and service users. The resource is available via the following link, with many of the articles open access: https://researchinvolvement.biomedcentral.com/ (accessed 4 June 2021).
- *NIHR INVOLVE.* Funded by the National Institute for Health Research to support active public involvement in the National Health Service, public health and social care research. It is one of the few government-funded programmes of its kind in the world. The resource is freely available at: www.invo.org.uk/ (accessed 4 June 2021).
- *UK Standards for Public Involvement.* Six core standards that are designed to improve the quality and consistency of public involvement in research. The standards are a description of what good public involvement looks

like and encourage approaches and behaviours that are the hallmark of good public involvement such as flexibility, sharing and learning, and respect for each other. The resource is freely available at: https://sites.google.com/nihr.ac.uk/pi-standards/home (accessed 4 June 2021).

- *Public Involvement Impact Assessment Framework (PiiAF)*. A framework and associated resources for exploring and assessing impact from PPI in research. The resource is freely available at: http://piiaf.org.uk/ (accessed 4 June 2021).
- *National Co-ordinating Centre for Public Engagement (NCCPE)*. Aligning with engagement as opposed to involvement, this organization provides training, resources and examples of best practice in undertaking public engagement. The resource is freely available at: www.publicengagement.ac.uk/ (accessed 4 June 2021).

References

Aluwé, M., Heyrman, E., Almeida, J.M., Babol, J., Battacone, G. *et al.* (2020) Exploratory survey on European consumer and stakeholder attitudes towards alternatives for surgical castration of piglets. *Animals* 10, 1758. DOI: 10.3390/ani10101758.

Belshaw, Z., Asher, L., Harvey, N.D. and Dean, R.S. (2015) Quality of life assessment in domestic dogs: an evidence-based rapid review. *The Veterinary Journal* 206(2), 203–212. DOI: 10.1016/j.tvjl.2015.07.016.

Byrne, B.M. (2013) *Structural Equation Modeling with Mplus: Basic Concepts, Applications, and Programming.* Routledge, New York.

Campbell, D.T. and Fiske, D.W. (1959) Convergent and discriminant validation by the multitrait-multimethod matrix. *Psychological Bulletin* 56(2), 81–105. DOI: 10.1037/h0046016.

Chalmers, I. and Glasziou, P. (2009) Avoidable waste in the production and reporting of research evidence. *The Lancet* 374(9683), 86–89. DOI: 10.1016/S0140-6736(09)60329-9.

Choi, B.C.K. and Pak, A.W.P. (2005) A catalog of biases in questionnaires. *Preventing Chronic Disease* 2(1), A13.

Christensen, T., Lawrence, A., Lund, M., Stott, A. and Sandøe, P. (2012) How can economists help to improve animal welfare? *Animal Welfare* 21(1), 1–10. DOI: 10.7120/096272812X13345905673449.

Connor, M., Currie, C. and Lawrence, A.B. (2018) Factors influencing the prevalence of animal cruelty during adolescence. *Journal of Interpersonal Violence* 36(7–8), 3017–3040. DOI: 10.1177/0886260518771684.

Cornish, A.R., Briley, D., Wilson, B.J., Raubenheimer, D., Schlosberg, D. *et al.* (2020) The price of good welfare: does informing consumers about what on-package labels mean for animal welfare influence their purchase intentions? *Appetite* 148, 104577. DOI: 10.1016/j.appet.2019.104577.

Costello, A.B. and Osborne, J. (2005) Best practices in exploratory factor analysis: four recommendations for getting the most from your analysis. *Practical Assessment, Research, and Evaluation* 10, 7.

Crowley, S.L., Cecchetti, M. and McDonald, R.A. (2020) Diverse perspectives of cat owners indicate barriers to and opportunities for managing cat predation of wildlife. *Frontiers in Ecology and the Environment* 18(10), 544–549. DOI: 10.1002/fee.2254.

de Olde, E.M., Carsjens, G.J. and Eilers, C.H.A.M. (2017) The role of collaborations in the development and implementation of sustainable livestock concepts in The Netherlands. *International Journal of Agricultural Sustainability* 15(2), 153–168. DOI: 10.1080/14735903.2016.1193423.

DeVellis, R.F. (2016) *Scale Development: Theory and Applications.* SAGE Publications, Thousand Oaks, California.

Fecher, B. and Friesike, S. (2014) Open science: one term, five schools of thought. In: Bartling, S. and Friesike, S. (eds) *Opening Science.* Springer, Cham, Switzerland, pp. 17–47.

Fernandes, J., Blache, D., Maloney, S.K., Martin, G.B., Venus, B. *et al.* (2019) Addressing animal welfare through collaborative stakeholder networks. *Agriculture* 9(6), 132. DOI: 10.3390/agriculture9060132.

Field, A. (2017) *Discovering Statistics Using IBM SPSS Statistics.* SAGE Publications, Thousand Oaks, California.

INVOLVE (2020) What is public involvement in research? National Institute for Health Research, UK. Available at: http://www.invo.org.uk/find-out-more/what-is-public-involvement-in-research-2/ (accessed 3 June 2021).

Klerkx, L. and Nettle, R. (2013) Achievements and challenges of innovation co-production support initiatives in the Australian and Dutch dairy sectors: a comparative study. *Food Policy* 40(3), 74–89. DOI: 10.1016/j.foodpol.2013.02.004.

Lund, V., Coleman, G., Gunnarsson, S., Appleby, M.C. and Karkinen, K. (2006) Animal welfare science—Working at the interface between the natural and social sciences. *Applied Animal Behaviour Science* 97(1), 37–49. DOI: 10.1016/j.applanim.2005.11.017.

Mullan, S. (2015) Assessment of quality of life in veterinary practice: developing tools for companion animal carers and veterinarians. *Veterinary Medicine: Research and Reports* 6, 203–210. DOI: 10.2147/VMRR.S62079.

National Institute for Health Research (2020) UK Standards for Public Involvement. Available at: https://sites.google.com/nihr.ac.uk/pi-standards/home (accessed 3 June 2021).

Neef, A. and Neubert, D. (2011) Stakeholder participation in agricultural research projects: a conceptual framework for reflection and decision-making. *Agriculture and Human Values* 28(2), 179–194. DOI: 10.1007/s10460-010-9272-z.

Pejman, N., Kallas, Z., Dalmau, A. and Velarde, A. (2019) Should animal welfare regulations be more restrictive? A case study in eight European Union countries. *Animals* 9(4), 195. DOI: 10.3390/ani9040195.

Pulido, M.A., Mariezcurrena-Berasain, M.A., Sepúlveda, W., Rayas-Amor, A.A., Salem, A.Z.M. *et al.* (2018) Hauliers' perceptions and attitudes towards farm animal welfare could influence the operational and logistics practices in sheep transport. *Journal of Veterinary Behavior* 23(3), 25–32. DOI: 10.1016/j.jveb.2017.10.004.

van Asselt, M., Ekkel, E.D., Kemp, B. and Stassen, E.N. (2016) Role of moral values in the trade-off between animal welfare and food safety risks in broiler husbandry. In: *Food Futures: Ethics, Science and Culture.* Wageningen Academic Publishers, Wageningen, The Netherlands, pp. 273–278.

van Dijk, L., Buller, H.J., Blokhuis, H.J., van Niekerk, T., Voslarova, E. *et al.* (2019) HENNOVATION: learnings from promoting practice-led multi-actor innovation networks to address complex animal welfare challenges within the laying hen industry. *Animals* 9(1), 24. DOI: 10.3390/ani9010024.

Vicente-Sáez, R. and Martínez-Fuentes, C. (2018) Open Science now: a systematic literature review for an integrated definition. *Journal of Business Research* 88(4), 428–436. DOI: 10.1016/j.jbusres.2017.12.043.

Weary, D.M., Ventura, B.A. and von Keyserlingk, M.A.G. (2016) Societal views and animal welfare science: understanding why the modified cage may fail and other stories. *Animal* 10(2), 309–317. DOI: 10.1017/S1751731115001160.

Wemelsfelder, F., Hunter, T.E.A., Mendl, M.T. and Lawrence, A.B. (2001) Assessing the 'whole animal': a free choice profiling approach. *Animal Behaviour* 62(2), 209–220. DOI: 10.1006/anbe.2001.1741.

Worthington, R.L. and Whittaker, T.A. (2006) Scale development research: a content analysis and recommendations for best practices. *The Counseling Psychologist* 34, 806–838.

6

Human Behaviour Change Models for Improving Animal Welfare

Grace A. Carroll[1]*, Jenny M. Groarke[2] and Lisa Graham-Wisener[2]

[1]Animal Behaviour Centre, School of Psychology, Queen's University Belfast, Belfast, UK; [2]Centre for Improving Health-Related Quality of Life (CIHRQoL), School of Psychology, Queen's University Belfast, Belfast, UK

6.1 Introduction

Understanding the reasons why pet owners, zookeepers or farmers act or fail to act in relation to animal welfare is a key consideration; it determines whether research findings are translated into real-world changes. Behaviour change frameworks were originally developed within the field of health psychology. These frameworks are applied to a number of human health-related behaviours such as smoking cessation, weight management and safe sexual health practices. While well-established within health psychology and related disciplines, in recent years there has been a growing interest in human behaviour change within the animal welfare community. Animal welfare studies aimed at changing human behaviour have traditionally focused on attitudes (Carroll and Groarke, 2019), with the idea that attitudes predict intentions, which, in turn, predict behaviour (Ajzen, 1985). Therefore, it is assumed that an intervention that alters attitudes will lead to a change in behaviour. However, psychologists have come to realize that what people intend to do does not always determine what they actually do (Sniehotta *et al.*, 2005). For example, zoo workers may intend to provide zoo animals with the best welfare standards possible. However, government regulations, management style or enclosure size constraints will also play an important role in determining whether a desired behaviour will be performed. A holistic approach is needed that considers both intrinsic and extrinsic factors affecting the behaviour of key animal welfare stakeholders.

*Corresponding author: G.Carroll@qub.ac.uk

 Bridging Research Disciplines to Advance Animal Welfare Science: A practical guide (ed. I. Camerlink)
DOI: 10.1079/9781789247893.0006

Attempts have been made within animal welfare science to develop and execute human behaviour change interventions. However, perhaps due to unfamiliarity with psychological methods, methodological flaws limit many of these. Glanville *et al.* (2020) reviewed behaviour change interventions reported in 47 animal care-related papers and found several issues. For example, the target behaviour to be changed was often too vaguely defined, making it difficult to determine whether the desired outcome had been achieved. In addition, intervention content was inadequately described, meaning that the aspects of the intervention that worked or did not work could not be identified. This, in turn, makes replication difficult. Moreover, only 16 of the 47 reviewed studies utilized a randomized controlled trial (RCT) design. Clearly, there is much work to be done in improving the methods that animal welfare scientists use when designing and evaluating human behaviour change interventions. Often, animal welfare scientists decide to assess the effectiveness of a particular training programme or information session without explicitly intending to carry out a behaviour change intervention. Indeed, many are unaware that an entire research discipline exists that is dedicated to the science of human behaviour change. It is therefore unsurprising that mistakes are made. These mistakes have already been made and overcome in other disciplines, which animal welfare science can now benefit from. This highlights the importance of interdisciplinary work as a way to avoid duplication of effort.

In recent years, psychologists have moved towards implementation of overarching frameworks of available theories and models of human behaviour change. In this chapter, three key frameworks will be outlined, as well as how they fit together. This will allow animal welfare scientists to develop human behaviour change plans from start to finish. These three frameworks are (i) the theoretical domains framework (TDF; Michie *et al.*, 2005); (ii) the behaviour change wheel (BCW; Michie *et al.*, 2011, 2014); and (iii) the behaviour change technique taxonomy (BCTT; Michie *et al.*, 2013). Each framework is described, and guidance is provided for applying these models to design effective human behaviour change interventions. Common pitfalls are outlined, and animal welfare examples are provided throughout to aid understanding.

6.2 Concepts

6.2.1 Behaviour Change Wheel

The BCW (Michie *et al.*, 2011) was developed from 19 frameworks of behaviour change, bringing together frameworks that focus on both internal factors, such as attitudes, and external factors, such as social environment. The main goal of the BCW is to support the development of theory-based interventions in a systematic way. It can also be used to help diagnose why

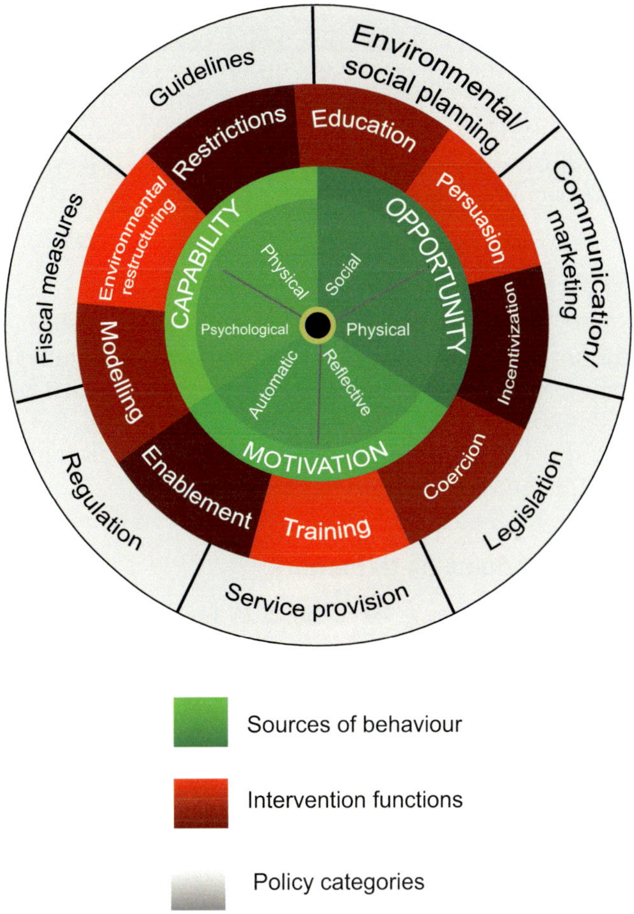

Fig. 6.1. The behaviour change wheel. From Michie *et al*. (2014). © S. Michie.

an intervention may have failed to achieve its goal (Michie *et al.*, 2014). The BCW is made up of three layers (COM-B, intervention functions and policy categories) (Fig. 6.1).

COM-B

The inner-most layer of the wheel identifies sources of behaviour that can be targeted for intervention. COM-B stands for 'capability opportunity motivation – behaviour'. For any *behaviour* to occur, the individual must have the *capability* to do it, the *opportunity* to perform the behaviour and the necessary *motivation* to carry it out (Michie *et al.*, 2014). The COM-B can be used to gain an understanding of the target behaviour, i.e. in terms of a person's capability, opportunity and motivation: what are the barriers to, or facilitators of, the behaviour that you want to change?

Intervention functions

The second layer of the wheel is made up of nine intervention functions: education, persuasion, incentivization, coercion, training, restrictions, environmental restructuring, modelling and enablement. Intervention functions are broad categories of methods that can be used to bring about changes to a target behaviour (Michie *et al.*, 2014).

Policy categories

The outer-most layer of the wheel is made up of seven policy categories; communication/marketing, guidelines, fiscal measures, regulation, legislation, environmental/social planning and service provision. The intervention under development may fall into any one of these policy categories, which reflect types of decisions made by authorities that help to support and sanction the intervention (Michie *et al.*, 2014).

6.2.2 Theoretical domains framework

The TDF is similar to the COM-B of the BCW but is more detailed (Michie *et al.*, 2014). There are 14 'domains' thought to influence behaviour: knowledge; skills; memory, attention and decision processes; behavioural regulation; social/professional role and identity; beliefs about capabilities; optimism; beliefs about consequences; intentions; goals; reinforcement; emotion; environmental context and resources; and social influences. These domains are a synthesis of several behaviour change theories. The TDF can be used to identify barriers to, and facilitators of, behaviour change by informing the schedule of questions included in questionnaires, focus groups and interviews. Each domain of the TDF is associated with a COM-B component. If it is not practicable to examine all 14 domains, the COM-B can be used as a screening tool to give an indication of which domains to focus on (Michie *et al.*, 2014).

6.2.3 Behaviour change technique

A behaviour change technique (BCT) is defined as an active element of an intervention designed to change behaviour (Michie *et al.*, 2014). BCTs should be observable and replicable.

6.2.4 Behaviour change technique taxonomy

The BCTT is a standardized catalogue of BCTs. This taxonomy specifies 93 different BCTs, which are organized into 16 overarching headings: goals and planning, feedback and monitoring, social support, shaping knowledge, natural consequences, comparison of behaviour, associations, repetition and substitution, comparison of outcomes, reward and threat, regulation, antecedents, identity, scheduled consequences, self-belief and covert learning (Michie *et al.*, 2013). Intervention designers select the BCTs that are most effective for addressing the target behaviour based on the available empirical evidence.

6.2.5 APEASE criteria

The APEASE criteria are: affordability, practicability, effectiveness and cost-effectiveness, acceptability, side-effects/safety and equity of a proposed intervention. They help intervention designers to evaluate intervention ideas. These criteria should be explored before the proposed intervention is implemented or scaled up. For example, if a farmer is operating on tight profit margins, an effective intervention that involves investment in precision livestock farming tools (see Chapter 15, this volume) may not be successful (affordability). Similarly, the public may oppose a ban on online puppy sales to tackle puppy farming (acceptability).

6.2.6 Randomized control trials

RCTs are common practice in clinical research and are viewed as the gold standard of empirical investigation. Participants are randomly assigned to the intervention (e.g. education session) or control group (e.g. no education session). The groups are followed up at the end of the study and scores on the outcome measure (e.g. depressive symptoms) are compared. RCTs provide rigorous testing of a treatment, intervention, hypothesis or causal mechanism. An example from animal welfare science might be that sheep farmers are randomly assigned to the intervention (e.g. expert demonstration of lameness detection) or a control group (e.g. no demonstration). The farmers are followed up at the end of the study and scores on the outcome measure (e.g. lameness prevalence in the flock) are compared. The use of randomization minimizes the risk that confounding factors will bias the results. This means that any difference in the outcome is due to the intervention and not because of baseline differences between the groups. An RCT is considered to offer reliable evidence for the effectiveness of an intervention.

6.3 Techniques

This section describes the four key stages in developing behaviour change interventions:

1. Understand the behaviour.
2. Identify intervention options.
3. Identify intervention content and implementation options.
4. Evaluate the intervention.

Within these stages, the BCW (Fig. 6.1), the TDF and the BCTT are applied.

6.3.1 Understand the behaviour

The first step in intervention design requires the intervention developer to fully understand the behaviour and what needs to happen for it to change. This involves defining the problem in behavioural terms (Michie *et al.*, 2014). For example, the researcher may want to reduce rates of obesity in dogs. However, obesity is an outcome and can be influenced by a range of behaviours. Therefore, consideration should be given to the specific behaviour or set of behaviours that need to change and the person or group of people who need to change it. In this example, a dog may have obesity because the animal is not getting enough exercise and because its diet is poor. Target behaviours should be identified for the dog's primary caregiver. Appropriate targets could be increasing the length and frequency of dog walks, and improving the dog's diet by purchasing food designed for weight management. Alternatively, intervention designers could focus on behaviours that should cease such as purchasing dog treats or feeding the dog scraps from the dinner table. One or two key target behaviours should be selected, rather than targeting several at once (Michie *et al.*, 2014). However, when developing an intervention, the intervention developer must first understand all the potential behaviours that come to bear on a specific problem. For example, is the person who walks the dog also the person responsible for buying the food that the dog eats? When the intervention designer has identified potential target behaviours, these should be narrowed down by considering the likely impact of changing the behaviour, how easy it will be to change, how changing this behaviour will affect other behaviours and how easy the behaviour will be to measure (Michie *et al.*, 2014). The intervention designer may choose to measure the success of the intervention by assessing the performance of the behaviour itself, such as increased step count in the pet owner, or by assessing an outcome of the behaviour, such as improved body condition score in the dog. This allows the success of behaviour change interventions to be determined on multiple levels. The intervention developer should then be specific in describing the target behaviour by considering who needs to perform the behaviour, what they need to do, when and where they should do it, how often and with whom (Michie *et al.*, 2014). If the target behaviour is not described in sufficient detail, it makes it difficult to determine the effectiveness of the intervention. For example, Rayner *et al.* (2020) assessed the effectiveness of an educational intervention on veterinarian knowledge and attitudes towards animal welfare and euthanasia. This study was carried out in India where, due to cultural or religious beliefs, there is often opposition to euthanizing animals. Rayner *et al.* (2020) found significant improvements in both knowledge and attitudes immediately post-intervention. However, knowledge and attitudes do not always translate into behaviour change (Sniehotta *et al.*, 2005), and therefore it is difficult to determine the success of this intervention, as its purpose was not sufficiently defined in behavioural terms. It would be useful to directly assess intervention success by specifying a target behaviour or outcome that can be measured. For example, the intervention designers could assess appropriate drug use when euthanizing animals (behaviour) or assess the number of animals within the veterinary practice euthanized pre- and post-intervention (outcome). The details can then be fleshed out using the 'who, what, when, where, how often, with whom' criteria (Michie *et al.*, 2014).

The intervention developer should then identify what specifically needs to change in order for the target behaviour to change (Michie *et al.*, 2014). This involves identifying barriers to, and facilitators of, the target behaviour or outcome. This can be done by reviewing the literature, and by consulting with key stakeholders via interviews, focus groups and questionnaires. The COM-B model (Michie *et al.*, 2011, 2014) and the TDF (Michie *et al.*, 2005) can assist in this process. For example, while surveying dog owners, it may be helpful to consider the dog owner's *capability*, *opportunity* and *motivation* to perform a desired *behaviour*, such as improving their dog's diet and increasing their physical activity. It may become apparent that there are several overarching issues. For example, it may be that dog owners do not have the *psychological capability* (i.e. the 'know-how') to distinguish nutritious dog food from unhealthy food. In addition, changing their current habits may not be a priority, possibly due to a perception that obesity does not harm dog health, i.e. the pet owner lacks the *reflective motivation* (i.e. the 'know-why') to change their dog-feeding behaviour. Some owners may have mobility issues and therefore lack the *physical capability* to provide an adequate amount of exercise for their pets. The TDF subdivides the components of the COM-B model further. For example, *reflective motivation* can be influenced by the domains of *social/professional role and identity*, *beliefs about capabilities*, *optimism*, *intentions*, *goals* and *beliefs about consequences*. These different influences on behaviour could then be delved into in more comprehensive interviews or focus groups. For example, do dog owners have sufficient *knowledge* to feed their dog a healthy diet? What are their specific *beliefs about consequences* of obesity in dogs? What roles do *environmental context and resources* play in the owners' ability to take their dog for regular walks? This process of systematically identifying barriers to, and facilitators of, behaviour change allows a comprehensive picture of the target behaviour to be identified and suggests potential avenues for changing behaviour through specific types of interventions and BCTs.

When the determinants of a behaviour or outcome are not considered, researchers are in a sense 'guessing' as to what might bring about change in the target behaviour. For example, Chilundo *et al.* (2020) designed an intervention to increase knowledge of pig production methods in order to improve livestock performance in a rural resource-poor community in Mozambique. However, while increasing knowledge (COM-B component: *psychological capability*) may well be an important facilitator of change, other key determinants were not explored. If farmers were interviewed using a script shaped by the COM-B model, the intervention developers would probably identify other key barriers to improvement in pig production methods. For example, given the poverty faced within these communities, lack of *opportunity* to perform desired target behaviours is likely to play a major role. Indeed, Chilundo *et al.* (2020) noted that some pig farmers became aware that they could improve their pig pens using inexpensive materials, and the number of farmers to do so was substantial. The exercises associated with 'understanding the behaviour' allow intervention developers to select the most appropriate intervention options.

6.3.2 Identify intervention options

Step 2 involves designing an appropriate intervention in light of the determinants of the target behaviour identified in step 1. Similar to how the TDF fits under the COM-B model, the COM-B model in turn sits under the *intervention functions* and *policy categories* of the BCW (Fig. 6.1). Depending on the determinants of behaviour, some intervention functions and policy categories will be more suitable for an intervention than others. Links between each layer of the BCW can be found in Michie *et al.* (2014; see p. 113 for links between COM-B and the TDF, p. 116 for COM-B and intervention functions, and p. 138 for links between intervention functions and policy categories). For example, an intervention aimed at increasing *psychological capability* could be achieved through the intervention functions of *education, training* or *enablement.* Using the 'obesity in dogs' example, an intervention with the function of *education* could involve an animal nutritionist educating owners on how to use the correct feeding schedule, food type and food quantity, considering a dog's age, breed and size. If for the general public, this intervention could fall into the policy category of *communication/marketing*, or the intervention developer may decide to use *service provision* to support and sanction the intervention in veterinary practices. Alternatively, an intervention with the function *persuasion* could involve using prompts to take the dog for a walk via a smart phone application, thereby increasing the dog owner's *reflective motivation* to engage in a greater number of walks per week. This would be facilitated through the policy category of *environmental planning*. Intervention developers will be restricted in the policy categories available to them. For example, while government bodies will have the ability to change behaviour via *legislation*, animal welfare scientists are more likely to use less authoritarian measures such as voluntary compliance via *guidelines* or persuasion via *communication/marketing*. Once the intervention function and policy category have been decided, the intervention content must be specified, and options for implementation are explored.

6.3.3 Identify intervention content and implementation options

In the final stage of intervention development, the intervention content and implementation options are selected. The BCTT (Michie *et al.*, 2013) can be used for this purpose. Appropriate BCTs are selected based on their successful application in previous research. The BCTT allows researchers to develop interventions in a systematic and rigorous way. This ensures accurate reporting of behaviour change interventions, thereby enabling future replication of the intervention. Any given intervention can have several 'active ingredients' that make up the content of the intervention (Michie *et al.*, 2013). If these are not clearly specified, it can be difficult to determine what worked and what did not work (see section 6.4). For example, the intervention could be described as: 'An animal nutritionist visits the dog owner's home and explains how an appropriate feeding schedule is developed. The animal nutritionist demonstrates how to measure out the appropriate amount of food and explains the potential health consequences of obesity in dogs.' The BCTT can be used to specify all the active ingredients within this intervention that could explain any changes seen in the target behaviour or outcome. This can be done by viewing the paper-based version of the BCTT (e.g. Michie *et al.*, 2014, p. 259). However, a smart phone application is also available, which can be found in major app stores by searching for 'BCT taxonomy'. Notes under each technique guide the intervention designer by suggesting other potential active ingredients that may be at play and help to clarify which criteria are, or are not, considered under each technique. For example, the intervention above is delivered by an animal nutritionist. It therefore falls under technique 9.1, 'Credible source', which is described as 'Present verbal or visual communication from a credible source in favour of or against a behaviour.' In the Notes section, it specifies the sources that can be considered credible. Health professionals are included as credible sources, suggesting that this intervention technique is suitable. Other BCTs are then suggested. For example, the text reads 'If information about health consequences, also code 5.1, Information about health consequences'. The intervention developer can then go to the page associated with this new technique. The Notes section suggests that consequences can be for any target and not just the individual receiving the intervention. Based on this information, it is decided that 'Information about health consequences' is also an active ingredient of the intervention, as the owner is receiving information about health consequences of obesity in their dog. After considering several different techniques, it is apparent that this behaviour change intervention has four active ingredients, namely, '4.1 Instruction on how to perform the behaviour', '5.1 Information about health consequences', '6.1 Demonstration of the behaviour' and '9.1 Credible source'. This process of mapping the intervention content to the BCTT makes interventions easier to replicate and allows different interventions to be compared within systematic reviews (Michie *et al.*, 2013). However, this step is often overlooked. For example, Khode *et al.* (2018) assessed the effectiveness of dairy farming training on knowledge levels in dairy farmers. However, the training procedure and content of the training was not described. It would therefore be impossible for other researchers to try to replicate these findings. Similarly, Dias Costa *et al.* (2017) assessed the effectiveness of a pet management programme but did not provide sufficient details of the programme to allow other researchers to identify its active ingredients. The detailed guidance provided by Michie *et al.* (2013) increases the accessibility of the BCTT, making it easier for animal welfare scientists to accurately map their intervention to its active components moving forward.

6.3.4 Evaluate the intervention

Once an intervention has been designed, it should be evaluated according to the APEASE criteria, which assess a number of intervention characteristics such as practicability, cost-effectiveness and acceptability (Michie *et al.*, 2014). Acceptability can, for example, be measured either quantitatively or qualitatively, by collecting feedback from relevant stakeholders. Depending on the outcome of this analysis, the intervention can be approved or amended prior to evaluating the effectiveness of the intervention. When the intervention development is complete, an important next step to consider is how to design a high-quality evaluation of the intervention. Much of the time, the interventions we develop are 'complex'; they involve a number of components or a number of outcomes. For example, an intervention to reduce lameness in dairy cattle may target the behaviour of farmers *and* their veterinarians. Outcomes might include recording lameness prevalence or observing the performance of specific management or treatment behaviours. After intervention development, a piloting phase is recommended where researchers test procedures, estimate recruitment/retention and determine the sample size needed for the main intervention study. A pilot study helps to ensure a more high-quality evaluation by assessing the feasibility of the proposed intervention before large-scale data collection begins. Ideally, a feasibility/pilot phase is followed by an RCT. An RCT will aim to assess whether the intervention was successful in changing the target behaviour. Sometimes, it is necessary to use less robust quasi-experimental designs. For example, it may not always be possible to randomly assign participants to a group (Glanville *et al.*, 2020).

6.3.5 Putting it all together

It may be difficult to visualize how these different frameworks come together. A complete example including all frameworks is provided below.

Imagine you are a researcher who wants to improve the welfare of gerbils. More specifically, you want to reduce the prevalence of stereotypic digging behaviour in gerbils living in the domestic household environment. You are aware that gerbils require a dark and narrow chamber to retreat to, and that this chamber must be reached via a tunnel from the surface if stereotypic behaviour is to be circumvented (Wiedenmayer, 1997). You interview gerbil owners, using a schedule of questions that were guided by the COM-B and TDF. From this exercise, you learn that gerbil owners do not know that their pet's corner digging is an abnormal behaviour, with many viewing it as an endearing or harmless 'quirk'. This lack of understanding (COM-B: psychological capability; TDF: knowledge) could be addressed by educating pet owners about appropriate housing of gerbils prior to purchase (intervention function: education; policy category: communication/marketing). In addition, pet owners will not be motivated to make improvements to the gerbils' environment if corner digging is viewed as a harmless behaviour. Thus, misinterpretation of gerbil behaviour (COM-B: reflective motivation; TDF: belief about consequences) could be addressed by playing a video about the welfare implications of stereotypic behaviour in gerbils, providing an example of how to avoid it (intervention function: modelling; policy category: communication/marketing). In addition, you learn that gerbil owners house their pets in whatever enclosure is available at the pet store. The provision of inappropriate housing (COM-B: physical opportunity; TDF: environmental context and resources) could be addressed by requiring pet shops to only stock enclosures that contain a tunnel and burrow (intervention functions: restriction, environmental restructuring). This could be delivered via establishing voluntary agreements with pet stores on the sale of pet enclosures (policy category: regulation). The behavioural target is 'getting owners to purchase two-level enclosures that have a chamber and burrow'. Therefore, there are two interventions, which you describe as:

- Intervention 1: An information day will be held for individuals intending to purchase a pet gerbil. Attendees will receive guidance from a rodent welfare expert on appropriate housing and how this relates to the gerbil's natural environment. The origin of stereotypic behaviour and its implications for subjective welfare in gerbils will be explained. Video clips will show stereotypic behaviour in gerbils and include a demonstration of how to avoid stereotypic behaviour by providing a suitable environment.
- Intervention 2: A voluntary agreement is signed by pet shops. The agreement states that pet shops will only stock enclosures that contain a tunnel and burrow to allow the gerbil access to a hiding place below the surface of the main enclosure.

Using the BCT Taxonomy smartphone app, the intervention developer maps intervention 1 to '4.1 Instruction on how to perform the behaviour', '5.1 Information about health consequences', '5.6 Information about emotional consequences', '6.1 Demonstration of the behaviour' and '9.1 Credible source'. The BCTT is then used to map intervention 2 to '1.8 Behavioural contact', '1.9 Commitment', '12.1 Restructuring the physical environment' and '12.5 Adding objects to the environment'.

Now that the intervention has been designed and mapped, the APEASE criteria can be used to evaluate the proposed interventions. Using intervention 2 as an example, the intervention is assessed for *affordability* (pet stores are keen to improve their reputation for high animal welfare standards and agree to change stock provider at no cost to the researchers); *practicability* (once the stock provider has been changed, no additional workload is placed on pet stores; pet owners shop as normal), *effectiveness and cost-effectiveness* (the intervention is effective as pet owners can only purchase an enclosure that meets the required welfare standards; there is low cost to the pet store but the new enclosures are more expensive for pet owners), *acceptability* (pet owners approve of welfare-friendly enclosures but may find the increased cost of enclosures unacceptable); *side-effects/safety* (no effects on safety) and *equity* (increased cost of enclosures may exclude individuals from lower socio-economic backgrounds from purchasing of enclosures). On completion of this analysis, it is decided that the intervention would only work if lower-cost alternatives are found. The intervention designer returns to earlier parts of the intervention design process and seeks solutions. Finally, an RCT is carried out to evaluate the effectiveness of the intervention in a real-world or applied setting with the target population.

6.4 Pitfalls

6.4.1 Failure to include a control group

A control group is a group of participants that receive no intervention, while an experimental group is one that is exposed to the intervention (Dempster and Hanna, 2015). Failure to include a control group means that you cannot reliably assess intervention effectiveness. For example, while it may be tempting to roll out trap–neuter–release programmes for feral cats in as many locations as possible, there needs to be an equivalent number of locations where this programme is not put in place. You can then measure differences in a target behaviour (e.g. homeowners trapping feral cats and getting them neutered) or outcome (e.g. a reduced number of feral cats in a certain location). If there is no control group for comparison, you cannot know for certain that any observed changes in the behaviours or outcomes of interest were the result of the intervention. For example, other factors such as time of year or food availability could explain any changes seen.

6.4.2 Lack of blinding

Blinding involves concealing group allocation from the participant, the researcher or both. The use of 'blinding' can reduce the risk of bias that may occur when participants and researchers are aware of what condition they have been assigned to. Their perceptions of the intervention or control may influence the results. It is not always possible to blind both the participants and the researchers (i.e. double-blind). If only the participants are kept unaware of group allocation, this is known as single-blind.

6.4.3 Failure to specify all active ingredients of an intervention

Interventions can be composed of several different mechanisms of change (Michie *et al.*, 2014). For example, after interviewing pet owners, you identify low motivation as one of the key barriers to timely spaying of female cats. Based on the available evidence, the function of the intervention will be *incentivization* (i.e. providing an incentive) in the form of a spaying discount voucher. The experimental group is informed that they will receive the voucher if they bring their cat to the vet for spaying in the next 30 days (BCT: goal-setting (behaviour)). The pet owners are also informed that a researcher will contact the vet to confirm that the procedure has been performed. While this intervention appears to be simply a 'financial reward' intervention, there are in actuality a number of components that may be influencing behaviour; the promise of a reward, the reward itself, setting and achieving a behavioural goal, or the 'feedback' and 'social reward' from the vet could all have influenced the pet owner's behaviour to varying extents.

6.4.4 Failure to evaluate the intervention

A failure to specify the target behaviour early on means that intervention success is hard to determine later. This links back to the beginning of the intervention design process. For example, Dias Costa *et al.* (2017) assessed the impact of a pet management programme that included neutering/spaying, lectures and free animal microchipping. The aim of the programme was to encourage responsible pet ownership and reduce the stray dog and cat population. However, rather than assessing the success of the programme in reducing the number of strays, Dias Costa *et al.* (2017) collected information on perceptions of stray dogs. Intervention effectiveness could therefore not be assessed in real terms. Clearly defining the target behaviour or outcome from the beginning allows the intervention developer to effectively assess the success of the intervention on completion.

6.4.5 Lack of collaboration

Before you begin designing your intervention, it is important to note that collaboration with psychologists or other behavioural scientists is recommended (Michie *et al.*, 2014). Combining your relative areas of expertise will enable you to develop a multidisciplinary intervention that is based on a deep understanding of the target behaviour, the context in which the behaviour occurs *and* the intervention design process.

6.5 Conclusions and Tips

Animal welfare cannot be achieved without gaining compliance from the humans who manage those animals. Therefore, designing interventions to change human behaviour is key. Intervention design can be a daunting task for those unfamiliar with the theories and frameworks surrounding human behaviour change. However, adopting a theory-based approach to intervention design can help in ensuring intervention effectiveness. To date, human behaviour change interventions have been designed by animal welfare scientists with limited success. By explicitly stating the common pitfalls to watch out for, this chapter will allow budding intervention designers to begin the process with an awareness of what constitutes good intervention design. The frameworks outlined within this chapter have been designed by experienced psychologists to be practical, accessible and usable (Michie *et al.*, 2014). Therefore, consulting with the recommended Further Reading is highly advisable. There is support for each phase of intervention design. The TDF can be used to identify barriers to, and facilitators of, behaviour change by informing the schedule of questions included in questionnaires, focus groups and interviews. The BCW then provides a useful way of linking sources of behaviour to functions of interventions to change that behaviour. These intervention functions in turn link to policy categories that facilitate behaviour change. Finally, the BCTT is used to specify intervention content. Once an intervention is designed, it needs to be trialled with the target population to determine its effectiveness. The worked examples provided in this chapter have been designed to allow the reader to place behaviour change frameworks within an animal welfare context. This should allow animal welfare researchers to see the relevance of these frameworks to their own work and should aid intervention design. However, it is always advisable to consult or collaborate with psychologists when developing any intervention.

Further Reading

In the reference list in bold, you will find some of the key texts outlining principles from the BCW, the TDF and the BCTT. Becoming familiar with these different frameworks, and how they fit together, can be confusing at first. Michie *et al.*'s (2014) book *The Behaviour Change Wheel: A Guide to Designing Interventions* expands on the key journal articles relating to these frameworks. The book is particularly accessible and provides a good starting point for researchers new to the field of human behaviour change by providing a step-by-step guide to intervention design. It may also aid your understanding to see examples of how these frameworks can be applied to animal welfare problems. Examples are provided of the application of these frameworks to tail biting in pigs (Carroll and Groarke, 2019), consumer food choices (Cornish *et al.*, 2019) and management of free-roaming cat populations (McLeod *et al.*, 2019). In addition, Glanville *et al.* (2020) provide a detailed review of limitations in existing human behaviour change research within animal welfare science.

References

Ajzen, I. (1985) From intentions to actions: a theory of planned behavior. In: Kuhl, J. and Beckmann, J. (eds) *Action Control.* SSSP Springer Series in Social Psychology. Springer, Berlin/Heidelberg, pp. 11–39.

Carroll, G.A. and Groarke, J.M. (2019) The importance of the social sciences in reducing tail biting prevalence in pigs. *Animals* 9, 591. DOI: 10.3390/ani9090591.

Chilundo, A.G., Mukaratirwa, S., Pondja, A., Afonso, S., Alfredo, Z. *et al.* (2020) Smallholder pig farming education improved community knowledge and pig management in Angónia district, Mozambique. *Tropical Animal Health and Production* 52(3), 1447–1457. DOI: 10.1007/s11250-019-02148-x.

Cornish, A., Jamieson, J., Raubenheimer, D. and McGreevy, P. (2019) Applying the behavioural change wheel to encourage higher welfare food choices. *Animals* 9(8), 524. DOI: 10.3390/ani9080524.

Dempster, M. and Hanna, D. (2015) *Research Methods in Psychology for Dummies.* Wiley, Chichester, UK.

Dias Costa, E., Martins, C.M., Cunha, G.R., Catapan, D.C., Ferreira, F. *et al.* (2017) Impact of a 3-year pet management program on pet population and owner's perception. *Preventive Veterinary Medicine* 139(Pt A), 33–41. DOI: 10.1016/j.prevetmed.2017.01.001.

Glanville, C., Abraham, C. and Coleman, G. (2020) Human behaviour change interventions in animal care and interactive settings: a review and framework for design and evaluation. *Animals* 10(12), 2333. DOI: 10.3390/ani10122333.

Khode, N., Singh, B.P., Chander, M., Bardhan, D., Verma, M.R. *et al.* (2018) Impact of training intervention on knowledge level of trained dairy animal owners: an application of propensity score matching method. *International Journal of Agricultural and Statistical Sciences* 14, 285–291.

McLeod, L.J., Hine, D.W. and Driver, A.B. (2019) Change the humans first: principles for improving the management of free-roaming cats. *Animals* 9(8), 555. DOI: 10.3390/ani9080555.

Michie, S., Johnston, M., Abraham, C., Lawton, R., Parker, D. *et al.* (2005) Making psychological theory useful for implementing evidence based practice: a

consensus approach. *Quality and Safety in Health Care* 14(1), 26–33. DOI: 10.1136/qshc.2004.011155.

Michie, S., van Stralen, M.M. and West, R. (2011) The behaviour change wheel: a new method for characterising and designing behaviour change interventions. *Implementation Science* 6, 42. DOI: 10.1186/1748-5908-6-42.

Michie, S., Richardson, M., Johnston, M., Abraham, C., Francis, J. *et al.* (2013) The behavior change technique taxonomy (v1) of 93 hierarchically clustered techniques: building an international consensus for the reporting of behavior change interventions. *Annals of Behavioral Medicine* 46(1), 81–95. DOI: 10.1007/s12160-013-9486-6.

Michie, S., Atkins, L. and West, R. (2014) *The Behaviour Change Wheel. A Guide to Designing Interventions*, 1st edn. Silverback Publishing, Sutton, UK.

Rayner, E.L., Airikkala-Otter, I., Bacon, H.J., Walters, H.M., Gamble, L. *et al.* (2020) Assessment of an educational intervention on the knowledge and attitudes of Indian national veterinarians to animal welfare and euthanasia. *Journal of Veterinary Medical Education* 47(2), 202–217. DOI: 10.3138/jvme.0518-063r.

Sniehotta, F.F., Scholz, U. and Schwarzer, R. (2005) Bridging the intention–behaviour gap: planning, self-efficacy, and action control in the adoption and maintenance of physical exercise. *Psychology & Health* 20(2), 143–160. DOI: 10.1080/08870440512331317670.

Wiedenmayer, C. (1997) Causation of the ontogenetic development of stereotypic digging in gerbils. *Animal Behaviour* 53(3), 461–470. DOI: 10.1006/anbe.1996.0296.

7

Applying Economic Models to Assess the Viability of Animal Welfare Interventions

Rachel S.E. Peden[1,2]* and Faical Akaichi[2]

[1]Newcastle University, Newcastle upon Tyne, UK; [2]Scotland's Rural College (SRUC), Edinburgh, UK

7.1 Introduction

Farm animals are sentient beings and are recognized as such by EU legislation (Treaty of Lisbon 2007/C 306/01, Article 13). Nevertheless, they are also resources involved in economic processes, whose value is in direct proportion to their productivity (McInerney, 1993). These 'sentient commodities' (Wilkie, 2005) have been reared under increasingly intensive production systems since the 1940s. Intensification in animal agriculture provides a number of human benefits (e.g. lower retail prices and higher availability of animal products), but arguably reduces the quality of animal life (Lusk and Norwood, 2012). Today, farms are much larger in size and highly specialized to grow and breed specific livestock species. Furthermore, advances in animal science and technology have driven an increase in livestock production traits (e.g. faster growth rates and increased litter sizes) (Thornton, 2010). As a result of these increased efficiencies and economies of scale, the costs of animal production have reduced, and consumers enjoy cheap, abundant meat (Appleby, 2005).

Although certain animal welfare standards are required to ensure productivity (e.g. feed, housing, health), animal welfare improvements are generally more expensive than the returns gained. For example, researchers have estimated the economic feasibility of animal welfare improvements in broilers (Gocsik *et al.*, 2013, 2015a, b) and laying hen production (Sumner *et al.*, 2011; Gocsik *et al.*, 2015a), in cattle production (Ahmed *et al.*, 2020), in extensive sheep production (Vosough Ahmadi *et al.*, 2010)

*Corresponding author: rachpeden91@hotmail.co.uk

 Bridging Research Disciplines to Advance Animal Welfare Science: A practical guide (ed. I. Camerlink)
DOI: 10.1079/9781789247893.0007

and in intensive pig production (Bornett *et al.*, 2003; Gocsik *et al.*, 2015b; d'Eath *et al.*, 2016), and all studies revealed a net cost to farmers.

Farmers are the primary stakeholders responsible for adopting farm animal welfare interventions in practice, and research into their behaviour regarding animal welfare has increased over recent years (reviewed by Balzani and Hanlon, 2020). Understanding farmers' behaviour and its determining factors is important for identifying targets for initiating a change in practice (also discussed in Chapters 5 and 6, this volume). We are now aware that farmers' behaviour regarding animal welfare is influenced by a wide range of internal (e.g. knowledge, empathy, personality) and external (e.g. economic viability, communication) factors (Carroll and Groarke, 2019; Balzani and Hanlon, 2020). Nevertheless, farmers must usually be confident that the costs of improving animal welfare will be covered by improvements in productivity, or by increases in food prices or subsidies (Bock and van Huik, 2007; Gocsik *et al.*, 2015a).

Despite general knowledge that, for farmers, the cost price of production is imperative, few studies have modelled the economic viability of animal welfare interventions. This is in direct contrast to the extensive research that has been conducted to identify specific changes to animal management, nutrition and genetics, which can improve the welfare of animals in agriculture (e.g. Dwyer *et al.*, 2016; Peden *et al.*, 2018). The lack of economically viable animal welfare interventions is likely to contribute to the limited uptake of animal welfare research in practice. In order to initiate real animal welfare improvements, it is essential that researchers better align animal welfare research with the financial constraints of farmers by identifying practical and economically viable solutions.

7.2 Concepts

Economies of scale

Economies of scale are cost savings that occur when the average cost of producing goods and/or services falls due to an increase in their production (i.e. costs are spread over a larger number of goods) (Silberston, 1972).

Economic viability

Economic viability is an evaluation of the various economic effects that may result from the implementation of a particular project. A project is economically viable if the economic benefits of the project exceed its economic costs.

Revealed preference approaches

Revealed preference (RP) approaches involve the exploration of individuals' preferences as revealed through their actions/choices in real market situations (i.e. the actions have actually occurred) (Hensher *et al.*, 2015). The travel cost method (Brown and Mendelsohn, 1984), hedonic price method (Costanigro *et al.*, 2011) and production function technique (Chakraborty *et al.*, 2002) are examples of RP approaches.

Stated preference approaches

Stated preference (SP) approaches involve asking individuals to state their preferences in hypothetical market situations. These hypothetical situations typically include alternatives/alternatives' attributes observed in actual markets as well as additional alternatives/alternatives' attributes not observed in actual markets (Hensher *et al.*, 2015). The contingent valuation (Venkatachalam, 2004) and discrete choice experiment method (Louviere *et al.*, 2000) are examples of SP approaches.

Partial budgeting

Partial budgeting (PB) is a planning and decision-making framework used to assess whether the changes in income and expenses that would result from implementing a specific production practice will increase or decrease the net benefit.

Stochastic partial budgeting

Instead of specifying a single point for each benefit and cost, as in PB, stochastic partial budgeting (SPB) attaches a probability of occurrence to each benefit or cost element, thereby generating the probability distribution of possible budget outcomes (Grové *et al.*, 2007; Dhoubhadel and Stockton, 2010).

7.3 Understanding Farmers' Preferences and Behaviour Using Revealed and Stated Preference Methods

The question of how to influence farmer decision-making behaviour has been the subject of a considerable number of research studies for several decades. Based on recommendations of researchers, extension workers and agricultural organizations, these studies have generally focused on providing farmers with information on the benefits of specific management practices, tools and policy measurements (e.g. practices related to sustainable intensification and improving farm animal welfare) and encouraging farmers to adopt them (e.g. Buddle *et al.*, 2011; Berckmans, 2014; d'Eath *et al.*, 2014). Efforts to change farmers' behaviour are unlikely to be successful without a good understanding of the behaviours determining factors (Rose *et al.*, 2018). To understand individuals' behaviour and its drivers, researchers and practitioners have used several qualitative and quantitative methods. These methods can be classified as RP or SP approaches (see section 7.2).

Both RP and SP approaches have advantages and disadvantages. The major advantage of the RP approach is its high reliability and face validity (i.e. similarity between what was observed to be chosen and what was actually chosen). This is because the RP data reflect individuals' real choices and take into account the effect of real market constraints on individual decisions (e.g. budget and time). In the real market, decision makers consider the internal costs and benefits of their decisions and experience their consequences, and this is reflected in RP data. By contrast, these constraints do not necessarily affect individuals' decision when they are asked to reveal their choices in a hypothetical market (as with SP data) (Louviere *et al.*, 2000).

SP methods are commonly criticized because of their hypothetical nature (i.e. based on what farmers said they would do and not what they actually did) (Loomis, 2011; Fifer *et al.*, 2014). Nevertheless, there are a number of compelling reasons why researchers and practitioners interested in understanding farmer decision-making behaviour should use SP approaches. First, SP methods are the only means for assessing farmers' preferences and values for products or services that are not actually used or traded explicitly in real markets (e.g. newly developed precision technology to monitor farm animals' welfare), or have public good characteristics that are generally used or consumed for free (e.g. animal welfare, reduction in greenhouse gas emissions). Hence, SP methods can be used to consider an array of alternatives that are fundamentally different from existing ones (Louviere *et al.*, 2000). Second, SP methods are based on hypothetical choices that can be specified in advance using statistical designs to minimize attributes' correlations (e.g. higher prices and higher quality), and hence allow a straightforward identification of all effects of interest. In a real marketplace, the attributes of products and services are highly collinear. This is probably the most common limitation of RP data (Ryan *et al.*, 2007). Third, ST preference methods allow the collection of large quantities of relevant data (including on individuals' attitudes and perception) at moderate cost. In contrast, the collection of RP data can be (perhaps prohibitively) costly, in terms of both time and money (Hensher *et al.*, 2015).

Despite the aforementioned divergence between RP and SP approaches in terms of validity and ability to consider non-traded or used goods and services, both approaches can provide data that is consistent with economic theory and from which similar econometric models can be estimated (Louviere *et al.*, 2000). Furthermore, because both approaches offer very attractive and complementary benefits, there is an increasing interest in the potential gain from combining RP and SP data (Hensher *et al.*, 1998; Whitehead *et al.*, 2008). One of the commonly used techniques to understand and predict farmers' decision making is the cost–benefit analysis (CBA). CBA generally relies on a combination of RP data (e.g. cost and benefits of actually used husbandry practices) and SP data (e.g. cost and benefits of newly developed and unused husbandry practices).

7.4 Techniques

Three well-established research techniques that can be utilized to identify economically viable animal welfare interventions are: (i) conducting a CBA to estimate the monetary consequences of animal welfare interventions; (ii) conducting discrete choice experiments to estimate farmers' preferences and willingness to pay for animal welfare interventions; and (iii) applying structural equation modelling (SEM) to evaluate the role of economic viability within the broader context of human behaviour change (for human behaviour changes, see Chapter 6, this volume). By combining these techniques, it is possible to identify animal welfare interventions that meet the demands of farmers, while also accounting for a wide range of factors that influence farmers' behaviour, and accounting for heterogeneity in the farmer population. Here, we provide examples of these techniques used in practice by drawing on recent research that utilized these techniques to identify economically viable interventions to reduce aggression among pigs.

7.4.1 Cost–benefit analysis

Implementation of an animal welfare intervention is likely to incur some costs and to yield some benefits. CBA involves trading off estimates of these relative costs and benefits in order to provide a measure of net benefit. Furthermore, the outcomes of different interventions can be compared to identify the one most likely to be beneficial, and thus the best allocation of limited resources (Dhoubhadel and Stockton, 2010). As it is possible to apply a monetary value to mortality and morbidity in animal production, CBA is a suitable tool for analysing the costs and benefits of animal welfare interventions (e.g. Groenendaal *et al.*, 2015). There are four main steps when conducting CBA to identify and compare the monetary consequences of animal welfare interventions.

Step 1: Identify relevant interventions

The first step is to identify all relevant, effective animal welfare interventions. These must be: (i) effective at improving the specific animal welfare problem under investigation; and (ii) able to be practically managed under the management system under investigation. Interventions should be identified by reviewing the relevant academic literature.

Step 2: Identify the costs and benefits likely to arise when implementing each intervention on the farm

Various costs and benefits may arise due to the impact of the intervention on production traits (e.g. on growth or mortality rates) or due to the costs of purchasing, installing and managing the intervention on the farm. A comprehensive list of all costs and benefits can be achieved by reviewing the relevant academic literature and by engaging with industry stakeholders, such as farmers.

Step 3: Estimate the monetary value of costs and benefits

The monetary value of each associated cost and benefit must then be estimated. Initial investment costs (e.g. for the purchasing or fitting of new materials) rely on business capital, while running costs (e.g. for labour for the replenishment of materials) mainly affect variable costs. Implementation of a strategy in practice may require only investment costs, only running costs or a combination of both. Nevertheless, in order to make these monetary values comparable, they must be described in a consistent and comparable unit of measurement (e.g. £/€/$ per animal produced, or £/€/$ per animal place on farm). Industry data can facilitate the estimation of monetary values, as can farmers' knowledge and experience of their own management systems. Involvement of farmers in monetary estimations is strongly recommended, as their estimates are likely to be more accurate than could be estimated by researchers alone.

Step 4: Analyse cost and benefit data

The net benefit of each animal welfare intervention can simply be calculated by subtracting the monetary costs from the monetary benefits as follows:

Costs of intervention – benefits of intervention = net benefit of intervention

This use of a single-point value for each cost or benefit element is known as PB. PB has been widely applied across a variety of academic fields, including for animal health and welfare interventions (Alvåsen *et al.*, 2017; Ahmed *et al.*, 2020). Where a net benefit is positive, the costs of the intervention are outweighed by the benefits, implying that the strategy is economically viable. Furthermore, the monetary consequences of multiple interventions can be compared to identify the best allocation of limited resources. However, farms vary widely in their structure, size, management approaches and skilled labour. As a result, the costs and benefits of animal welfare interventions vary widely among farms. Therefore, the use of single-point values renders the analysis deterministic and likely to be inaccurate. Instead, the use of SPB is recommended due to the approach's ability to account for variation in costs and benefits. SPB allows the analyst to use a range of the estimated costs and benefits (e.g. minimum, mean and maximum values) and attach to them probabilities of occurrence. A combination of the estimated values of costs and revenues and their probability distributions are used to determine the range and probability of the change's final possible outcomes (Grové *et al.*, 2007; Dhoubhadel and Stockton, 2010). For ease of presentation and interpretation, the final outcome of the stochastic analysis is generally graphed as a cumulative distribution function.

Example: CBA of interventions to reduce aggression between pigs

As an example, the stages described above were followed to estimate the economic viability of strategies to reduce aggression among pigs. Aggression among pigs was used as an example because there is a large literature base from which to draw interventions, and the problem has persisted for decades without resolution (Peden *et al.*, 2018). Specifically, Peden *et al.* (2021) conducted CBA of the following three strategies to reduce aggression among pigs: (i) pre-weaning socialization of piglets; (ii) housing pigs in large social groups; and (iii) exposing pigs to synthetic maternal pheromones. These strategies had the overall aim of determining intervention viability under commercial conditions. The appropriate interventions to reduce aggression were identified based on a stepwise selection process whereby only effective interventions were included, based on greatest consensus in the literature that they effectively reduce aggression (step 1). Furthermore, only interventions that could realistically be implemented under commercial farming conditions were included; for example, where they are dependent on bought-in products, these products are commercially available.

The costs and benefits of implementing each intervention under commercial conditions were identified and estimated (in £ per pig produced) using a combined approach involving review of the academic and industry literature and a survey of 16 pig farmers (steps 2 and 3). This process revealed that pre-weaning socialization required only small running costs associated with the removal of barriers between adjacent farrowing pens when litters were 3 weeks old. Housing pigs in large social groups and exposing pigs to synthetic maternal pheromones required greater running and investment costs due to the slowed growth rates with increasing group size (Turner *et al.*, 2003). Similarly, the purchase price of pheromones resulted in higher costs. All strategies were linked to monetary benefits as a result of alleviating injuries and stress caused by aggression (McGlone, 1985; Turner *et al.*, 2006). SPB of the monetary data revealed that the strategy most likely to be economically viable for farmers was allowing litters to mix prior to weaning (i.e. pre-weaning socialization; step 4). When implementing this strategy, the cumulative distribution function line intersects the *y*-axis at approximately the 62% level (Fig. 7.1a). This indicates that farmers are likely to obtain a neutral or positive net effect 38% of the time, and a negative net effect 62% of the time. The maximum net benefit is £0.22 per pig produced and the maximum net cost is £0.39 per pig produced. Both exposing pigs to synthetic maternal pheromones and converting existing buildings in order to house pigs in large social groups resulted in 100% probability of obtaining a negative effect. Specifically, for converting existing buildings to house pigs in large social groups, the net costs ranged between –£2.72 and –£1.20 per pig produced (Fig. 7.1b). When applying synthetic maternal pheromones, the net costs ranged between –£0.45 and –£0.06 per pig produced (Fig. 7.1c). The authors concluded that allowing litters to mix prior to weaning should be a fundamental component of campaigns promoting the control of aggression in the industry.

7.4.2 Discrete choice experiments

Improving animal welfare generally comes at a cost to producers. However, animal welfare interventions with net costs are not necessarily economically non-viable. Farmers' willingness to adopt animal welfare interventions beyond the minimum required by legislation is determined not only by the economic value derived from returns in productivity, but also by the value of improved animal welfare independent of any returns in productivity (e.g. McInerney, 2004; Hansson and Lagerkvist, 2016). Farmers are not a homogeneous group; their businesses differ in size, structure, location and management practices, farmers differ in personal attitudes and perceptions, and all of these factors can influence decisions with regard to animal welfare (Bock and van Huik, 2007; Kauppinen *et al.*, 2012). As a result, some farmers are willing to pay a price premium to improve animal welfare beyond profitability goals, and this should be accounted for when conducting an economic evaluation of animal welfare interventions. This can be achieved by estimating farmers' preferences and willingness to pay for animal welfare interventions through SP methods.

Discrete choice experiments are generally accepted as a valid method for preference elicitation and are typically implemented in surveys. Respondents are asked to indicate their preferred option among a number of different hypothetical alternatives with assigned prices. Preferences are measured via systematic variations of product characteristics ('attributes') in an experimental design. Based on respondent choices, it is possible to estimate the relative importance of the different attributes (Ryan and Farrar, 2000; Breidert *et al.*, 2006). Discrete choice experiments can also be used to estimate willingness to pay for both changes in individual

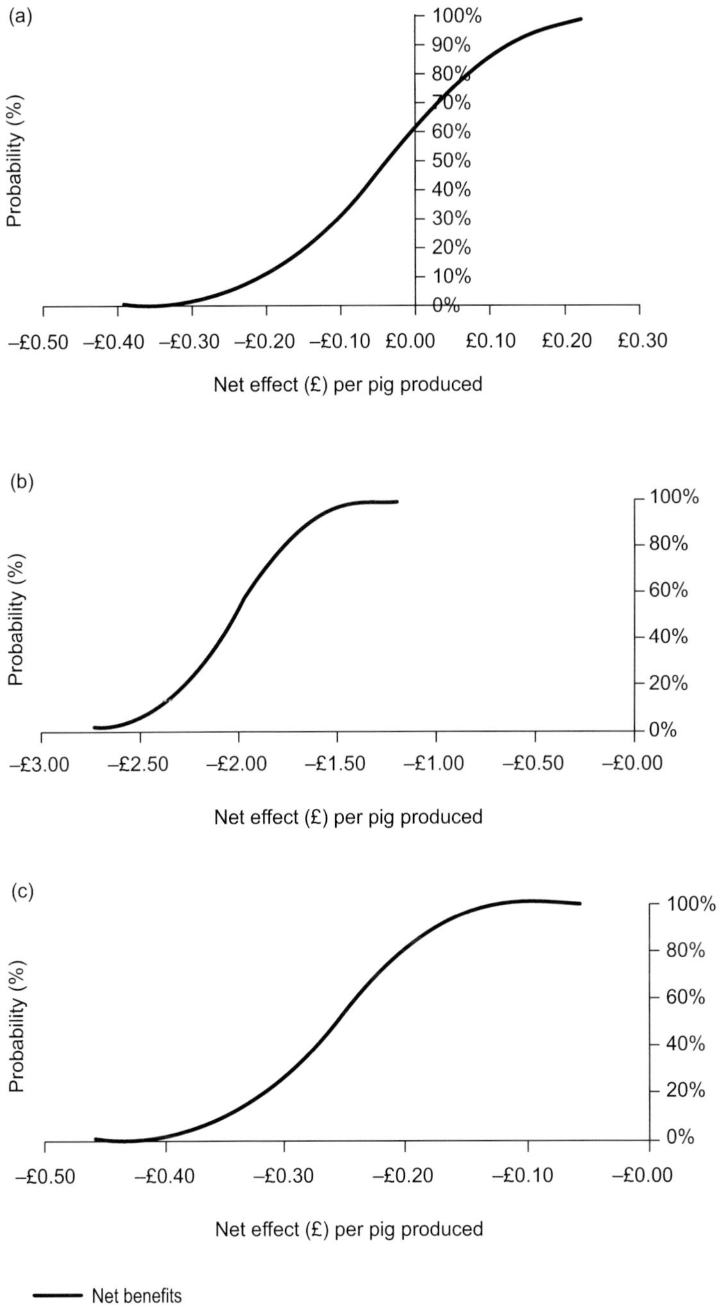

Fig. 7.1. The cumulative distribution function for (a) pre-weaning socialization of litters, (b) housing pigs in large social groups, and (c) applying synthetic maternal pheromones, as modelled through stochastic partial budgeting. Modified from Peden *et al*. (2021).

attributes and changes in any combination of attributes. Furthermore, heterogeneity in these preferences can be understood by employing latent class modelling, which identifies different classes of respondents based on their demographic characteristics, attitudes and preferences (Greene and Hensher, 2003).

Example: Farmers' willingness to pay for interventions to reduce aggression among pigs

Peden *et al.* (2019a) utilized a discrete choice experiment to model 82 pig farmers' decisions to invest in an intervention to reduce aggression among pigs. The pig farm owners and managers were asked to choose between hypothetical aggression control strategies described in terms of four attributes: (i) installation cost; (ii) ongoing cost; (iii) impact on skin lesions from aggression (i.e. reduction in aggression); and (iv) impact on growth rate. If they did not like any of the strategies, they could opt to keep their current farm practice. Systematic variations in product attributes allowed farmers' preferences and willingness to pay to be estimated. Furthermore, farmers answered questions on their demographic characteristics and their practices, attitudes and perceptions regarding animal welfare. Latent class modelling accounted for heterogeneity in responses. Specifically, the results revealed three independent subgroups of farmers, each with different preferences and willingness to pay for animal welfare and business goals. The first class (18% of respondents) would not invest in an aggression control strategy as they were unlikely to regroup unfamiliar pigs, and thus were unlikely to experience high levels of aggression. The second class (32% of respondents) were willing to invest in an aggression control strategy but were unwilling to pay anything to reduce aggression specifically. They were only interested in the extent to which the strategy improved growth rates. Finally, the third class (50% of respondents) were motivated to reduce aggression for welfare as well as production benefits, and were willing to pay £0.11 per pig place (installation cost) and £0.03 per pig produced (running cost) for each 1% reduction in lesions as a result of aggression. This equates to £0.77 per pig produced for a realistic reduction in aggression of 25% (Peden *et al.*, 2021). When the willingness to pay of these farmers was taken into account, both pre-weaning socialization and synthetic maternal pheromones were likely to be economically viable options, even when the minimum net effect is incurred (see section 7.4.1.5). Converting existing buildings to increase group size to 100+ pigs remained beyond the realms of farmers' willingness to pay. Thus, the authors concluded that synthetic maternal pheromones should also be included in campaigns promoting the control of aggression in the industry; however, they should be targeted specifically at farmers who are motivated to improve animal welfare for reasons beyond an expected benefit for production.

7.4.3 Structural equation modelling

As discussed previously, farmers' willingness to adopt animal welfare interventions is influenced not only by economic factors but also by a range of other internal and external factors (Carroll and Groarke, 2019; Balzani and Hanlon, 2020). In this section, SEM is introduced as a powerful tool for estimating the relationships between these factors in order to identify targets for initiating a change in practice. Specifically, SEM tests the causal links specified in a theoretical model and describes the nature and magnitude of the relationship between these factors (Schumacker and Lomax, 2016).

The recommended theoretical model is the behaviour change wheel (BCW) (see Fig. 6.1, Chapter 6, this volume) because it uses a simple but comprehensive framework that can be applied to any human behaviour (Michie *et al.*, 2011). It has also previously been emphasized as a useful tool for developing interventions to encourage the uptake of animal welfare research in practice (Carroll and Groarke, 2019; Chapter 6, this volume). Furthermore, unlike traditional behaviour change models such as the theory of planned behaviour (Ajzen, 1985), the BCW accounts for the social, political and economic context of the behaviour. At the centre of the wheel, *capability*, *opportunity* and *motivation* are proposed as interacting determinants of behaviour. Capability may be physical (e.g. affordability) or psychological (e.g. awareness), opportunity may be physical (e.g. availability) or social (e.g. social norms or pressure), and motivation may be reflective (e.g. personal goals) or automatic (e.g. habit). The middle layer of the BCW outlines the ways that sources of behaviour can be influenced through nine possible intervention functions. The outer layer outlines seven policy categories. For more information on the BCW, see Michie *et al.* (2011). Multiple survey or interview questions should be carefully designed in order to correspond to each of the six central BCW components, as well as farmers' behaviour regarding the animal welfare issue under investigation. The SEM process will then describe the nature and magnitude of the relationship between variables specified in the model and farmers' behaviour regarding animal welfare. Specifically, when conducting SEM, the two-step approach proposed by Anderson and Gerbing (1988) is recommended:

- *Step 1: The measurement model.* Following data collection, the first stage of conducting SEM analyses requires confirmatory factor analysis to confirm that the observed variables (i.e. responses to individual survey or interview questions) load on to the relevant latent variables (i.e. the six central BCW components) in a manner that supports the definition of the latent variable. This step is referred to as the 'measurement model'.
- *Step 2: The structural model.* The second step involves assigning the relevant relationships among latent variables to build and test the

structural model. The structural model is typically displayed in a path diagram with regression coefficients presented for each relationship, thus describing the magnitude and direction of the relationship between latent variables. The term SEM refers to the full model, which consists of both the measurement model and the structural model.

Once the source of behaviour has been identified, it can then be linked to suitable interventions using the middle and outer layers of the BCW (see Fig. 6.1, Chapter 6, this volume). Depending on the behavioural determinants, there are nine possible intervention functions (see the middle layer of the wheel). These reflect broad categories of means by which an intervention can change behaviour. Having selected the intervention type most likely to be effective, these can then be linked to policy categories that support the delivery of the behaviour change intervention (see the outer section of the wheel). For examples of where SEM has been used to explore the factors that determine farmers' behaviour regarding animal welfare, see Kauppinen *et al.* (2010) and Peden *et al.* (2019b).

7.4.4 Other stakeholder groups

This chapter has focused on the economic viability of animal welfare interventions from the farmers' perspective. However, it is important to note that change in animal welfare is piecemeal, and other stakeholder groups also have the capacity to influence and pay for animal welfare standards. For example, consumers have the capacity to cover the additional costs of improved animal welfare. Under current animal welfare and organic quality assurance schemes, the additional costs of certain animal welfare improvements can be passed to consumers and retailers through product labelling (Akaichi *et al.*, 2019). For example, Royal Society for the Prevention of Cruelty to Animals (RSPCA)-assured pigs benefit from freedom of movement during farrowing and increased space allowances, and consumers of these products pay a price premium to cover the additional costs (RSPCA, 2016). However, animal welfare concerns do not influence consumer buying behaviour of the vast majority of consumers (Te Velde *et al.*, 2002; Schröder and McEachern, 2004; Miele and Evans, 2010). In general, consumers accept and pay for the minimum standards of animal welfare as guaranteed by legislation.

7.5 Pitfalls

This chapter takes the position that farmers, as animal caregivers, are ultimately responsible for making the necessary husbandry and management changes required to implement aggression control strategies in practice. Nevertheless, it is important to note that a change in animal welfare is piecemeal, and other stakeholder groups play important roles. Consumers and retailers have the capacity to influence animal welfare by creating a market pull towards animal-friendly production.

7.5.1 Choice experiments

Despite the advantages outlined above, the use of the choice experiment has some potential limitations that have to be taken into account when designing and implementing it. One major concern is the inability of respondents to make consistent choices if they are provided with a high number of choice sets. This may incentivize respondents to choose the opt-out alternative to avoid making cognitively challenging choices. This concern may be overcome by keeping the number of attributes and attribute levels low for each choice set. However, the robustness of the output of choice experiments (i.e. data and results) relies heavily on the accuracy and completeness of the attributes used to describe the alternatives of the product/service under study. Any omission of relevant attributes is likely to result in biased estimates. Therefore, the exclusion of attributes needs to be weighed against task complexity due to too many attributes. Typically, the final design of the choice experiment is preceded by rigorous and iterative piloting to get the balance right. The best practices to design and implement a choice experiment have been well described by Hensher *et al.* (2015).

7.5.2 Cost–benefit analysis

Costs and benefits are highly estimated and are likely to vary widely in the industry; therefore, there is likely to be a high degree of uncertainty. It is best to include stakeholders in the estimation of costs and benefits and to tailor these estimates for specific farms.

7.6 Conclusions and Tips

The economic consequences of animal welfare interventions are a primary determinant of farmers' willingness to improve animal welfare in practice. The economic viability of animal welfare interventions can be identified by conducting cost–benefit analyses. Nevertheless, even interventions that come at a cost to producers may be economically viable if farmers are willing to pay the additional cost, and willingness to pay can be estimated using discrete choice experiments. Despite knowledge that economic factors are key determinants of whether effort will be made to improve animal welfare, these have rarely been estimated. Animal welfare researchers should collaborate with economists more frequently to identify economically viable strategies to improve animal welfare. Nevertheless, farmers' decisions are not only influenced by economic viability, but also influenced by a wide range of internal and external factors. SEM assesses the role of economic viability within the broader context of human behaviour change. By combining these techniques, it is possible to identify animal welfare interventions that meet the demands of farmers, while also accounting for the wide range of factors that influence farmers' behaviour, and accounting for heterogeneity in the farmer population.

Further Reading

For more guidance on the use of the described methods, we refer to the references given in bold in the reference list.

References

Ahmed, H., Alvåsen, K., Berg, C., Hansson, H., Hultgren, J. *et al.* (2020) Assessing economic consequences of improved animal welfare in Swedish cattle fattening operations using a stochastic partial budgeting approach. *Livestock Science* 232(2), 103920. DOI: 10.1016/j.livsci.2020.103920.

Ajzen, I. (1985) From intentions to actions: a theory of planned behavior. In: Kuhl, J. and Beckmann, J. (eds) *Action Control.* SSSP Springer Series in Social Psychology. Springer, Berlin/Heidelberg, pp. 11–39.

Akaichi, F., Glenk, K. and Revoredo-Giha, C. (2019) Could animal welfare claims and nutritional information boost the demand for organic meat? Evidence from non-hypothetical experimental auctions. *Journal of Cleaner Production* 207(7), 961–970. DOI: 10.1016/j.jclepro.2018.10.064.

Alvåsen, K., Hansson, H., Emanuelson, U. and Westin, R. (2017) Animal welfare and economic aspects of using nurse sows in Swedish pig production. *Frontiers in Veterinary Science* 4, 204. DOI: 10.3389/fvets.2017.00204.

Anderson, J.C. and Gerbing, D.W. (1988) Structural equation modeling in practice: a review and recommended two-step approach. *Psychological Bulletin* 103(3), 411–423. DOI: 10.1037/0033-2909.103.3.411.

Appleby, M.C. (2005) The relationship between food prices and animal welfare. *Journal of Animal Science* 83, E9–E12.

Balzani, A. and Hanlon, A. (2020) Factors that influence farmers' views on farm animal welfare: a semi-systematic review and thematic analysis. *Animals* 10(9), 1524–1525. DOI: 10.3390/ani10091524.

Berckmans, D. (2014) Precision livestock farming technologies for welfare management in intensive livestock systems. *Revue Scientifique et Technique de l'OIE* 33(1), 189–196. DOI: 10.20506/rst.33.1.2273.

Bock, B.B. and van Huik, M.M. (2007) Animal welfare: the attitudes and behaviour of European pig farmers. *British Food Journal* 109(11), 931–944. DOI: 10.1108/00070700710835732.

Bornett, H.L.I., Guy, J.H. and Cain, P.J. (2003) Impact of animal welfare on costs and viability of pig production in the UK. *Journal of Agricultural and Environmental Ethics* 16(2), 163–186. DOI: 10.1023/A:1022994131594.

Breidert, C., Hahsler, M. and Reutterer, T. (2006) A review of methods for measuring willingness-to-pay. *Innovative Marketing* 2, 8–32.

Brown, G. and Mendelsohn, R. (1984) The hedonic travel cost method. *The Review of Economics and Statistics* 66(3), 427–433. DOI: 10.2307/1924998.

Buddle, B.M., Denis, M., Attwood, G.T., Altermann, E., Janssen, P.H. *et al.* (2011) Strategies to reduce methane emissions from farmed ruminants grazing on pasture. *The Veterinary Journal* 188(1), 11–17. DOI: 10.1016/j.tvjl.2010.02.019.

Carroll, G.A. and Groarke, J.M. (2019) The importance of the social sciences in reducing tail biting prevalence in pigs. *Animals* 9, 591. DOI: 10.3390/ani9090591.

Chakraborty, K., Misra, S. and Johnson, P. (2002) Cotton farmers' technical efficiency: stochastic and nonstochastic production function approaches. *Agricultural and Resource Economics Review* 31(2), 211–220. DOI: 10.1017/S1068280500004019.

Costanigro, M., McCluskey, J.J. and Lusk, J.L. (2011) Hedonic price analysis in food markets. In: Lusk, J., Roosen, J. and Shogren, J.F. (eds) *The Oxford Handbook of the Economics of Food Consumption and Policy*. Oxford University Press, Oxford, UK, pp. 152–180.

d'Eath, R.B., Arnott, G., Turner, S.P., Jensen, T., Lahrmann, H.P. *et al.* (2014) Injurious tail biting in pigs: how can it be controlled in existing systems without tail docking? *Animal* 8(9), 1479–1497. DOI: 10.1017/S1751731114001359.

d'Eath, R.B., Niemi, J.K., Vosough Ahmadi, B., Rutherford, K.M.D., Ison, S.H. *et al.* (2016) Why are most EU pigs tail docked? Economic and ethical analysis of four pig housing and management scenarios in the light of EU legislation and animal welfare outcomes. *Animal* 10(4), 687–699. DOI: 10.1017/S1751731115002098.

Dhoubhadel, S.P. and Stockton, M. (2010) Stochastic partial budgeting: a new look at an old tool. *Cornhusker Economics* 424. Available at: https://core.ac.uk/download/pdf/17210893.pdf (accessed 9 June 2021).

Dwyer, C.M., Conington, J., Corbiere, F., Holmøy, I.H., Muri, K. *et al.* (2016) Invited review: improving neonatal survival in small ruminants: science into practice. *Animal* 10(3), 449–459. DOI: 10.1017/S1751731115001974.

Fifer, S., Rose, J. and Greaves, S. (2014) Hypothetical bias in stated choice experiments: is it a problem? And if so, how do we deal with it? *Transportation Research Part A: Policy and Practice* 61, 164–177.

Gocsik, E., Lansink, A. and Saatkamp, H.W. (2013) Mid-term financial impact of animal welfare improvements in Dutch broiler production. *Poultry Science* 92(12), 3314–3329. DOI: 10.3382/ps.2013-03221.

Gocsik, E., Lansink, A., Voermans, G. and Saatkamp, H.W. (2015a) Economic feasibility of animal welfare improvements in Dutch intensive livestock production: a comparison between broiler, laying hen, and fattening pig sectors. *Livestock Science* 182, 38–53. DOI: 10.1016/j.livsci.2015.10.015.

Gocsik, E., van der Lans, I.A., Lansink, A. and Saatkamp, H.W. (2015b) Willingness of Dutch broiler and pig farmers to convert to production systems with improved welfare. *Animal Welfare* 24(2), 211–222. DOI: 10.7120/09627286.24.2.211.

Greene, W.H. and Hensher, D.A. (2003) A latent class model for discrete choice analysis: contrasts with mixed logit. *Transportation Research Part B: Methodological* 37(8), 681–698. DOI: 10.1016/S0191-2615(02)00046-2.

Groenendaal, H., Zagmutt, F.J., Patton, E.A. and Wells, S.J. (2015) Cost–benefit analysis of vaccination against *Mycobacterium avium* ssp. *paratuberculosis* in dairy cattle, given its cross-reactivity with tuberculosis tests. *Journal of Dairy Science* 98(9), 6070–6084. DOI: 10.3168/jds.2014-8914.

Grové, B., Taljaard, P.R. and Cloete, P.C. (2007) A stochastic budgeting analysis of three alternative scenarios to convert from beef-cattle farming to game ranching. *Agrekon* 46(4), 514–531. DOI: 10.1080/03031853.2007.9523784.

Hansson, H. and Lagerkvist, C.J. (2016) Dairy farmers' use and non-use values in animal welfare: determining the empirical content and structure with anchored best-worst scaling. *Journal of Dairy Science* 99(1), 579–592. DOI: 10.3168/jds.2015-9755.

Hensher, D., Louviere, J. and Swait, J. (1998) Combining sources of preference data. *Journal of Econometrics* 89(1–2), 197–221. DOI: 10.1016/S0304-4076(98)00061-X.

Hensher, D.A., Rose, J.M. and Greene, W.H. (2015) *Applied Choice Analysis: a Primer*, 2nd edn. Cambridge University Press, Cambridge, UK.

Kauppinen, T., Vainio, A., Valros, A., Rita, H. and Vesala, K.M. (2010) Improving animal welfare: qualitative and quantitative methodology in the study of farmers' attitudes. *Animal Welfare* 19, 523–536.

Kauppinen, T., Vesala, K.M. and Valros, A. (2012) Farmer attitude toward improvement of animal welfare is correlated with piglet production parameters. *Livestock Science* 143(2–3), 142–150. DOI: 10.1016/j.livsci.2011.09.011.

Loomis, J. (2011) What's to know about hypothetical bias in stated preference valuation studies? *Journal of Economic Surveys* 25(2), 363–370. DOI: 10.1111/j.1467-6419.2010.00675.x.

Louviere, J.J., Hensher, D.A. and Swait, J.D. (2000) *Stated Choice Methods: Analysis and Applications*. Cambridge University Press, Cambridge, UK.

Lusk, J.L. and Norwood, F.B. (2012) Speciesism, altruism and the economics of animal welfare. *European Review of Agricultural Economics* 39(2), 189–212. DOI: 10.1093/erae/jbr015.

McGlone, J.J. (1985) A quantitative ethogram of aggressive and submissive behaviors in recently regrouped pigs. *Journal of Animal Science* 61(3), 556–566. DOI: 10.2527/jas1985.613556x.

McInerney, J.P. (1993) Animal welfare: an economic perspective. *Paper presented at the Annual Conference of the Agricultural Economics Society*, Oxford, UK.

McInerney, J.P. (2004) Animal welfare, economics and policy. Report on a study undertaken for the Farm and Animal Health Economics Division of Defra. Available at: https://webarchive.nationalarchives.gov.uk/20110318142209/http://www.defra.gov.uk/evidence/economics/foodfarm/reports/documents/animalwelfare.pdf (accessed 9 June 2021).

Michie, S., van Stralen, M.M. and West, R. (2011) The behaviour change wheel: a new method for characterising and designing behaviour change interventions. *Implementation Science* 6, 42. DOI: 10.1186/1748-5908-6-42.

Miele, M. and Evans, A. (2010) When foods become animals: ruminations on ethics and responsibility in care-*full* practices of consumption. *Ethics, Place & Environment* 13(2), 171–190. DOI: 10.1080/13668791003778842.

Peden, R.S.E., Turner, S.P., Boyle, L.A. and Camerlink, I. (2018) The translation of animal welfare research into practice: the case of mixing aggression between pigs. *Applied Animal Behaviour Science* 204(3), 1–9. DOI: 10.1016/j.applanim.2018.03.003.

Peden, R.S.E., Akaichi, F., Camerlink, I., Boyle, L.A. and Turner, S.P. (2019a) Pig farmers' willingness to pay for management strategies to reduce aggression between pigs. *PLoS One* 14(11), e0224924. DOI: 10.1371/journal.pone.0224924.

Peden, R.S.E., Akaichi, F., Camerlink, I., Boyle, L. and Turner, S. (2019b) Factors influencing farmer willingness to reduce aggression between pigs. *Animals* 9(1), 6. DOI: 10.3390/ani9010006.

Peden, R.S.E., Turner, S.P., Camerlink, I. and Akaichi, F. (2021) An estimation of the financial consequences of reducing pig aggression. *PLoS One* 16(5), e0250556. DOI: 10.1371/journal.pone.0250556.

Rose, D.C., Keating, C. and Morris, C. (2018) Understanding How to Influence Farmers' Decision-making Behaviour: A Social Science Literature Review. Agriculture and Horticulture Development Board, Kenilworth, UK. Available at: https://ueaeprints.uea.ac.uk/id/eprint/67271/1/FarmersDecisionMaking_2018_09_18.pdf (accessed 9 June 2021).

RSPCA (Royal Society for the Prevention of Cruelty to Animals) (2016) RSPCA Welfare Standards for Pigs. Available at: https://science.rspca.org.uk/documents/1494935/9042554/RSPCA+welfare+standards+for+pigs+%28PDF+5.72MB%29.pdf/8b2d5794-9a10-cd1f-f27d-e3642c0c1945?t=1557668440116 (accessed 9 June 2021).

Ryan, M. and Farrar, S. (2000) Using conjoint analysis to elicit preferences for healthcare. *Education and Debate* 320, 1530–1533.

Ryan, M., Gerard, K. and Amaya-Amaya, M. (eds) (2007) *Using Discrete Choice Experiments to Value Health and Health Care,* Vol. 11. Springer Science and Business Media, Berlin, Germany.

Schröder, M.J.A. and McEachern, M.G. (2004) Consumer value conflicts surrounding ethical food purchase decisions: a focus on animal welfare. *International Journal of Consumer Studies* 28(2), 168–177. DOI: 10.1111/j.1470-6431.2003.00357.x.

Schumacker, R.E. and Lomax, R.G. (2016) *A Beginner's Guide to Structural Equation Modeling,* 4th edn. Routledge, New York.

Silberston, A. (1972) Economies of scale in theory and practice. *The Economic Journal* 82(325), 369–391. DOI: 10.2307/2229943.

Sumner, D.A., Gow, H., Hayes, D., Matthews, W., Norwood, B. *et al.* (2011) Economic and market issues on the sustainability of egg production in the United States: analysis of alternative production systems. *Poultry Science* 90(1), 241–250. DOI: 10.3382/ps.2010-00822.

te Velde, H., Aarts, N. and Van Woerkum, C. (2002) Dealing with ambivalence: farmers' and consumers' perceptions of animal welfare in livestock breeding. *Journal of Agricultural and Environmental Ethics* 15(2), 203–219. DOI: 10.1023/A:1015012403331.

Thornton, P.K. (2010) Livestock production: recent trends, future prospects. *Philosophical Transactions of the Royal Society B: Biological Sciences* 365(1554), 2853–2867. DOI: 10.1098/rstb.2010.0134.

Turner, S.P., Allcroft, D.J. and Edwards, S.A. (2003) Housing pigs in large social groups: a review of implications for performance and other economic traits. *Livestock Production Science* 82(1), 39–51. DOI: 10.1016/S0301-6226(03)00008-3.

Turner, S.P., Farnworth, M.J., White, I.M.S., Brotherstone, S., Mendl, M. *et al.* (2006) The accumulation of skin lesions and their use as a predictor of individual aggressiveness in pigs. *Applied Animal Behaviour Science* 96(3–4), 245–259. DOI: 10.1016/j.applanim.2005.06.009.

Venkatachalam, L. (2004) The contingent valuation method: a review. *Environmental Impact Assessment Review* 24(1), 89–124. DOI: 10.1016/S0195-9255(03)00138-0.

Vosough Ahmadi, B., Dwyer, C.M., Erhard, H.W., Morgan-Davies, C. and Waterhouse, A. (2010) Impacts of labour on interactions between economics and animal welfare in extensive sheep farms. *Paper presented at the 84th Annual Conference of the Agricultural Economics Society*, Edinburgh, UK, 29–31 March 2010.

Whitehead, J.C., Pattanayak, S.K., Van Houtven, G.L. and Gelso, B.R. (2008) Combining revealed and stated preference data to estimate the nonmarket value of ecological services: an assessment of the state of the science. *Journal of Economic Surveys* 22(5), 872–908. DOI: 10.1111/j.1467-6419.2008.00552.x.

Wilkie, R. (2005) Sentient commodities and productive paradoxes: the ambiguous nature of human–livestock relations in northeast Scotland. *Journal of Rural Studies* 21(2), 213–230. DOI: 10.1016/j.jrurstud.2004.10.002.

8 Application of Psychometrics to Assess Quality of Life in Animals

Jacky Reid[1]*, Andrea Nolan[2] and Marian Scott[3]

[1]NewMetrica Ltd, Glasgow, UK; [2]Edinburgh Napier University, Edinburgh, UK; [3]School of Mathematics and Statistics, University of Glasgow, Glasgow, UK

8.1 Introduction

Quality of life (QoL) is an abstract construct that has been recognized and widely employed in human medicine using formally assessed structured questionnaire instruments, developed using psychometric methodology, to measure the impact of chronic disease (Streiner *et al.*, 2015). The application of the term 'quality of life' to animals is less common, and the veterinary and scientific focus has been on 'welfare' and welfare measurement. Some use the terms QoL and welfare interchangeably, but the authors believe there to be a distinction between the two as follows: 'Welfare (and similarly well-being) is a complex construct that combines both subjective and objective aspects of the conditions of life for animals' (Scott *et al.*, 2003), for example, as described in the Five Freedoms of the Farm Animal Welfare Council (see Chapter 1, this volume). Traditional methods of welfare assessment ensure that minimum standards of care are provided and so ensure that animals do not feel too bad, rather than helping them feel good or very good. It is now widely recognized that how an animal feels is important to its welfare, and if we accept that a conceptualization and definition of QoL for animals is similar to that for people, then consideration of QoL in animal science makes clear the importance of the individual's perspective. This significantly shifts the focus from the avoidance of poor QoL to the attainment of good and even excellent QoL (Scott *et al.*, 2007).

*Corresponding author: jacky.reid@newmetrica.com

 Bridging Research Disciplines to Advance Animal Welfare Science: A practical guide (ed. I. Camerlink)
DOI: 10.1079/9781789247893.0008

In the field of human health, the World Health Organization defines QoL as an 'individual's' perception of their position in life in the context of the culture and value systems in which they live and in relation to their goals, expectations, standards and concerns' (https://www.who.int/tools/whoqol; accessed 11 June 2021). However, no widely accepted definition for QoL in animals exists. We have proposed the following definition, intended to be relevant in any circumstances (including those of ill health): QoL is 'the subjective and dynamic evaluation by the individual of its circumstances (internal and external) and the extent to which these meet its expectations (that may be innate or learned and that may or may not include anticipation of future events), which results in, or includes, an affective (emotional) response to those circumstances (the evaluation may be a conscious or an unconscious process, with a complexity appropriate to the cognitive capacity of the individual)' (Wiseman-Orr *et al.*, 2006, p. 1826).

For people, QoL 'is a broad-ranging concept incorporating, in a complex way, the person's physical health, psychological state, level of independence, social relationships, personal beliefs, and relationship to salient features of the environment' (WHOQOL Group, 1994, p. 28). Similarly, for animals, it will be affected by health state, relationships with other animals and people, husbandry and the environment. However, within the field of healthcare (human or animal), it is impossible to encompass every aspect of life, and for this reason it is often easier to explore health-related quality of life (HRQL), which for people is defined as 'how well a person functions in their life and his or her perceived wellbeing in physical, mental, and social domains of health' (Hays and Reeve, 2010, p. 195). In our work, carried out over many years on the impact of pain on the lives of animals, we have focused on HRQL. While in the literature QoL and HRQL are often used interchangeably, each has its own meaning as explained above. QoL is a broader concept that covers all aspects of life, while HRQL is concerned with the effects of illness and the impact treatment may have on QoL.

Instruments developed to quantify HRQL can be used to measure the difference between patients at a point in time (discriminative purposes) or differences within a patient over time (evaluative purposes). They can be specific, focusing on particular conditions (disease specific), or they can be generic, designed for use in a variety of contexts. Disease-specific instruments may be more sensitive to clinical change, but generic instruments are useful to quantify a range of impacts related to disease and its treatment, and may be the only choice when a patient has more than one condition, a situation encountered commonly in veterinary medicine (Mattin *et al.*, 2014; Noble *et al.*, 2019). Instruments either generate a single index score, which indicates that a patient is better or worse (Brazier *et al.*, 2017), or a profile of scores, which offers more information and may be more sensitive to group differences and to changes in health status over time (Streiner *et al.*, 2015).

8.1.1 Measuring the 'unmeasurable'

As QoL is an entirely subjective experience that is uniquely personal to the individual, self-reports are regarded as the 'gold standard' in assessing a person's QoL. However, animals are incapable of verbal self-report and so must rely on an observer, often referred to as a 'proxy', to assess their subjective experiences. A proxy-reported outcome measure is defined as a report by someone who is not the patient responding as if that person were the patient (US Department of Health and Human Services FDA Center for Drug Evaluation and Research *et al.*, 2006). In the veterinary context, this can be very misleading. For example, an owner, if asked 'How is your dog's quality of life?' may think of those things that the owner considers important, such as a warm bed and two good walks a day, good nutrition and immediate veterinary attention if required, and so respond 'My dog has a good quality of life.' In this case, the owner's focus is on the dog's circumstances, rather than the dog's perception of those circumstances, which is the essence of QoL. Circumstances may be quite differently perceived by different canine subjects. For example, a long walk on a windswept beach with lots of intriguing things to eat is likely to be perceived in one way by an active young Labrador and may be perceived entirely differently by an elderly Cavalier King Charles Spaniel that enjoys spending most of its time resting on its owner's knees. Accordingly, the only true measure of QoL is how the individual feels about its circumstances.

Accessing that subjective experience in animals is very challenging, and although we have chosen to interpret how animals feel using measures of HRQL, others have used trained observers to conduct qualitative behaviour assessment, mainly in farm species and horses. This approach summarizes the different aspects of an animal's dynamic style of interaction with its environment to enable the identification of the main dimensions of mood states (Wemelsfelder, 2007; Minero *et al.*, 2009; Mendl *et al.*, 2010).

It has been suggested that non-verbal pain behaviour in infants is a form of self-report (Anand and Craig, 1996), and behavioural disturbances have long been recognized as potential indicators of the presence of pain in animals. This recognition of the value of behavioural cues has, in the last decade, led to a focus on developing HRQL structured questionnaire instruments, designed to measure pain and QoL in companion animals using observer-reported outcome measures where the owner, who is best placed to report subtle behaviour changes, is the observer (Wiseman-Orr *et al.*, 2004). Similarly, for food producing animals, the farmer or stockperson fulfils that role best (Wiseman-Orr *et al.*, 2011a, b).

In attempting to access an animal's subjective perception of its circumstances using behavioural observation, it is important to bear in mind that each species may manifest its own unique behaviours or disturbances of these, which may have been established in the evolutionary process (e.g.

selection pressures may have ensured that prey species do not 'advertise' weakness to predators) and therefore these cannot be translated to another non-prey species. Even within a species, it is possible that breed variations will result in differences such that poor QoL may be more quickly reflected in reduced appetite in dogs that are highly motivated to eat, such as Labradors, compared with other breeds who may 'pick' at their food. Consequently, instruments designed to measure HRQL must be developed for each species and, in some cases, must be sensitive to breed differences.

8.1.2 The process of developing a measurement tool for HRQL

Some decades ago, psychologists and psychiatrists developed psychometric scales to measure abstract, multiple-attribute constructs such as intelligence and personality. Over the last three decades, these same approaches have been adopted in the medical field to develop instruments to measure the abstract, subjective construct of QoL. Defined steps lead to the creation of a structured questionnaire instrument with a formal scoring mechanism. The validity of the instrument is established first through the methods used during its construction, and then evidence for various kinds of validity is sought by testing the instrument (see section 8.2.2). Finally, development and testing are fully reported. It is important to appreciate that this process is iterative, such that newly developed instruments are refined and retested with new populations in new contexts and for new purposes. An example of this is the VetMetrica Canine tool (Fig. 8.1), originally developed as the 109 item Glasgow University Veterinary School Questionnaire (GUVQuest) to assess the impact of chronic pain on HRQL in dogs (Wiseman-Orr *et al.*, 2004, 2006). A cycle of iterations led to the creation of a 46-item tool (Reid *et al.*, 2013) and subsequently a 22-item owner-based questionnaire (Reid *et al.*, 2018), which has been shown to be valid and reliable in a range of populations and contexts such as obesity (Yam *et al.*, 2016) and dogs suffering from a range of chronic diseases (Rodger *et al.*, 2021).

8.2 Concepts

8.2.1 Concept of measurement

Stevens (1951, p. 1) noted that 'measurement is the assignment of numerals to objects or events according to rules'. The adherence to rules is essential for the measurement to have meaning, even if these rules are not stated explicitly. However, there are clear differences between the measurements of certain kinds of properties. For example, height, weight or speed are all properties that can be measured directly. Other properties are not observable, for example latent traits such as depression or pain, and attributing a quantifying number to a level of this property or attribute becomes

The following are some guidelines to help you interpret the results.

- 70% of healthy pets will score above the grey threshold line
- You are likely to see fluctuations in the HRQL scores over time, and to help you interpret these we recommend that you consider an improvement of around 7 or above on the 0–70 scale to indicate a clinically meaningful change

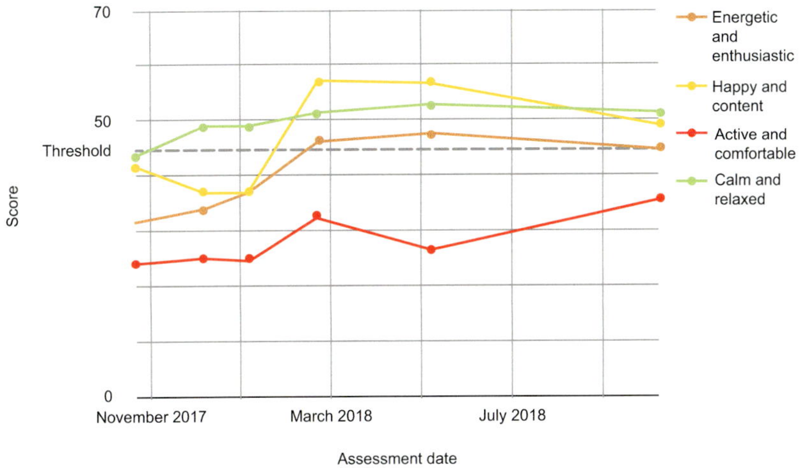

Fig. 8.1. The VetMetrica HRQL 22-item tool for dogs. This graph shows the profile of scores (from the output of the VetMetrica web platform) in four domains of QoL (energetic/enthusiastic, happy/content, active/comfortable, calm/relaxed) for a dog with elbow dysplasia treated with stem cell therapy. The first assessment is before treatment. A score of 50 represents the score for the age-related average healthy dog, and 70% of healthy dogs will score above the grey dotted threshold line. The minimum important difference (MID) is 7 in all domains. © NewMetrica Ltd.

more problematic. It is hence necessary to postulate a theory to relate that attribute to some other observable feature(s) and measure that, be it behavioural signs or an individual's own report.

The measurement process is focused on quantification, to assign a numerical value to the measure of interest, in this case QoL. We want to have confidence in the value, and we want it to have certain properties (discussed later). The resulting values can be categorized into four levels of measurement, classified as nominal, ordinal, interval and ratio (Stevens, 1958).

We will not discuss the nominal scale in the context of QoL but rather start with the *ordinal* scale, so here we might describe (measure) QoL in terms of low or high, or mild, moderate or severe. We thus have relative information, high has more QoL than low, but we do not know how much higher. *Interval* scales assume another property, where measurement is continuous, and where the spacing between points is meaningful and constant. For example, if $a=3$, $b=6$ and $c=9$ on an interval scale, this implies that the difference between a and b and b and c is equal (akin to a 'ruler'). The highest level of measurement is provided by *ratio* scales, which assume that the zero point on a scale is meaningful (i.e. the property of interest is not present). An interval scale, such as temperature in Celsius, implies an arbitrary zero point, whereas temperature on a Kelvin scale is a ratio scale, as zero degrees Kelvin is absolute freezing.

Ordinal and interval level measurements are both practicable and desirable for the assessment of QoL. Ordinal scales have been the most frequently created, yet they present difficulties in analysis and interpretation. They may offer the precision required, although their sensitivity and responsiveness to change is compromised if the ordered categories are broad. There must be careful consideration of the number and definition of ordered levels. Interval level measurement is more demanding to create but provides more precise measurement: the rank ordering of the individuals is known and the distance separating the individuals on the attribute is also known.

8.2.2 Key properties

For any measurement scale there are several key properties that the scale should possess: these are validity, reliability and responsiveness (Fayers and Machin, 2007). There is considerable literature on these properties, so here we briefly summarize them in the context of QoL measurement.

Validity

Validity provides evidence that the QoL tool is able to measure QoL. There are three kinds of validity: criterion, content (face validity, which is related to content validity, may or may not be sought) and construct validity (Streiner, 1993; Coste *et al.*, 1995; Johnston, 1998; Jensen, 2003). If there is an existing gold standard of QoL measurement, then a new tool would have criterion validity if the agreement of the new tool with the 'gold standard' is demonstrated. As there is no gold standard for QoL, we must instead provide evidence for content and construct validity. Content validity is the extent to which attribute(s) of QoL are assessed by the QoL tool's items and the appropriateness of each of the items to QoL: it is largely established through the methodology used to collect and choose the items to be included in an instrument, but is often formally assessed by an independent group of 'experts' (Brod *et al.*, 2009; Streiner *et al.*, 2015). It is commonly the case that the items to be considered are suggested through expert elicitation, focus groups and surveys.

Face validity would describe a QoL tool where the items appear 'on the face of it' to be measuring what the tool is intended to measure. This kind of validity does not improve the psychometric properties of an instrument, but it generally increases the instrument's acceptability to the respondent.

QoL is usually inferred from a variety of observations, so is hypothetical. Factorial validity is one kind of construct validity that requires the statistical analysis of correlations between responses to the items of a QoL instrument. Groupings of items revealed by such analysis (that are also related on clinical or other grounds) are termed 'factors' and provide evidence for a factor structure underlying the data generated by the instrument. If this underlying factor structure fits the construct upon which the instrument was developed, then some evidence has been provided for the validity of the instrument and also for that hypothetical construct (Feinstein, 1987; Johnston, 1998). Evidence for the construct validity of an instrument is also provided when the scores obtained for different known groups or within groups over time can be predicted by that construct (Guyatt *et al.*, 1993; Streiner, 1993; Johnston, 1998).

Reliability

Reliability is whether the QoL instrument can measure accurately and repeatedly what it is intended to measure, so that 'measurements of individuals on different occasions, or by different observers, or by similar or parallel tests, produce the same or similar results' (Streiner *et al.*, 2015, p. 9). If an instrument is to be used by an independent observer, then inter-rater reliability is sought. Alternatively, an instrument's reliability can be estimated by examining the stability of responses when scores are not expected to change between administrations; this is called 'test–retest reliability'. If an instrument is valid, then it is likely also to be reliable, but it may be highly reliable yet lack validity because it is measuring something other than that which it was intended to measure (Fallowfield, 1990).

Responsiveness

While reliability is an important attribute of an instrument, it is possible for an instrument to be reliable yet be unresponsive to clinical change. The ability of an instrument to capture changes that are important (statistically and clinically) has been termed its 'responsiveness'. This is an essential requirement of evaluative instruments – those designed principally to measure change over time (Guyatt *et al.*, 1987).

Utility

A useful clinical instrument must not only be valid, reliable and responsive but also be 'practical and easy to administer, score and interpret' (Landgraf and Abetz, 1996). Even if a measure is valid and reliable, it will not be used if it requires lengthy training, is time-consuming to administer or if scoring is complex (Streiner, 1993).

8.3 Techniques

8.3.1 Psychometric approach to instrument design

The methods used to develop psychometric instruments for the measurement of abstract, multiple-attribute constructs such as QoL, originally developed by psychiatrists and psychologists, are now well established, and there is a large body of literature covering this subject. In the interest of brevity, this description is based on the collective experience of the authors, with references to a small number of key texts and articles to facilitate a broader appreciation of the topic.

The psychometric process is composed of three phases:

- phase 1: item generation and construct validity;
- phase 2: questionnaire design; and
- phase 3: field testing for validity and reliability plus, in the case of an evaluative instrument, responsiveness.

Phase 1 involves the specifying of measurement goals, the identification of the patient population and the development of a pool of potential items for inclusion in the instrument. Semi-structured interviews along with item-generating questionnaires with the intended end users (key informants) define the pool of items. The number of interviews required during this phase is determined by sampling to redundancy, where additional interviews cease to elicit new information. Open-ended and closed questions are worded carefully to avoid response bias. In phase 2, suitable items are selected from the item pool and that selection is subjected to expert validation in order to ensure adequacy of content relevance and coverage. The validated collection of items is then incorporated into an instrument, with suitable consideration given to layout, response options provided for items, instructions to respondent and other details of administration. The resulting prototype is then pre-tested with the end user to ensure that the target respondent can use the instrument correctly. Phase 3 involves field testing the instrument, in order to evaluate its psychometric properties.

8.3.2 Choosing response options

Each item in a questionnaire instrument is accompanied by an answer option or options. If item responses are likely to lie on a continuum, the opportunity for respondents to answer in this way should be embraced to ensure minimum loss of information (Streiner and Norman, 1995). This enhances the sensitivity of the instrument and the precision of the measure

(Bowling, 1991). There are a wide variety of response options. For example, for the direct estimation of continuous variables, numerical rating scales, visual analogue scales, verbal rating scales and Likert scales (where the respondent rates agreement with a series of statements on an agree–disagree continuum) have all been used (see Chapter 5, this volume). Where a number of response options are offered, such as in a Likert scale, between five or seven options appear to result in good reliability in scales in which people are asked to discriminate (Cicchetti *et al.*, 1985; Preston and Colman, 2000).

8.3.3 Constructing a composite indicator or combined score

In phase 2, the multiple items must be combined in some way to generate the composite indicator or score. Thus, a scaling model must be chosen that allows weights to be devised for each item response reflecting the level of the attribute of interest; how these weights are developed defines the model.

A variety of scaling models exist, including direct or indirect estimation models (from classical test theory) and item response theory (IRT). There are two main types of classical scaling models: direct or subjective estimation techniques, and indirect or discriminant techniques (Nunnally and Bernstein, 1994). The direct or subjective estimation techniques are based on the developers' best subjective estimate of the weights that should be assigned to the items. In indirect or discriminant techniques, the weight for each item is derived from experimental observations. Commonly used indirect scaling models include the equally weighted and paired comparison models.

Latent trait models (including factor analysis models) and IRT provide an alternative to classical scaling models. Latent trait models assume that the responses to the multiple items can be described by a smaller number of factors. An item should 'load' on to a particular factor, which is hidden or 'latent'. Each factor consists of clusters of items that are correlated among themselves. Factor analysis also plays an important role in evaluation of instrument validity (Nunnally and Bernstein, 1994).

In classical psychometrics, tests generally require their validity and reliability to be re-established when using the scale on samples from differing populations. With IRT, different sets of items can be used on differing occasions. In particular, classical test theory approaches produce scales that are population specific, whereas in IRT the item characteristics are not dependent on the group used to develop the scale (Yen and Edwardson, 1999). IRT adopts an approach where both items and subjects are scaled with respect to the underlying trait value (Embretson and Reise, 2000).

8.3.4 Methods to optimize interpretability of scales

Interpretability is an important feature of an effective measurement scale, but currently there is no agreement in human healthcare as to how HRQL scores should best be presented for easy interpretation by the user. Users must be able to understand what an individual score produced by an instrument means, for example by comparison with a healthy population. They also need to know when any change in scores is meaningful: is the treatment effective or is the disease getting worse over time?

Normalization of scores

Various methods have been described to improve interpretability of human HRQL instruments including enabling the comparison of scores from instruments that are measuring the same thing but on different metrics (Streiner *et al.*, 2015). One common approach to equating scores reported using different metrics is to transform the raw scores to normalized scores (T-scores). This 'transforms' each scale to have a mean of 50 and a standard deviation of 10, rendering them comparable (Carey and Delaney, 2010). A further approach to facilitate interpretation relates the HRQL scores to those of a specific population, enabling judgement as to whether an observed score is typical of what would be expected for that population (norm-based scoring). Norm-based comparisons can be related to the general population, to subpopulations with shared demographics such as age or gender, or to a population with a particular disease (Lam *et al.*, 1999). In a veterinary context, we have combined these two techniques to normalize HRQL scores to the healthy population, such that a score of 50 represents the average healthy cat or dog, and to create a threshold score above which 70% of healthy animals will score (Davies *et al.*, 2019, 2021).

Calculation of minimal important difference

Schünemann *et al.* (2006) described how clinicians can use in-built thresholds to interpret HRQL scores by referring to a score above which a patient will be deemed to have improved or by associating a change in score with improvement or worsening. An extension of this empirical approach is to quantify a score change by calculating the minimum important difference (MID), defined as 'the smallest difference in score in the outcome of interest that informed patients or informed proxies perceive as important, either beneficial or harmful, and which would lead the patient or clinician to consider a change in the management' (Schünemann *et al.*, 2005, p. 82).

The MID can be established with distributional and/or anchor-based techniques. Distribution-based methods rely on the statistical properties of the instrument with no reference to an external impression of change and include effect size (Kazis *et al.*, 1989), normalized response mean

(Liang *et al.*, 1990) and the modified normalized response mean (Guyatt *et al.*, 1987). Anchor-based techniques use an external criterion such as the patient's perspective of meaningful improvement or worsening to identify the change on the HRQL scale that corresponds to the MID (Lydick and Epstein, 1993). A more objective approach uses receiver operating characteristic (ROC) curves, which are often used to determine the precision of diagnostic tests (Zweig and Campbell, 1993). Deyo and Centor (1986) suggested that scales could be viewed as 'diagnostic tests' for discriminating between patients that had improved and those that had not and that ROC curves could be used to describe a scale's ability to detect improvement. To use ROC curves, the characteristics of the test, namely sensitivity and specificity, which describe how well the test discriminates between two groups, are calculated. The ROC curve plots true positive rate (sensitivity) against false positive rate (1 – specificity), and its shape is used to make decisions about the most appropriate MID. A test with perfect discrimination (no overlap in the diseased and non-diseased distributions) would have a ROC curve that passes through the upper left corner of the ROC graph, providing 100% sensitivity and 100% specificity. Accordingly, the closer the ROC curve is to the upper left corner of the graph, the higher the overall accuracy of the test (Zweig and Campbell, 1993). In a veterinary context, the owner's impression of change (improved or unchanged) and the corresponding change in HRQL score are used to compute a series of sensitivity and 1 – specificity pairs, which then make up the points on the ROC curve. Each point on the ROC curve can be converted back to a value, which in this case is a change in score. A point on the curve is then chosen as the MID, taking into account the clinical implications of that choice.

8.4 Pitfalls

With respect to observations made by an observer on behalf of an animal it is important to be aware of the distinction between causal and indicator variables for QoL. Causal variables such as vomiting in dogs or tail biting in pigs have an impact on QoL but do not tell us anything about how the animal feels, while indicator variables like playfulness and lethargy tell us how the animal 'feels' and therefore inform QoL (Fayers and Hand, 2002; Scott *et al.*, 2007). Some researchers have confused the measurement of health status and the measurement of QoL, and in some cases this has resulted in the inclusion in the instrument of both causal and indicator variables. However, these variables relate to HRQL in different ways and so should contribute in different ways to its measurement (Fayers and Hand, 2002). The confounding of causal and indicator variables has important implications for the analysis of data obtained in such studies, and suitable models have been proposed for that purpose (Fayers and Machin, 2000).

A number of tools intended to measure HRQL in animals have been developed in recent years, and range in complexity from a single visual analogue scale or numerical rating scale on which to record a global score to scientifically robust composite instruments developed using the psychometric methods described above. The latter are costly and time-consuming to develop, with recruitment of client-owned animals to a field test, designed to ensure validity and reliability, being a significant rate-limiting factor. Additionally, several publications (Noble *et al.*, 2019; Rodger *et al.*, 2021) have highlighted the danger of relying on an owner's impression of their pet's health status and this should be considered when embarking on the development of a new tool.

According to Streiner *et al.* (2015), generating a plethora of measurement tools can impede research because of the difficulty of generalizing the findings of one with another. Researchers contemplating a new scale should therefore always be aware of existing scales that might suit the purpose and apply criteria to judge their usefulness as follows:

- What does the instrument intend to measure, and what is its validity for this purpose (e.g. consider how measurement goals were defined, how items were generated and selected, and with what population(s) and in what contexts the instrument has been tested)?
- Do the response options seem appropriate and is the method of generating instrument scores from item responses fully explained and justified?
- What evidence has testing of the instrument provided for its: criterion or construct validity, its reliability, its responsiveness (for evaluative instruments) and its utility?

If no suitable instrument exists and a decision is made to devise a new instrument for a particular purpose, care should be taken to use best practices in instrument development and testing (and these should be published) before the novel instrument is used for clinical or research purposes.

8.5 Conclusions and Tips

The measurement of QoL/HRQL enhances the field of animal welfare by making explicit the importance of how the animal 'feels', and in doing so it addresses the increasing need for measuring subjective experiences and enhancing positive welfare in animals. It is important to realize that validity and reliability are not inherent properties of a scale that are determined by a single value. Developing HRQL instruments is an iterative process, in which instruments are refined and retested with new populations in new contexts and for new purposes. Their development is a time-consuming, challenging and costly undertaking, but the importance of the psychometric approach to such instrument development is widely accepted (Cook *et al.*, 2003). By adopting rigorous methods to construct HRQL instruments that ensure their validity and reliability, veterinary practitioners can be more confident in the management and treatment of chronic disease in their patients, secure in the knowledge that they uphold their promise made upon entry to the veterinary profession: '…my constant endeavour will be to ensure the welfare of animals committed to my care' (RCVS, 2020).

Further Reading

For more information on QoL, please see the references in bold in the reference list.

References

Anand, K.J. and Craig, K.D. (1996) New perspectives on the definition of pain. *Pain* 67(1), 3–6. DOI: 10.1016/0304-3959(96)03135-1.

Bowling, A. (1991) *Measuring Health: A Review of Quality of Life Measurement.* Open University Press, Buckingham, UK.

Brazier, J., Ratcliffe, J., Saloman, J. and Tsuchiya, A. (2017) *Measuring and Valuing Health Benefits for Economic Evaluation.* Oxford University Press, New York.

Brod, M., Tesler, L.E. and Christensen, T.L. (2009) Qualitative research and content validity: developing best practices based on science and experience. *Quality of Life Research* 18(9), 1263–1278. DOI: 10.1007/s11136-009-9540-9.

Carey, J.J. and Delaney, M.F. (2010) T-scores and Z-scores. *Clinical Reviews in Bone and Mineral Metabolism* 8(3), 113–121. DOI: 10.1007/s12018-009-9064-4.

Cicchetti, D.V., Shoinralter, D. and Tyrer, P.J. (1985) The effect of number of rating scale categories on levels of interrater reliability: a Monte Carlo investigation. *Applied Psychological Measurement* 9(1), 31–36. DOI: 10.1177/014662168500900103.

Cook, K.F., Monahan, P.O. and McHorney, C.A. (2003) Delicate balance between theory and practice: health status assessment and item response theory. *Medical Care* 41(5), 571–574. DOI: 10.1097/01.MLR.0000064780.30399.A4.

Coste, J., Fermanian, J. and Venot, A. (1995) Methodological and statistical problems in the construction of composite measurement scales: a survey of six medical and epidemiological journals. *Statistics in Medicine* 14(4), 331–345. DOI: 10.1002/sim.4780140402.

Davies, V., Reid, J., Wiseman-Orr, M.L. and Scott, E.M. (2019) Optimising outputs from a validated online instrument to measure health-related quality of life (HRQL) in dogs. *PLoS One* 14(9), e0221869. DOI: 10.1371/journal.pone.0221869.

Davies, V., Reid, J. and Scott, E.M. (2021) Optimisation of scores generated by an online feline health-related quality of life (HRQL) instrument to assist the veterinary user interpret its results. *Frontiers in Veterinary Science* 7, 1120. DOI: 10.3389/fvets.2020.601304.

Deyo, R.A. and Centor, R.M. (1986) Assessing the responsiveness of functional scales to clinical change: an analogy to diagnostic test performance. *Journal of Chronic Diseases* 39(11), 897–906. DOI: 10.1016/0021-9681(86)90038-X.

Embretson, S.E. and Reise, S.P. (2000) *Item Response Theory for Psychologists.* L. Erlbaum Associates, Mahwah, New Jersey.

Fallowfield, L. (1990) *The Quality of Life: The Missing Measurement in Health Care.* Souvenir Press, London.

Fayers, P.M. and Machin, D. (2000) Principles of measurement scales. In: *Quality of Life: Assessment, Analysis and Interpretation.* Wiley, New York, pp. 28–42.

Fayers, P.M. and Hand, D.J. (2002) Causal variables, indicator variables and measurement scales: an example from quality of life. *Journal of the Royal Statistical Society: Series A* 165(2), 233–253. DOI: 10.1111/1467-985X.02020.

Fayers, P.M. and Machin, D. (2007) Scores and measurements: validity, reliability, sensitivity. In: *Quality of Life: Assessment, Analysis and Interpretation,* 2nd edn. Wiley, New York, pp. 77–108.

Feinstein, A.R. (1987) *Clinimetrics.* Yale University Press, New Haven, Connecticut.

Guyatt, G., Walter, S. and Norman, G. (1987) Measuring change over time: assessing the usefulness of evaluative instruments. *Journal of Chronic Diseases* 40(2), 171–178. DOI: 10.1016/0021-9681(87)90069-5.

Guyatt, G.H., Feeny, D.H. and Patrick, D.L. (1993) Measuring health-related quality of life. *Annals of Internal Medicine* 118(8), 622–629. DOI: 10.7326/0003-4819-118-8-199304150-00009.

Hays, R.D. and Reeve, B.B. (2010) Measurement and modeling of health-related quality of life. In: Killewo, J., Heggenhougen, H.K. and Quah, S.R. (eds) *Epidemiology and Demography in Public Health.* Academic Press, San Diego, California, pp. 195–205.

Jensen, M.P. (2003) Questionnaire validation: a brief guide for readers of the research literature. *The Clinical Journal of Pain* 19(6), 345–352. DOI: 10.1097/00002508-200311000-00002.

Johnston, C.C. (1998) Psychometric issues in the measurement of pain. *Progress in Pain Research and Management* 10, 5–20.

Kazis, L.E., Anderson, J.J. and Meenan, R.F. (1989) Effect sizes for interpreting changes in health status. *Medical Care* 27 (Suppl), S178–S189.

Lam, C.L.K., Lauder, I.J., Lam, T.P. and Gandek, B. (1999) Population based norming of the Chinese (HK) version of the SF-36 health survey. *Hong Kong Practitioner* 21, 460–470.

Landgraf, J.M. and Abetz, L.N. (1996) Measuring health outcomes in pediatric populations: issues in psychometrics and application. In: Spilker, B. (ed.) *Quality of Life and Pharmacoeconomics in Clinical Trials.* Lippincott-Raven, Philadelphia/New York, pp. 793–802.

Liang, M.H., Fossel, A.H. and Larson, M.G. (1990) Comparisons of five health status instruments for orthopedic evaluation. *Medical Care* 28(7), 632–642. DOI: 10.1097/00005650-199007000-00008.

Lydick, E. and Epstein, R.S. (1993) Interpretation of quality of life changes. *Quality of Life Research* 2(3), 221–226. DOI: 10.1007/BF00435226.

Mattin, M., O'Neill, D., Church, D., McGreevy, P.D., Thomson, P.C. *et al.* (2014) An epidemiological study of diabetes mellitus in dogs attending first opinion practice in the UK. *Veterinary Record* 174(14), 349. DOI: 10.1136/vr.101950.

Mendl, M., Burman, O.H.P. and Paul, E.S. (2010) An integrative and functional framework for the study of animal emotion and mood. *Proceedings of the Royal Society B: Biological Sciences* 277(1696), 2895–2904. DOI: 10.1098/rspb.2010.0303.

Minero, M., Tosi, M.V., Canali, E. and Wemelsfelder, F. (2009) Quantitative and qualitative assessment of the response of foals to the presence of an unfamiliar human. *Applied Animal Behaviour Science* 116(1), 74–81. DOI: 10.1016/j.applanim.2008.07.001.

Noble, C.E., Wiseman-Orr, L.M., Scott, M.E., Nolan, A.M. and Reid, J. (2019) Development, initial validation and reliability testing of a web-based, generic feline health-related quality-of-life instrument. *Journal of Feline Medicine and Surgery* 21(2), 84–94. DOI: 10.1177/1098612X18758176.

Nunnally, J.C. and Bernstein, I.H. (1994) *Psychometric Theory*, 3rd edn. McGraw-Hill, New York.

Preston, C.C. and Colman, A.M. (2000) Optimal number of response categories in rating scales: reliability, validity, discriminating power, and respondent preferences. *Acta Psychologica* 104(1), 1–15. DOI: 10.1016/S0001-6918(99)00050-5.

RCVS (Royal College of Veterinary Surgeons) (2020) Code of Professional Conduct for Veterinary Surgeons. Available at: www.rcvs.org.uk/setting-standards/advice-and-guidance/code-of-professional-conduct-for-veterinary-surgeons/ (accessed 14 June 2021).

Reid, J., Wiseman-Orr, M.L., Scott, E.M. and Nolan, A.M. (2013) Development, validation and reliability of a web-based questionnaire to measure health-related quality of life in dogs. *Journal of Small Animal Practice* 54(5), 227–233. DOI: 10.1111/jsap.12059.

Reid, J., Wiseman-Orr, L. and Scott, M. (2018) Shortening of an existing generic online health-related quality of life instrument for dogs. *Journal of Small Animal Practice* 59(6), 334–342. DOI: 10.1111/jsap.12772.

Rodger, S., Scott, E.M., Nolan, A., Wright, A.K. and Reid, J. (2021) Effect of age, breed, and sex on the health-related quality of life of owner assessed healthy dogs. *Frontiers in Veterinary Science* 8, 603139. DOI: 10.3389/fvets.2021.603139.

Scott, E.M., Fitzpatrick, J.L., Nolan, A.M., Reid, J. and Wiseman, M.L. (2003) Evaluation of welfare state based on interpretation of multiple indices. *Animal Welfare* 12, 457–468.

Scott, E.M., Nolan, A.M., Reid, J. and Wiseman-Orr, M.L. (2007) Can we really measure animal quality of life? Methodologies for measuring quality of life in people and other animals. *Animal Welfare* 16, 17–24.

Schünemann, H.J., Puhan, M., Goldstein, R., Jaeschke, R. and Guyatt, G.H. (2005) Measurement properties and interpretability of the Chronic Respiratory Disease questionnaire (CRQ). *COPD* 2(1), 81–89. DOI: 10.1081/COPD-200050651.

Schünemann, H.J., Akl, E.A. and Guyatt, G.H. (2006) Interpreting the results of patient reported outcome measures in clinical trials: the clinician's perspective. *Health and Quality of Life Outcomes* 4, 62. DOI: 10.1186/1477-7525-4-62.

Stevens, S.S. (1951) Mathematics, measurement and psychophysics. In: Stevens, S.S. (ed.) *Handbook of Experimental Psychology*. Wiley, Oxford, UK, pp. 1–49.

Stevens, S.S. (1958) Measurement and man. *Science* 127(3295), 383–389. DOI: 10.1126/science.127.3295.383.

Streiner, D.L. (1993) A checklist for evaluating the usefulness of rating scales. *The Canadian Journal of Psychiatry* 38(2), 140–148. DOI: 10.1177/070674379303800214.

Streiner, D.L. and Norman, G.R. (1995) *Health Measurement Scales: A Practical Guide to Their Development and Use*, 2nd edn. Oxford University Press, New York.

Streiner, D.L., Norman, G.R. and Cairney, J. (2015) *Health Measurement Scales: A Practical Guide to Their Development and Use.* Oxford University Press, New York.

US Department of Health and Human Services FDA Center for Drug Evaluation and Research, US Department of Health and Human Services FDA Center for Biologics Evaluation and Research and US Department of Health and Human Services FDA Center for Devices and Radiological Health (2006) Guidance for industry: patient-reported outcome measures: use in medical product development to support labeling claims: draft guidance. *Health and Quality of Life Outcomes* 4, 79. DOI: 10.1186/1477-7525-4-79.

Wemelsfelder, F. (2007) How animals communicate quality of life: the qualitative assessment of behaviour. *Animal Welfare* 16, 25–31.

WHOQOL Group (1994) Development of the WHOQOL: rationale and current status. *International Journal of Mental Health* 23(3), 24–56. DOI: 10.1080/00207411.1994.11449286.

Wiseman-Orr, M.L., Nolan, A.M., Reid, J. and Scott, E.M. (2004) Development of a questionnaire to measure the effects of chronic pain on health-related quality of life in dogs. *American Journal of Veterinary Research* 65(8), 1077–1084. DOI: 10.2460/ajvr.2004.65.1077.

Wiseman-Orr, M.L., Scott, E.M., Reid, J. and Nolan, A.M. (2006) Validation of a structured questionnaire as an instrument to measure chronic pain in dogs on the basis of effects on health-related quality of life. *American Journal of Veterinary Research* 67(11), 1826–1836. DOI: 10.2460/ajvr.67.11.1826.

Wiseman-Orr, M.L., Scott, E.M. and Nolan, A.M. (2011a) Development and testing of a novel instrument to measure health-related quality of life (HRQL) of farmed pigs and promote welfare enhancement (Part 1). *Animal Welfare* 20, 535–548.

Wiseman-Orr, M.L., Scott, E.M. and Nolan, A.M. (2011b) Development and testing of a novel instrument to measure health-related quality of life (HRQL) of farmed pigs and promote welfare enhancement (Part 2). *Animal Welfare* 20, 549.

Yam, P.S., Butowski, C.F., Chitty, J.L., Naughton, G., Wiseman-Orr, M.L. *et al.* (2016) Impact of canine overweight and obesity on health-related quality of life. *Preventive Veterinary Medicine* 127, 64–69. DOI: 10.1016/j.prevetmed.2016.03.013.

Yen, M. and Edwardson, S.R. (1999) Item-response theory approach in scale development. *Nursing Research* 48(4), 234–238. DOI: 10.1097/00006199-199907000-00008.

Zweig, M.H. and Campbell, G. (1993) Receiver-operating characteristic (ROC) plots: a fundamental evaluation tool in clinical medicine. *Clinical Chemistry* 39(4), 561–577. DOI: 10.1093/clinchem/39.4.561.

9

Behavioural Tests for Learning and Cognition in Humans and Animals

Rebecca E. Nordquist*

Faculty of Veterinary Medicine, Utrecht University, Utrecht, The Netherlands

9.1 Introduction

Anyone who has spent time with animals – be it professionally, as pets or simply through observing animals outdoors – has little doubt about their capacity to learn. For quite a while throughout history, however, humans were considered unique in many cognitive abilities that we now realize non-human animals may also possess. Tool use is a well-known example, where what was first considered as obviously unique to humans was later demonstrated in primates, famously by Jane Goodall (Goodall, 1986), and is now widely accepted also to occur in other species, including crows and possibly dolphins (Seed and Byrne, 2010). There is an increasing body of knowledge that many animals can perform complex learning tasks that can sometimes surprise us. Sheep can recognize other sheep from photographs (Kendrick *et al.*, 2001). Pigs turned out to be capable of learning complex cognitive tasks (Nordquist *et al.*, 2017). Indeed, the more research is conducted in animal cognition, the more capabilities are uncovered in species previously deemed 'lower' species; see, for instance, recent reviews on the cognitive capacities of fish (Salena *et al.*, 2021) and cephalopods such as octopus and squid (Schnell *et al.*, 2021).

In this chapter, I will discuss the importance of animal cognition to the field of animal welfare and will put the development of the field of animal cognition into historical perspective. The practicality of testing animal cognition will be illustrated with several tests frequently used in both humans and non-human animals, and I will end with some tips and recommendations for those wishing to embark on cognitive testing in relation to animal welfare.

*r.e.nordquist1@uu.nl

 Bridging Research Disciplines to Advance Animal Welfare Science: A practical guide (ed. I. Camerlink)
DOI: 10.1079/9781789247893.0009

9.1.1 The importance of learning and cognition in animal welfare

In a field so broad as animal welfare, one can ask why it is important to study cognition in animals. After all, if an animal is healthy and seems happy, why is it important what its cognitive capacities are? Cognition is essential to both the animals themselves and to humans studying animals; this is visualized in Fig. 9.1. Cognition is essential to animals for two main reasons. First is the practical aspect: animals must have sufficient cognitive abilities to thrive in their own environments. Second, an animal's cognition is part of

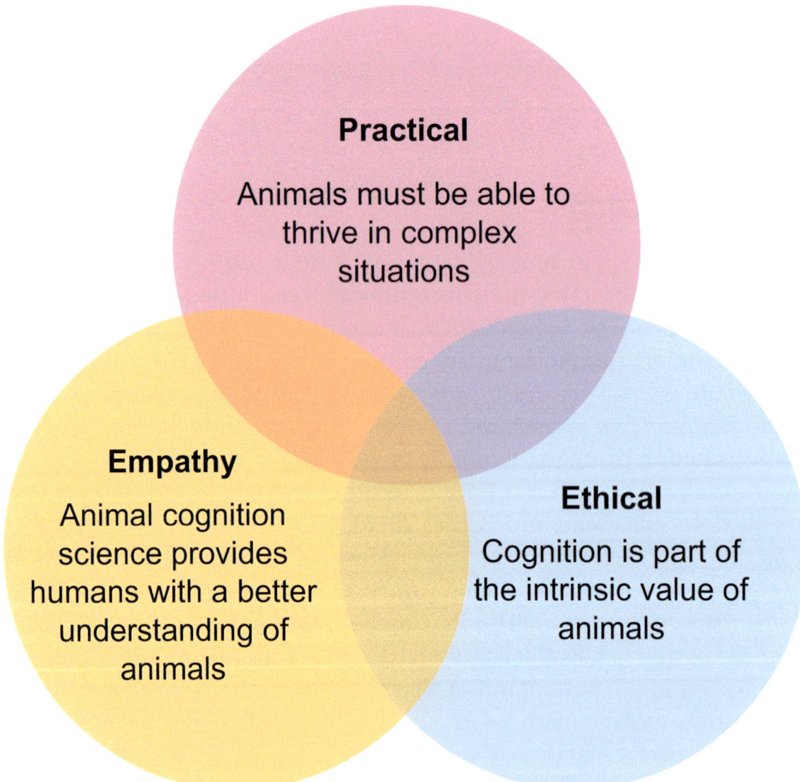

Fig. 9.1. The importance of animal cognition within the field of animal welfare. There are three main reasons to study cognition in human-kept animals: (i) practical: humans keep animals in situations that can be complex for them; for example, farms are becoming increasingly complex, and we need to make sure that human-kept animals' intelligence is on a par with the challenges on farms; (ii) ethical: similar to other species-specific behaviours, the cognitive capacity of animals is part of their intrinsic value; if our breeding or management systems are changing the cognitive capacities of animals, this poses ethical issues; and (iii) empathy: by better understanding what animals are capable of, we as humans will be better animal stewards.

its intrinsic value. Intrinsic value refers to the value that an animal has as a living being, regardless of the value that humans may place on animals as a monetary commodity, a source of food or clothing, or a companion. This means that humans have ethical responsibilities regarding animal cognition. Just as other species-specific behaviours need to be taken into account when humans deal with animals, the cognitive capacities of animals must be taken into account. That ethical responsibility links the importance of cognition to the field of animal welfare research.

There are a number of reasons why an understanding of animal cognition is important to the commitment of the field to improve animal welfare. First, better understanding of cognitive capacities allows humans to be better animal stewards by creating housing and management systems that match the cognitive capacities of animals: both by not overtaxing their cognition and also by providing sufficient challenge to their cognitive capacities. Second, a better understanding of animals leads to more societal concern for their well-being. It is not yet established that a better understanding of the cognitive capacities of animals leads to more empathy for those animals, but studies do point to the reverse situation being true: people tend to discount the cognitive capacities of animals that are to be consumed, compared with those that are not (Bastian *et al.*, 2012). Finally, cognition is an important proxy for internal state. Happy animals perform well, while compromised animals (i.e. animals in pain) perform more poorly (see section 9.2.3 on bias tasks below).

9.1.2 Historical perspective of associative learning and cognition in animals

The focus of the present chapter is on cognition research at the overlap between human and animal research. I will therefore outline the history of cognitive science from a psychological research standpoint, which focuses on understanding learning and the mind at a fundamental level. This type of research has been conducted in humans and in other animals. This is a different starting point compared with research from an ethological standpoint, which is geared more to understanding the animal itself. I will focus here on the history of cognitive research in animals that came from psychology, but this is not to dismiss the work of ethologists such as Niko Tinbergen (see Chapter 14, this volume), Jane Goodall, Irene Pepperberg or any of the other important contributions from ethology to understanding animal cognition.

Cognitive science within psychology was initially focused almost solely on understanding the human cognition. Psychology as a field started with early scientists such as Wilhelm Wundt, a German scientist in the late 1800s, whose primary interest was in understanding how learning takes place, and who studied this process in humans. He described (among other things) the process of 'apperception', in which past experiences colour a subject's current perception (Mandler, 2011). Thus, from the start, the importance of internal states was recognized, with the potential for experience to alter cognitive abilities – much the same as current judgement bias tasks (see section 9.3.3). However, from the 1930s to the 1960s, the seminal work of B.F. Skinner on associative learning processes gave rise to 'behaviourism'. In behaviourism, (nearly) all behaviour is seen as the result of an animal receiving a reward or punishment for an action, and increasing or decreasing that behaviour accordingly. This extensively influential work was instrumental in understanding many fundamental learning processes but left little room for examining motivations or internal states of animals as conscious beings, despite the fact that this detached view of animals does not seem to have been Skinner's intention (O'Donohue and Ferguson, 2001).

In line with this shift to more 'hard' science in the psychology of learning, in the mid-20th century, psychology turned increasingly to quantitative measures such as latency times and correct choices in learning tasks to show how animals (including humans) learn, rather than qualitative reports. This corresponded with the rise of the analysis of variance (ANOVA) as a statistical method in psychology to parse out experimental effects, with all of the requirements the test entails in terms of independence of groups, ratio-level measures and normality of data distribution. This helped to lend credibility to the field of psychology as a 'real' science, which may not be a spectacularly good reason to adopt a methodology but was important to the maturation of the field nonetheless. The wider use of statistics in general, but ANOVAs specifically, was also instrumental in the development of ways to reliably test hypotheses about behaviour and cognition. The history of statistics in psychology is very readably discussed in Michael Cowles' book on this subject (Cowles, 2005).

The peak of the quantification movement in psychological research in animal learning was perhaps the introduction of the Rescorla–Wagner model. This model quantifies and predicts classical conditioning responses through the formula (Rescorla and Wagner, 1972):

$$\Delta V = \alpha\beta(\lambda - \Sigma V)$$

which specifies that the amount of learning (the change, Δ, in the predictive value of a stimulus, V) depends on the amount of surprise (the difference between what actually happens, represented by λ, and what is expected to happen, ΣV), with the learning rate included as α and the unconditioned stimulus used as β. While this view of learning as reducible to a formula was extremely helpful in understanding basic learning processes, and sparked many new areas of research (including having a strong influence on the development of artificial intelligence; Gershman, 2015; Hollis, 2019), reduction of learning processes to mathematical terms led to a distancing of behavioural psychologists from their subjects, the animals, and was perhaps not helpful for the development of empathy of researchers for animals, or for viewing learning from an animal's perspective.

Starting in the 1990s, the role of individual variation in behaviour began to regain attention. In particular, the role of coping styles (Coppens *et al.*, 2010) made a strong contribution to examination of animals as individuals. It became clear that some animals show problem-solving strategies that are stable across time and correlate with, for instance, fear reactivity. As more stable traits were developed that differed among individuals within a species, the term 'personality' came more into use in animal research (Gosling, 2001; Stamps and Groothuis, 2010; Kaiser and Müller, 2021), and it has been observed that in some cases personality as defined by 'temporally stable behavioural traits' can also correlate with cognitive performance (Carere and Locurto, 2011).

From this background, with individual variation and cognitive performance only relatively recently regaining attention in animal cognition research, it is perhaps not surprising that it has taken some time to recognize that cognition is important to animal welfare as viewed from the perspective of the animal. Until recently, the study of animal learning and cognitive processes was mainly in the service of humans in order to uncover fundamental processes of learning, or in animal models of disease or disorder, but there was very little interaction between this large body of research, going back over a century, and animal welfare research. Fortunately, this is changing, and in the past decades, learning and cognition are increasingly seen as an integral part of animal welfare research.

9.2 Concepts

Animal welfare

There are many definitions of animal welfare in use today (as described in Chapter 1, this volume). Most modern definitions in the field include adaptation or flexibility to adapt, and/or positive welfare indicators. For the current chapter, I will use the following definition, from Ohl and van der Staay (2012, p. 17):

An individual is in a positive welfare state when it has the freedom *adequately to react to*

- hunger, thirst or incorrect food;
- thermal and physical discomfort;
- injuries or diseases;
- fear and chronic stress; and thus,
- the freedom to display normal behavioural patterns that *allow the animal to adapt to the demands of the prevailing environmental circumstances and enable it to reach a state that it perceives as positive.*

Cognition

As with animal welfare scientists and defining animal welfare, cognitive scientists similarly struggle with a definition of 'cognition'. Indeed, there have been arguments from cognitive science to simply not define cognition and move on (Allen, 2017), or to accept multiple definitions as this does not inhibit the work of cognitive scientists (Akagi, 2018). For the sake of pragmatism, in the current chapter, I will use a broad definition of cognition. I refer to a definition relevant for comparative cognition (Buckner, 2015, p. 314), which poses that a 'general characteristic that nearly every test for cognition is meant to elicit is behavioural flexibility'. In this definition, responses in tests that contain a cognitive component contain at least a few of the following properties: context sensitivity, speed (with cognitive responses presumed faster than trial and error), class formation, higher-order and abstract learning, multi-modality transfer, inhibition, monotonic integration (i.e. ordering by size), and expectation generation and monitoring. For further explanation of these properties, see Buckner (2015).

Bias task

A bias task is a task where *affect* (how an animal is 'feeling') has an effect on performance of the task. Examples are judgement bias, where an animal has to choose between an expected positive or negative outcome of an action, and attention bias, where strong emotional stimuli, such as the threat of a predator, can cause animals to perform a task worse than without the predator.

9.3 Methodologies

Below, I will outline three types of tests that are used in both humans and animals, to serve as examples of tests that may tell us something about the fundamental cognitive capacities of animals related to animal welfare, or where cognitive performance may serve as a proxy to show the internal state of the animal. These examples were chosen because they cover different aspects of cognition and have been conducted in many species, but they are by no means more or less important than other tasks in investigation of animal cognition.

Delayed match tasks

Delayed match tasks are used to examine working (or short-term) memory. In these tasks, the subject is given a cue, followed by a delay. Following the delay, the subject must give a response based on the cue, so the subject must keep the cue in their short-term memory for the duration of the delay. The cues can be based on a sample (i.e. respond to a green light versus a white light) or based on a location (i.e. respond to the left if the cue was on the left), and the task can be performed as a match (respond to the left if the cue on the left is given) or non-match (respond to the right if the cue on the left is given).

Delayed match tasks are used in humans to study the basic neurobiology of learning, for instance the role of the hippocampus in working memory using magnetic resonance imaging (O'Neil and Lee, 2019). Working memory as measured by delayed match tasks has been shown to be impaired by pharmacological interventions in healthy human volunteers (Jäkälä *et al.*, 1999). Humans born with low birth weight also show impaired performance in the task in adulthood (Daniel *et al.*, 2016; Mathewson *et al.*, 2020). This task has been used widely and extensively in laboratory animals to investigate fundamental learning processes and neurobiological substrates of learning using electrophysiology and psychopharmacology (Daniel *et al.*, 2016).

More recently, Teutsch and Kätzel (2019) developed a delayed match-to-place task (DMTP) in mice based on the five-choice serial reaction time set-up, which is considerably more complex than the typical two-lever DMTP set-up. This has the advantage of being able to more reliably exclude the animals using, for instance, body position as a marker to remember which cue requires a response. Given the increased complexity, however, the task has the distinct disadvantage that it takes several months of training before the mice are able to perform reliably. This trade-off is frequently seen in cognitive testing, where more detail requires longer and more intensive training, which is not always possible or feasible.

Set-shifting tasks

Set-shifting tasks include moving from rule learning based on one characteristic or modality, to another. The classic test of set shifting in humans is the Wisconsin Card Sorting Test, which was originally developed in the late 1940s (Berg, 1948). In this task, human subjects are given cards with one to four geometric figures (star, circle, square or cross) printed in one of four colours. Originally, these were physical cards, but currently the test is more often performed on a computer. Subjects are told to sort the cards, but not what the rule for sorting is, only 'correct' or 'incorrect'. The rule might be sorting by colour, by number of symbols (i.e. geometric figures) on the card, or by type of symbol; the subject is to discover the rule by trial and error. Once the subject has correctly discovered the rule, the sorting rule is changed without warning. For instance, if the sorting is first based on number of symbols, it is then based on colour. The ability of a subject to change the characteristic used to sort is taken as a measure of behavioural flexibility. Human subjects with a number of disorders thought to hamper their cognitive flexibility, including schizophrenia (Everett *et al.*, 2001;

Prentice *et al.*, 2008) and autism (Yerys *et al.*, 2009), score poorly on the parameters that measure set shifting in the Wisconsin Card Sorting Task.

In animals, the intradimensional–extradimensional set shift task investigates similar underlying constructs, namely the ability of an animal to learn to solve a task based on one modality, and then to change to another modality. In the most commonly used protocol, rodents are taught that a reward is in one of two dishes, covered with substrate. The rodents learn that the reward is always covered with either a specific substrate (coarse versus fine) or in a dish that contains a specific odour; thus, there is a task-relevant modality (i.e. tactile information from the substrate) and a task-irrelevant modality (olfactory cue). In this example, the animal needs to learn to respond to the correct substrate (tactile information given as the task-relevant modality) and disregard the olfactory cue (olfactory information as the task-irrelevant modality). After the first association has been learned to a performance criterion, the relevant and irrelevant modalities are switched. In this example, the rodent must now respond to the previously disregarded olfactory information and disregard the tactile information. In rodents, as recently reviewed by Tait *et al.* (2018), performance on set shifting has been shown to be impaired by a number of manipulations that would be expected to influence set shifting based on human literature, such as treatment with valproic acid (McKinnell *et al.*, 2021) and early-life lead exposure (Neuwirth *et al.*, 2019). Along with the wide use in more traditional laboratory animals, this task has recently also been adapted for use in sheep (Sorby-Adams *et al.*, 2021).

Bias tasks as a proxy for measuring affect

Assessment of subjective affect (i.e. how an animal 'feels') is notoriously difficult in animals. One cannot simply ask how they are feeling, of course. At the same time, subjective affect is an integral part of some definitions of animal welfare. For example, in the definition I use in this chapter, good welfare includes 'the freedom to display normal behavioural patterns that allow the animal to adapt to the demands of the prevailing environmental circumstances and enable it to reach a state that it perceives as positive' (Ohl and van der Staay, 2012, p. 17). When applying this definition to animal affect, there is an obvious component of subjectivity and individual response included, which is difficult to measure. The use of cognition as a proxy to reflect internal state is one approach to try to elucidate the inner emotional state of animals.

Judgement bias tasks are intended to investigate effects of affect on cognitive function, thus using cognitive performance as a proxy for subjective affect. Figure 9.2 provides a more detailed explanation of a judgement bias task. Basically, a subject is taught that one stimulus predicts a positive outcome (e.g. a high-frequency tone predicts a large reward) and a different stimulus predicts a negative outcome (e.g. a low-frequency tone

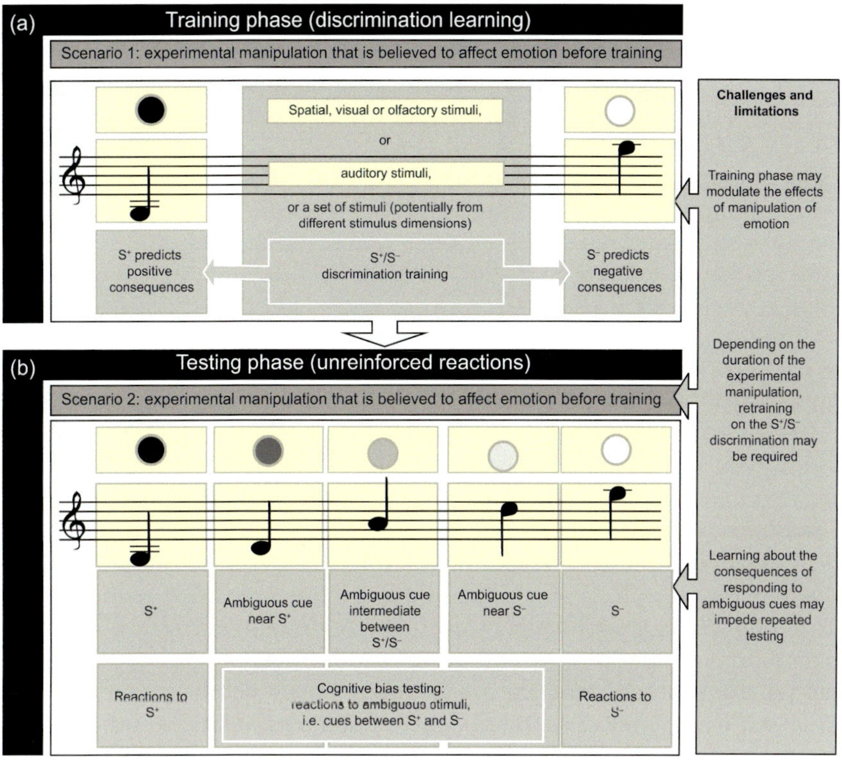

Fig. 9.2. Judgement bias training and testing. Schematic representation of judgement bias training and testing using visual, olfactory, spatial or auditory cues, or a combination of cues from different stimulus dimensions (inspired by Bateson *et al.*, 2011; Mendl *et al.*, 2011). The experimental manipulation that is believed to affect emotion precedes the training phase (a; scenario 1) or the testing phase (b; scenario 2). Refreshment of the discrimination acquired during the training phase may be necessary, if the experimental manipulation preceding phase (b) lasts for a longer time period. An example of scenario 1 is studying the effects of growing up in different housing systems, whereas scenario 2 might be applied in a study assessing the effects of shorter-lasting experimental manipulations, such as confinement, on emotion. Phase (b) may be repeated multiple times to test the effects of different experimental manipulations in the same animal. Specific challenges and limitations may be connected to the different phases. From Roelofs *et al.* (2016); reprinted under the terms of Creative Commons Attribution License (CC BY).

predicts a small reward). The important response is then: what happens when the animal is given a stimulus in the middle (i.e. a mid-frequency tone)? Does the subject respond to the stimulus as if it expects a positive outcome or a negative outcome? In essence, is the glass half full or half empty? Figure 9.2 shows an example of the specific contingencies connected with responding to positive, negative and ambiguous cues. In humans, judgement bias is seen to be affected by long-term negative affect, such as people suffering from depression (Mineka and Sutton, 1992). Judgement bias has been regularly used within the field of animal welfare as a proxy for measuring the subjective state of animals. It has now been tested in the context of animal welfare in a wide variety of animals, including, among others, pigs, sheep, chickens, rodents and hummingbirds. Recent reviews have examined the utility of the judgement bias paradigm for examining animal welfare and have covered the scientific literature quite extensively (see reviews by Roelofs *et al.*, 2016; Watters and Krebs, 2019; Mendl and Paul, 2020; Neville *et al.*, 2020).

Although there is certainly potential for this test to be useful, a number of hurdles are still to be overcome before the test can be practically implemented in animal welfare research. Judgement bias protocols require relatively long and labour-intensive training in learning tasks, making them impractical for use in more natural and/or farm settings. Recent efforts have also been made to design faster and more practical judgement bias paradigms (e.g. Hintze *et al.*, 2018), although the test will still need a great deal of refinement to be useful on a larger scale. Aside from practicality, the long training phase may be detrimental to interpretation of the results. The long training sessions may also provide enrichment with both relief from boredom and positive interactions with humans. This may mask effects of, for instance, housing conditions, as an animal that spends all day, every day in an impoverished environment will respond differently from an animal housed in an impoverished environment that is trained for an hour or two per day in an operant task.

Attention bias is similar to judgement bias in that it uses a proxy to elucidate emotional state. Where in judgement bias an operant task is used, which (by definition) requires training, in the attention bias task the amount of attention given to a threat versus a neutral or positive stimulus is measured. In general, humans and other animals will show a bias in attention towards a threat, compared with attention towards a neutral or positive stimulus. This bias is probably composed of both increased attention to the threat and decreased ability to disengage attention to the threat (Cisler *et al.*, 2009). In humans, anxiety disorders have been shown to increase the attentional bias towards threats compared with subjects without anxiety disorders (Bar-Haim *et al.*, 2007; Mobini and Grant, 2007), an effect seen already at an early age in humans (Puliafico and Kendall, 2006).

A common paradigm for attention bias in animals is the provision of a primary reward, such as food or water for bathing (in birds), in the presence of a predator or conspecific alarm calls (recently reviewed by Crump *et al.*, 2018). In primates, showing pictures of positively valanced stimuli (e.g. food) or negatively valanced stimuli (e.g. snakes) disrupts performance of a complex task more than neutral stimuli (Hopper *et al.*, 2021). Sheep paid less attention to a predator in the form of a dog, and were more likely to feed in the presence of the stimuli when given the anxiolytic diazepam prior to testing (Lee *et al.*, 2016), supporting the influence of anxiety on task performance. In pigs, current housing conditions (but not early-life housing conditions) affect performance on attention bias tasks (Luo *et al.*, 2019).

9.4 Common Pitfalls in the Integration of Cognitive Research with Animal Welfare

The field of cognitive research in both humans and animals is vast and has a long history. Cognition research in animals was initially propelled primarily by basic questions about how cognition is fundamentally organized. At present, cognitive research in animals is most strongly in use in animal models for human disease with cognitive deficits, and this is likely to continue. The importance of cognition to animal welfare is starting to take hold in the field of animal welfare. This recognition is encouraging, but animal welfare researchers still have quite a way to go. There are a number of issues that generally impede the advancement of integration between broad cognitive research and cognitive research geared towards animal welfare.

The first hurdle is communication between fields. This is discussed at length in Chapter 2 (this volume), but I will address a few cognition-specific issues here. There is at present very little interaction between scientists working on cognition in animals as models for human disease and scientists working on animal cognition in animal welfare. Animal welfare research is a vastly smaller field than animal models of human disease. Animal welfare journals, consequently, generally have lower impact factors compared with those focused on human disease. This is a serious hurdle to scientists in the field of human disease who may want to work with animal welfare researchers or on animal welfare topics. Low-impact-factor publications will not help (and may even hurt) their careers at tenure review or other major milestones, and changing this viewpoint will require a more widespread change in valuation of scientific output. From the other side, animal welfare researchers can be strong in beliefs about animal welfare and mistrusting of the use of animals in models of human disease. This can be a hindrance to the openness needed to learn from another field. Furthermore, the huge volume of research in cognition in which animals are used as models can be daunting to dive into. More dialogue between the two areas could produce a more constructive discourse.

Second is the complexity of translation of tests from humans to animals, specifically for cognition. Cognitive tests in humans tend to rely on language, making translation to non-verbal animals difficult. One area of missed potential is in collaboration with scientists working on pre-verbal humans (young children). Dating back to the seminal work of Jean Piaget, cognitive testing in developmental psychology has been aimed at testing of pre-verbal humans. There is much overlap in the difficulties of scientists researching neonatal abilities and the clinicians aiming to directly improve the welfare of neonates, and the science and practice of animal welfare. In one recent example where developmental psychology was taken to animal behaviour, teasing was reviewed in animals (Eckert *et al.*, 2020). Teasing in children is fairly well described developmentally, and requires the ability to understand what the to-be-teased partner is expecting in order to mislead them. There are a few examples, but at present there is too little interaction between these two fields, while there is much to be gained by both sides from the cross-pollination.

Finally, cognitive testing in animals often relies on extensive training, elaborate set-ups and relatively subtle differences across many trials, and traditionally requires standardized set-ups with little variation in distractions such as background noise. All of these aspects make the use of cognitive testing outside laboratories challenging. This often leads to protests about the relevance of cognitive testing to animal welfare. If we cannot test cognition in relevant, real-world settings, what use are the results to improve animal welfare? In part, cognitive testing is about understanding the basic biology of the species we keep for companions, for food and for work in order to better meet their basic needs in practice. To some extent, cognitive testing will remain a laboratory discipline, geared towards better understanding of the cognitive abilities of animals. Better understanding then leads to more compassion on the one hand and opportunities to develop better management practices on the other. There are opportunities, however, to bring cognitive testing out of the laboratory. In conservation studies, the importance of animal cognition in human–animal interactions in the wild is gaining traction (Goumas *et al.*, 2020). In a recent example where cognition testing has made its way into ornithology of wild birds, wild collared flycatchers (*Ficedula albicollis*) were studied in an ingenious nest-box designed as a behavioural set-up (Morinay *et al.*, 2020). Similarly, squirrels have been tested in their own habitats (Chow *et al.*, 2021). More cognitive tests that can be performed in the home area, with animals freely deciding to participate or not, will certainly increase the utility of cognitive testing in natural and/or farm settings.

9.5 Conclusions and Tips

On a final note, I provide a few conclusions and practical tips for integrating cognitive research in animal welfare work. Cognitive testing takes time if you want results that can be interpreted. Involve a specialist to think through necessary controls and possible interpretations of results. Do not underestimate things like habituation to a test set-up or building in all of the steps needed to train an animal in a cognitive task. Habituation time can be considerably shortened by testing in a home environment where possible, and testing in pairs when feasible. Faster learning of responses can be facilitated by making use of species-specific behaviours, such as rooting in pigs, digging in rodents or pecking in chickens, as responses. If using cues, make sure they are in a modality that is attractive to the species being tested. Even if a test is constructed optimally, the fact remains that cognitive testing in animals is nearly always time-intensive. The exact time required will differ among test and species, but cognitive testing takes easily a week and can take months for complex tasks. This needs to be taken into consideration when planning experiments.

A final issue is the need for recognition of the importance of cognition to animal welfare. Cognition is a much less obvious need of an animal compared with freedom from disease or a place to root (for a pig), scratch (for a chicken) or dig (for a rabbit). Within the field of animal welfare, there is increased attention to the need to examine cognition, as detailed at the beginning of the present chapter. There is, however, a lot of work to be done to educate non-researchers, such as the general public and policy makers, on why we should care about cognition in animals.

Further Reading

References that are recommended for further information are given in bold in the reference list.

References

Akagi, M. (2018) Rethinking the problem of cognition. *Synthese* 195, 3547–3570.

Allen, C. (2017) On (not) defining cognition. *Synthese* 194, 4233–4249.

Bar-Haim, Y., Lamy, D., Pergamin, L., Bakermans-Kranenburg, M.J. and van IJzendoorn, M.H. (2007) Threat-related attentional bias in anxious and nonanxious individuals: a meta-analytic study. *Psychological Bulletin* 133(1), 1–24. DOI: 10.1037/0033-2909.133.1.1.

Bastian, B., Loughnan, S., Haslam, N. and Radke, H.R.M. (2012) Don't mind meat? The denial of mind to animals used for human consumption. *Personality & Social Psychology Bulletin* 38(2), 247–256. DOI: 10.1177/0146167211424291.

Bateson, M., Desire, S., Gartside, S.E. and Wright, G.A. (2011) Agitated honeybees exhibit pessimistic cognitive biases. *Current Biology* 21(12), 1070–1073. DOI: 10.1016/j.cub.2011.05.017.

Berg, E.A. (1948) A simple objective technique for measuring flexibility in thinking. *The Journal of General Psychology* 39, 15–22. DOI: 10.1080/00221309.1948.9918159.

Buckner, C. (2015) A property cluster theory of cognition. *Philosophical Psychology* 28, 307–336.

Carere, C. and Locurto, C. (2011) Interaction between animal personality and animal cognition. *Current Zoology* 57, 491–498.

Chow, P.K.Y., Clayton, N.S. and Steele, M.A. (2021) Cognitive performance of wild eastern gray squirrels (*Sciurus carolinensis*) in rural and urban, native, and non-native environments. *Frontiers in Ecology and Evolution,* in press.

Cisler, J.M., Bacon, A.K. and Williams, N.L. (2009) Phenomenological characteristics of attentional biases towards threat: a critical review. *Cognitive Therapy and Research* 33(2), 221–234. DOI: 10.1007/s10608-007-9161-y.

Coppens, C.M., de Boer, S.F. and Koolhaas, J.M. (2010) Coping styles and behavioural flexibility: towards underlying mechanisms. *Philosophical Transactions of the Royal Society of London. Series B, Biological Sciences* 365(1560), 4021–4028. DOI: 10.1098/rstb.2010.0217.

Cowles, M. (2005) *Statistics in Psychology: An Historical Perspective,* 2nd edn. Psychology Press, New York.

Crump, A., Arnott, G. and Bethell, E.J. (2018) Affect-driven attention biases as animal welfare indicators: review and methods. *Animals* 8, 136. DOI: 10.3390/ani8080136.

Daniel, T.A., Katz, J.S. and Robinson, J.L. (2016) Delayed match-to-sample in working memory: a BrainMap meta-analysis. *Biological Psychology* 120, 10–20. DOI: 10.1016/j.biopsycho.2016.07.015.

Eckert, J., Winkler, S.L. and Cartmill, E.A. (2020) Just kidding: the evolutionary roots of playful teasing. *Biology Letters* 16(9), 20200370. DOI: 10.1098/rsbl.2020.0370.

Everett, J., Lavoie, K., Gagnon, J.F. and Gosselin, N. (2001) Performance of patients with schizophrenia on the Wisconsin Card Sorting Test (WCST). *Journal of Psychiatry & Neuroscience* 26(2), 123–130.

Gershman, S.J. (2015) A unifying probabilistic view of associative learning. *PLOS Computational Biology* 11, e1004567. DOI: 10.1371/journal.pcbi.1004567.

Goodall, J. (1986) *The Chimpanzees of Gombe: Patterns of Behavior.* Harvard University Press, Cambridge, Massachusetts.

Gosling, S.D. (2001) From mice to men: what can we learn about personality from animal research? *Psychological Bulletin* 127(1), 45–86. DOI: 10.1037/0033-2909.127.1.45.

Goumas, M., Lee, V.E., Boogert, N.J., Kelley, L.A. and Thornton, A. (2020) The role of animal cognition in human–wildlife interactions. *Frontiers in Psychology* 11, 589978. DOI: 10.3389/fpsyg.2020.589978.

Hintze, S., Melotti, L., Colosio, S., Bailoo, J.D., Boada-Saña, M. *et al.* (2018) A cross-species judgement bias task: integrating active trial initiation into a spatial Go/No-go task. *Scientific Reports* 8(1), 5104. DOI: 10.1038/s41598-018-23459-3.

Hollis, G. (2019) Learning about things that never happened: a critique and refinement of the Rescorla–Wagner update rule when many outcomes are possible. *Memory & Cognition* 47(7), 1415–1430. DOI: 10.3758/s13421-019-00942-4.

Hopper, L.M., Allritz, M., Egelkamp, C.L., Huskisson, S.M., Jacobson, S.L. *et al.* (2021) A comparative perspective on three primate species' responses to a pictorial emotional stroop task. *Animals* 11(3), 588. DOI: 10.3390/ani11030588.

Jäkälä, P., Sirviö, J., Riekkinen, M., Koivisto, E., Kejonen, K. *et al.* (1999) Guanfacine and clonidine, alpha2-agonists, improve paired associates learning, but not delayed matching to sample, in humans. *Neuropsychopharmacology* 20(2), 119–130. DOI: 10.1016/S0893-133X(98)00055-4.

Kaiser, M.I. and Müller, C. (2021) What is an animal personality? *Biology and Philosophy* 36, 1.

Kendrick, K.M., da Costa, A.P., Leigh, A.E., Hinton, M.R. and Peirce, J.W. (2001) Sheep don't forget a face. *Nature* 414(6860), 165–166. DOI: 10.1038/35102669.

Lee, C., Verbeek, E., Doyle, R. and Bateson, M. (2016) Attention bias to threat indicates anxiety differences in sheep. *Biology Letters* 12(6), 20150977. DOI: 10.1098/rsbl.2015.0977.

Luo, L., Reimert, I., de Haas, E.N., Kemp, B. and Bolhuis, J.E. (2019) Effects of early and later life environmental enrichment and personality on attention bias in pigs (*Sus scrofa domesticus*). *Animal Cognition* 22(6), 959–972. DOI: 10.1007/s10071-019-01287-w.

Mandler, G. (2011) *A History of Modern Experimental Psychology: From James and Wundt to Cognitive Science*. MIT Press, Cambridge, Massachusetts.

Mathewson, K.J., Maurer, D., Mondloch, C.J., Saigal, S., Van Lieshout, R.J. *et al.* (2020) Visual configural processing in adults born at extremely low birth weight. *Developmental Science* 23(2), e12890. DOI: 10.1111/desc.12890.

McKinnell, Z.E., Maze, T., Ramos, A., Challans, B. and Plakke, B. (2021) Valproic acid treated female Long-Evans rats are impaired on attentional set-shifting. *Behavioural Brain Research* 397, 112966. DOI: 10.1016/j.bbr.2020.112966.

Mendl, M. and Paul, E.S. (2020) Animal affect and decision-making. *Neuroscience and Biobehavioral Reviews* 112, 144–163. DOI: 10.1016/j.neubiorev.2020.01.025.

Mendl, M., Paul, E.S. and Chittka, L. (2011) Animal behaviour: emotion in invertebrates? *Current Biology* 21(12), R463–R465. DOI: 10.1016/j.cub.2011.05.028.

Mineka, S. and Sutton, S.K. (1992) Cognitive biases and the emotional disorders. *Psychological Science* 3, 65–69.

Mobini, S. and Grant, A. (2007) Clinical implications of attentional bias in anxiety disorders: an integrative literature review. *Psychotherapy: Theory, Research, Practice, Training* 44, 450–462.

Morinay, J., Cauchard, L., Bize, P. and Doligez, B. (2020) The role of cognition in social information use for breeding site selection: experimental evidence in a wild passerine population. *Frontiers in Ecology and Evolution* 8, 559690.

Neuwirth, L.S., Masood, S., Anderson, D.W. and Schneider, J.S. (2019) The attention set-shifting test is sensitive for revealing sex-based impairments in executive functions following developmental lead exposure in rats. *Behavioural Brain Research* 366, 126–134. DOI: 10.1016/j.bbr.2019.03.022.

Neville, V., Nakagawa, S., Zidar, J., Paul, E.S., Lagisz, M. *et al.* (2020) Pharmacological manipulations of judgement bias: a systematic review and meta-analysis. *Neuroscience and Biobehavioral Reviews* 108, 269–286. DOI: 10.1016/j.neubiorev.2019.11.008.

Nordquist, R.E., Meijer, E., van der Staay, F.J. and Arndt, S.S. (2017) Chapter 39 – Pigs as model species to investigate effects of early life events on later behavioral and neurological functions. In: Conn, P.M. (ed.) *Animal Models for the Study of Human Disease*, 2nd edn. Academic Press, London, pp. 1003–1030.

O'Donohue, W. and Ferguson, K.E. (2001) *The Psychology of B.F. Skinner*. SAGE Publications, Thousand Oaks, California.

Ohl, F. and van der Staay, F.J. (2012) Animal welfare: at the interface between science and society. *Veterinary Journal* 192, 13–19.

O'Neil, E.B. and Lee, A.C.H. (2019) Examining the representational content of perirhinal cortex and posterior ventral visual pathway regions when maintenance of visual information is interrupted. *Cortex* 121, 16–26. DOI: 10.1016/j.cortex.2019.08.008.

Prentice, K.J., Gold, J.M. and Buchanan, R.W. (2008) The Wisconsin Card Sorting impairment in schizophrenia is evident in the first four trials. *Schizophrenia Research* 106(1), 81–87. DOI: 10.1016/j.schres.2007.07.015.

Puliafico, A.C. and Kendall, P.C. (2006) Threat-related attentional bias in anxious youth: a review. *Clinical Child and Family Psychology Review* 9(3–4), 162–180. DOI: 10.1007/s10567-006-0009-x.

Rescorla, R. and Wagner, A.R. (1972) A theory of Pavlovian conditioning: variations in effectiveness of reinforcement and nonreinforcement. In: Black, A.H. and Pokasy, W.F. (eds) *Classical Conditioning II: Current Research and Theory*. Appleton-Centry-Crofts, New York, pp. 64–99.

Roelofs, S., Boleij, H., Nordquist, R.E. and van der Staay, F.J. (2016) Making decisions under ambiguity: judgment bias tasks for assessing emotional state in animals. *Frontiers in Behavioral Neuroscience* 10, 119. DOI: 10.3389/fnbeh.2016.00119.

Salena, M.G., Turko, A.J., Singh, A., Pathak, A., Hughes, E. *et al.* (2021) Understanding fish cognition: a review and appraisal of current practices. *Animal Cognition* 24(3), 395–406. DOI: 10.1007/s10071-021-01488-2.

Schnell, A.K., Amodio, P., Boeckle, M. and Clayton, N.S. (2021) How intelligent is a cephalopod? Lessons from comparative cognition. *Biological Reviews* 96(1), 162–178. DOI: 10.1111/brv.12651.

Seed, A. and Byrne, R. (2010) Animal tool-use. *Current Biology* 20(23), R1032–R1039. DOI: 10.1016/j.cub.2010.09.042.

Sorby-Adams, A.J., Schneider, W.T., Goncalves, R.P., Knolle, F. and Morton, A.J. (2021) Measuring executive function in sheep (*Ovis aries*) using visual stimuli in a semi-automated operant system. *Journal of Neuroscience Methods* 351, 109009. DOI: 10.1016/j.jneumeth.2020.109009.

Stamps, J. and Groothuis, T.G.G. (2010) The development of animal personality: relevance, concepts and perspectives. *Biological Reviews* 85(2), 301–325. DOI: 10.1111/j.1469-185X.2009.00103.x.

Tait, D.S., Bowman, E.M., Neuwirth, L.S. and Brown, V.J. (2018) Assessment of intradimensional/extradimensional attentional set-shifting in rats. *Neuroscience and Biobehavioral Reviews* 89, 72–84. DOI: 10.1016/j.neubiorev.2018.02.013.

Teutsch, J. and Kätzel, D. (2019) Operant assessment of DMTP spatial working memory in mice. *Frontiers in Behavioral Neuroscience* 13, 193. DOI: 10.3389/fnbeh.2019.00193.

Watters, J.V. and Krebs, B.L. (2019) Assessing and enhancing the welfare of animals with equivocal and reliable cues. *Animals* 9, 680. DOI: 10.3390/ani9090680.

Yerys, B.E., Wallace, G.L., Harrison, B., Celano, M.J., Giedd, J.N. *et al.* (2009) Set-shifting in children with autism spectrum disorders: reversal shifting deficits on the Intradimensional/Extradimensional Shift Test correlate with repetitive behaviors. *Autism* 13(5), 523–538. DOI: 10.1177/1362361309335716.

10

Using Neuroscience to Better Understand Animals and their Welfare

Irene Camerlink*

Institute of Genetics and Animal Biotechnology, Polish Academy of Sciences, Jastrzebiec, Poland

10.1 Introduction

Neuroscience is a broad interdisciplinary science which evolved rapidly from various disciplines, including, among others, physiology, molecular biology and computer science. As the nervous system is such a complex and vast system (see also Chapter 1, this volume, on complex systems), multiple subdisciplines within neuroscience evolved rapidly, such as behavioural neuroscience, cognitive neuroscience and neuroethology, to name but a few, with the latter focusing on the neural basis underlying animals' natural behaviour (Bullock, 1990). In many of the branches of neuroscience, animals are used for research, mainly as model species for humans. The application of neuroscience techniques to animals primarily involves rats, mice and other small rodents (Keifer and Summers, 2016). However, it is increasingly recognized that other species may form more suitable models for understanding human neurobiology. Larger mammals provide more, and more easily accessible, brain tissue to work with. Other species' brains may also have better similarity to the human brain. While primates form the best substitute for human research, there are strong ethical and moral reasons not to use non-human primates for research, and in many countries strict regulations on their use for research have been put in place. Alternative species that are less constrained by regulations, but are more similar to humans than mice and rats, are being actively explored (Hale, 2019). While such alternatives include *Drosophila*, zebrafish, reptiles and birds (Keifer and Summers, 2016), they also include species that are of interest to animal welfare researchers, such as dogs and pigs. Dogs have

*i.camerlink@igbzpan.pl

DOI: 10.1079/9781789247893.0010

been proposed as promising models for comparative neuroscience studies, partly due to their shared social environment with humans (Bunford *et al.*, 2017). Pigs are physiologically very similar to humans, and this also holds true for some aspects of the brain (Lind *et al.*, 2007; Sauleau *et al.*, 2009; Gieling *et al.*, 2011). Sheep have recently also been suggested as a model species for neurodegenerative research (McBride and Morton, 2018). As such, pet and farm animals have increasingly been used as models for human neurobiology and cognition. The diversification in use of species has been acknowledged as a strength for neuroscience, by creating more valid and biologically relevant models (Keifer and Summers, 2016). With the gradual change in research species, neurobiology and animal welfare are gradually coming closer together.

The use of neuroscience for animal welfare was suggested by Broom in 1987, and his review shows that the studies then were mainly on stereotypic behaviour of farm animals and the related brain dopamine metabolism. Broom and colleagues have over the years explored this topic further and provided the early literature for the synthesis between neuroscience and animal welfare (e.g. Zanella *et al.*, 1996, on brain opioid receptors in relation to stereotypies in sows; Broom and Zanella, 2004). However, only from around 2014 did the research in this area gradually take off (Fig. 10.1). This may be partly due to the requirement for technical expertise, skills, and expensive

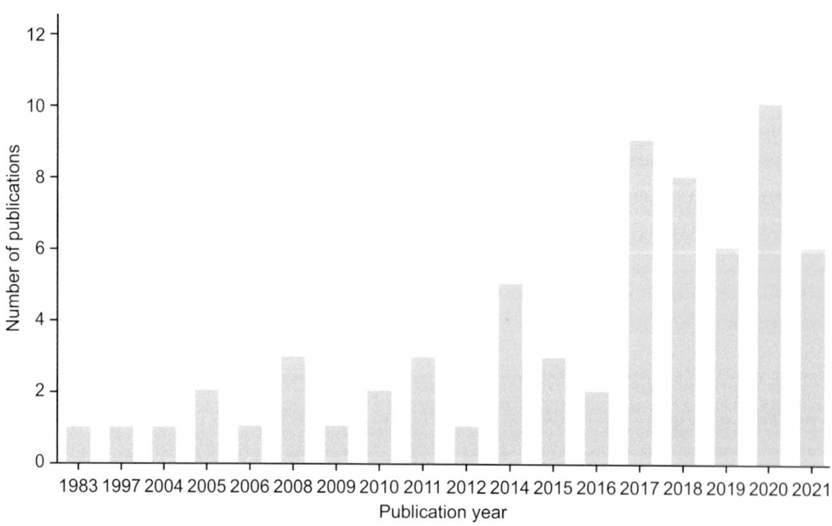

Fig. 10.1. Neuroscience in animal welfare studies. An overview of the number of publications on animal welfare and neuroscience per year. The search was performed in PubMed by searching for 'animal welfare' AND 'neuroscience' in the title and abstract. When searching for studies that apply neuroscience techniques in relation to animal behaviour, there are approximately 100 more results.

and not widely available equipment. Equipment and resources have become more available with the increased (cross-disciplinary) collaboration between research institutes and research departments. Collaborations could often be found within the facilities of the same institute or even campus, or in collaboration with local veterinary clinics or human medical facilities. As the integration of neuroscience and animal welfare is still in its infancy, the use of facilities is sometimes compromised, e.g. researchers are reliant on when equipment is available and the equipment is often less powerful compared with the newer models for human medical research (e.g. magnetic resonance imaging (MRI) scanners with a 1.5 T magnet or less are used for animal welfare studies whereas in human clinical work nowadays 3.0 T is used, and for medical research 7.0–11 T is available). Despite the limitations that researchers may face, it is a promising start for a new cross-disciplinary direction in animal welfare research.

While the initial application of neuroscience to study stereotypies is still relevant (e.g. stereotypies in horses: McBride and Hemmings, 2009), neuroscience opens up opportunities to study the background of behaviours and mental states that are less obvious to the human observer (as advocated in a review by Broom and Zanella, 2004). This has become particularly relevant with the increasing interest in affective states and positive animal welfare. Measures to indicate animal suffering are generally quite accessible, as threats to survival will relate to a profound behavioural and physiological response by the animal in its attempts to cope. The behavioural and physiological response to a positive event is often more nuanced, short lived and difficult to grasp. As positive experiences do not form a threat to survival for the individual or social group, there is less need to communicate it openly or to instantly warn conspecifics. In other words, although all emotions are internal experiences, there is less evolutionary relevance in openly communicating positive states as compared with negative states.

Measures of positive mental states are currently limited, and mainly relate to play behaviour and test paradigms of affective states. The application of neuroscience can change this (Panksepp *et al.*, 2017), as its techniques may provide insight in animal cognition (Huber and Lamm, 2017), affective states (Poirier *et al.*, 2019; Paul *et al.*, 2020) and measures to assess positive emotions (humans: Alexander *et al.*, 2021; pigs: Lawrence *et al.*, 2018). Already it has provided much insight in positive social behaviour (Carter and Keverne, 2002), but application to domestic animals is still in its infancy. Neuroscience also opens up new opportunities for testing hypotheses on animal consciousness (Paul *et al.*, 2020). Some intriguing neuroscience studies from outside the field of animal welfare science have revealed that mammals dream when they sleep (reviewed by Manger and Siegel, 2020) and that rats memorize in their sleep what they have learned in a maze during the day (Wilson and McNaughton, 1994; Ji and Wilson, 2007), thus providing a glimpse of what happens in an animal's brain.

Studies that originated from animal welfare research are, for example, studies on how environmental enrichment alters brain gene expression in pigs (Brown *et al.*, 2017, 2018), and how brain monoamines such as serotonin affect feather-pecking behaviour in hens (Kops *et al.*, 2017).

The aim of this chapter is to provide a brief overview of how neuroscience techniques can be of use for animal welfare research and highlight some of the relevant literature. It is by no means a complete overview of all techniques or all related literature, but can be used as a starting point for developing ideas.

10.2 Concepts

There are many concepts, terms and abbreviations in neuroscience that may be foreign to animal welfare researchers. Below is just a handful of terms that you are likely to come across in presentations and publications. A good glossary of neuroscience terms can be found in the guidebook of Carter and Shieh (2015).

- *Neuropeptides* are small proteins produced by neurons that modulate a postsynaptic response to a neurotransmitter. They thereby influence many functions, including social behaviour. Oxytocin and vasopressin are examples of neuropeptides that have been studied in relation to animal behaviour and welfare (reviewed by Rault *et al.*, 2017).
- *Neurotransmitters* are polypeptide molecules that transmit a signal from a neuron across the synapse to a target cell. The main neurotransmitters are acetylcholine, dopamine, γ-aminobutyric acid (GABA), glutamate, histamine, norepinephrine and serotonin; of these, dopamine, norepinephrine and serotonin have received particular interest in animal behaviour and welfare research (e.g. pigs: Arroyo *et al.*, 2016; Stracke *et al.*, 2017; poultry: Kops *et al.*, 2017, reviewed by de Haas and van der Eijk, 2018).
- *Neurotrophins* are secreted proteins that enhance different activities in the nervous system, such as neuronal survival and synaptic plasticity. The most abundant neurotrophin in the brain is brain-derived neurotrophic factor (BDNF). BDNF was found to be higher in pigs from enriched environments, and could have potential as an animal welfare indicator (Rault *et al.*, 2018).
- *Gene expression* is the process by which the information encoded in a gene is used to direct protein synthesis and produce the structures of the cell. Gene expression is assessed in terms of up- and downregulated genes, whereby the cell either increases or decreases the quantity of a cellular component (e.g. RNA or protein) in response to an external stimulus. In gene-expression studies, often a select number of genes of interest are investigated based on their potential relationship to the treatment. Gene expression studies have been carried out alongside behavioural

observations and physiological measures to obtain a total picture of how they interrelate (e.g. poultry: Gualtieri *et al.*, 2019).

- *MRI* is a non-invasive imaging technique to assess the structure of the brain or body. *Functional MRI* (fMRI) uses the principles of MRI to indirectly measure neural activity in the nervous system over time. *Blood oxygen level-dependent imaging* (BOLD) is a method used in fMRI studies to observe the flow of oxygen-rich blood to neurons after an increase of neural activity. Brain imaging with MRI or fMRI is applied to various studies that inform about animals' perception to external stimuli, either under anaesthesia (e.g. Coquery *et al.*, 2018) or when awake (e.g. Cook *et al.*, 2018).
- *Electroencephalography* (EEG) is a non-invasive technique used to measure the electrical brain activity on the surface of the brain (i.e. the skull) (see section 10.3.1 for various applications to animal welfare studies).

10.3 Techniques

Neuroscience has a wide array of techniques, and their potential for animal welfare has in many cases yet to be explored. Various techniques are briefly mentioned below, based on the guidebook of Carter and Shieh (2015), with a selection related to the current use in animal welfare research. The techniques are not explained in terms of how they function, as this would go beyond the scope of this chapter, and has been well described by Carter and Shieh (2015) and in the cited publications. As well as neuroscience techniques contributing to animal welfare research, animal welfare research also influences neuroscience, as there is growing attention for the welfare of (laboratory) animals used for neuroscience studies (van der Staay *et al.*, 2009).

10.3.1 Whole-brain imaging

Whole-brain imaging can be categorized into structural imaging, which looks at the anatomical structure of the brain, and functional imaging, which indicates the physiological processes related to neural activity. Structural techniques include computed tomography (CT), MRI and diffusion MRI. CT scans are commonly used in animals, for example to assess production performances such as meat quality (sheep: Lambe *et al.*, 2003). The use of CT scanners for studying animal brains in relation to behaviour or welfare is still largely unexplored (but see Sauleau *et al.*, 2009, for CT scans of pig brains). MRI has been used to show, for example, structural changes in the brains of rats in response to pain (Seminowicz *et al.*, 2009). Rats that experienced long-term pain had reduced frontal cortex volumes, and this reduction coincided with anxiety-like behaviours. Welfare conditions that result in long-term profound behavioural responses (e.g. stereotypic behaviour) would be good candidates for investigation with structural imaging.

Functional techniques include fMRI, positron emission tomography (PET), electroencephalography (EEG) and optical imaging. fMRI is actively used in research on animal cognition, behaviour and emotion. In an MRI/fMRI scanner, the individual has to lie completely still in order for a good quality scan to be made. As animals commonly cannot be asked to lie still, they are either restrained or scanned under anaesthesia. Under anaesthesia, the response to sensory stimuli can still be tested, e.g. the response to flavour (pigs: Coquery *et al.*, 2018). In Hungary (Cook *et al.*, 2018) and Austria (Huber and Lamm, 2017), groups are training dogs to voluntarily lie still in the scanner in order to carry out tests during the scans and observe the changes in neural activity in the brain. In Hungary, for example, Cook *et al.* (2018) used fMRI to study the amygdala in awake dogs when exposed to a situation that may trigger jealousy. The training associated with scanning awake animals is time-consuming but feasible (Karl *et al.*, 2019). For some species, imaging awake but restrained animals is possible, as shown by Yee *et al.* (2016) in prairie voles. A similar immobilization of the animal in the fMRI, using a custom-built holder, was also done with minipigs in China to study visual centres (Fang *et al.*, 2006). However, for most species of interest, anaesthesia would be needed to make the procedure ethically acceptable.

PET measures the changes in metabolic processes in the brain, including blood flow and acute changes in neurotransmitter systems. In a methodological study, Villadsen *et al.* (2018) first used MRI scans to construct an atlas of the pig brain, and then related PET scans to the MRI scans to create a PET template. With this technique, they assessed PET radiotracers with different kinetics and spatial distributions. This can be used for further pig brain studies without the need for an MRI scanner. PET was also used in (anaesthetized) pigs to study brain responses to flavours with hedonic values (Clouard *et al.*, 2012). In a study with pharmacological interventions, Jørgensen *et al.* (2018) investigated the cerebral extracellular serotonin levels in pigs. Serotonin is actively studied in relation to animal behaviour and welfare (e.g. pigs: Stracke *et al.*, 2017; poultry: de Haas and van der Eijk, 2018), and such studies can thus inform further work on this.

EEG records electrical brain activity on the surface (i.e. skull) of the brain. The term electrocorticogram refers to data recorded directly from the cortex, and electrosubcorticogram when recorded from subcortical structures (e.g. via implanted electrodes). A review by Klemm (1992) suggested the use of EEG for assessing mental states and welfare in animals but received little attention (maybe because of naming behavioural observations highly subject to anthropomorphism, which highlights one of the potential conflicts that may arise in collaboration between neuroscientists and animal welfare researchers). EEG has so far mainly been used to study pain (reviewed by Murrell and Johnson, 2006), for example during tail docking in pigs (Kells *et al.*, 2017), and consciousness during euthanasia (reviewed by Verhoeven *et al.*, 2015). Resting-state EEG profiles in horses have been associated with their welfare scores, with fewer gamma waves in the left brain hemisphere in horses with good welfare (Stomp *et al.*, 2021). Measuring EEG will produce a huge amount of data that is difficult to interpret. Moreover, the output will be influenced by muscle movements if measured on the skull, creating noise in the data. Rault *et al.* (2020), in collaboration with researchers in medicine, elegantly overcame this by using wireless epidural (under the skull) EEG in pigs, which also allowed the study of positive human–animal interactions (Rault *et al.*, 2019). However, epidural EEG is invasive and may not be feasible or allowed in every laboratory.

10.3.2 Electrophysiology

Electrophysiological techniques inform about the electrical activity of living neurons and the processes related to their signalling. From the many types of electrophysiological techniques, electromyography (EMG) may have some potential for animal welfare studies. EMG records biomedical electrical signals obtained from neuromuscular activities (reviewed by Gohel and Mehendale, 2020). It can be used to quantify the activity in individual muscles, and has been applied, among other examples, for assessing startle responses, for gait analysis (including in dogs and horses) and for analysing feeding behaviour in goats (Campos *et al.*, 2019). It can also be applied to study facial expressions (Gohel and Mehendale, 2020), which in animals is emerging as a method to assess emotional state.

10.3.3 Microscopy and visualization of structures and functions

Microscopy techniques that are used in neuroscience include fluorescence microscopy and electron microscopy. Fluorescence microscopy allows the labelling of different structures and proteins with fluorophores (fluorescent chemical components that can absorb and emit light at specific wavelengths), so that each component in an image can be distinguished clearly by a colour (Fig. 10.2). Fluorescence microscopy can also be used to map brain activity related to animal behaviour (Guzowski *et al.*, 2005). This was applied to detect immediate-early genes (IEGs) (Guzowski *et al.*, 2005), which provide a static measure of activity and function of neural activity. IEGs are a group of genes that transiently and rapidly activate during periods of high neural activity. IEGs provide information about certain proteins, and through staining methods, it is possible to see which regions of the brain were active just before the brain was fixed. A classic staining method is Golgi staining, which is used to label a small subset of neurons. It visualizes the morphology of neurons and their circuits throughout the brain, and allows, for example, visualization of dendrite branching.

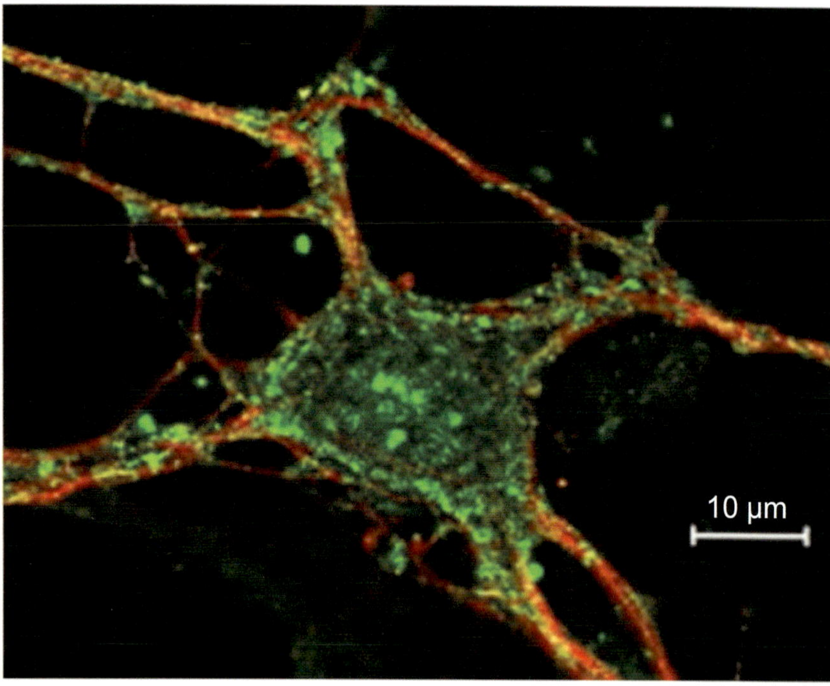

Fig. 10.2. Image of a cultured hippocampal neuron stained with neurexin 3 (green) and β-actin (red). Image reproduced with permission from S.M. Brown.

There is also the application of micro-microscopes (also called micro-endoscopes), which can be (surgically) attached to the skull of the animal, either with or without a wire (reviewed for mice: Malvaut *et al.*, 2020). This enables brain activity to be correlated directly to animal behaviour, but it requires a different approach from common animal welfare research (e.g. invasive procedures in singly housed animals under highly controlled conditions).

10.3.4 Techniques to identify genes and gene expression

The expression of genes between one or more tissue samples can be measured and compared using various techniques. Using a microarray, van der Staay *et al.* (2010) investigated gene expression in the frontal cortex and hippocampus of tethered versus loose-housed sows. Older tethered sows had increased β-globin mRNA expression in the hippocampus. From their findings, combined with behavioural observations and physiological parameters, the authors concluded that chronically stressed pigs develop depression-like symptoms.

Gene expression can also be investigated by using polymerase chain reaction (PCR) methods for the sample analysis. Real-time quantitative PCR (qPCR) is a method to amplify and quantify a specific DNA fragment. Transcription is the process of synthesizing RNA from a DNA template; reverse transcriptase PCR (RT-PCR) first produces the RNA from the DNA template and then amplifies it. Brown *et al.* (2017) looked at gene expression in the frontal cortex of piglets that had been exposed to environmental enrichment or not. They used a qPCR with a set of target genes including BDNF and insulin-like growth factor 1 (IGF-1), and found that the latter was upregulated in the frontal cortex of piglets from the environmental enrichment group. A recent study on outdoor laying hens (Armstrong *et al.*, 2020) used qPCR to assess proliferation and neuronal differentiation in the hippocampus and related this to individual differences in tonic immobility, showing that the tonic immobility response and the use of outdoor areas related to differences in hippocampal plasticity. Kanitz *et al.* (2009) used real-time RT-PCR to study gene expression in the hypothalamus, hippocampus and amygdala of pigs that had experienced social isolation. They found that social isolation resulted in neuroendocrine and behavioural changes and differently expressed genes, predominantly in the hypothalamus.

10.3.5 Biochemical assays and intracellular signalling

Chromatography is a method to separate biomolecules (e.g. proteins) from a solution. Arroyo *et al.* (2016) used this technique to determine dopamine and serotonin levels in the brains of pigs that underwent stressful handling at the slaughterhouse. They assessed four brain regions (amygdala, prefrontal cortex, hippocampus and hypothalamus) and found a decrease in serotonin in the amygdala in response to stressful handling.

10.4 Pitfalls

Cross-disciplinary collaboration is not always easy, and researchers in animal welfare have for quite some time taken the lead themselves in borrowing techniques from other disciplines, thereby avoiding the need for collaboration. This 'independent' approach is less suitable for neuroscience, as most techniques and the interpretation of the resulting data require training, expertise and experience. The brain is a fragile tissue and is easily damaged by improper handling or storage, and different brain regions can easily be misidentified. To attempt to analyse the brain yourself without adequate experience risks the loss of precious tissue for which animals may have been sacrificed. Collaboration will thus contribute not only to the expansion of research networks and the quality of the research, but also to the 3Rs (replacement, reduction and refinement; see Blakemore *et al.*, 2012 for a perspective on the 3Rs in neuroscience). The trend in animal welfare research to use increasingly large sample sizes and to study animals in their commercial settings (which do not fulfil the requirements for housing conditions for research animals) makes studies more applied and statistically powerful, but is distinctly different from the laboratory set-ups in which neuroscience studies mostly take place. Researchers thus have to adjust their way of thinking when undertaking the research design. Neuroscience studies are often more invasive (e.g. surgery or implantations) and stressful (e.g. singly housed animals, use of restraint, social isolation) than animal welfare studies. Moreover, techniques may require the sacrifice of the animal, which can then not be studied further. The ethical aspects of euthanizing animals for knowledge generation (see Chapter 4, this volume) is something to take into consideration, and sample sizes need to be carefully calculated. As animal welfare researchers may be even more under the scrutiny of society (and animal protection organizations), it is extremely important that knowledge on animal welfare is applied to refine the experimental protocols as much as possible. For example, the enormous amount of literature on the negative impact of social isolation on social species such as sheep and pigs should not be ignored in order to follow a standardized protocol that was developed for rodents. Thus, rather than *applying* neuroscience techniques, it is important to *integrate* the techniques, by using animal welfare expertise throughout the project, including its neuroscience aspects.

Studies that are carried out on awake animals in an MRI scanner take time and labour investment, and although it has been shown to work with a clear protocol (Karl *et al.*, 2019), the success rate may depend on the animals' background and the skills of the trainer. There are still many techniques that have not yet been explored, and alternatives that can retrieve the required information from anaesthetized animals may be available. Moreover, not all techniques can be transferred to animals of other species, the other sex or even other breeds without modification. For example, studies based on brain structures require a brain atlas (a three-dimensional graphical representation of the brain), and if an atlas already exists, it may need to be adjusted for the breed, sex and age of the animal (e.g. minipigs and domestic pigs differ in their brain structure; Sauleau *et al.*, 2009). To be fully aware of the possible alternatives and required adjustments, it is important to gain advice from neuroscientists before deciding on the experimental designs. Collaboration therefore ideally starts at the proposal-writing stage.

Collaboration across disciplines is not an easy feat, as also described in the other chapters in this book. In order to establish a satisfactory collaboration, animal welfare researchers need to invest time in gaining some basic understanding of the background, terminology and techniques, so that the right questions can be asked. Animal welfare researchers will also need to communicate clearly which aspects of the protocols may need to be reconsidered in order to make the experimental design less stressful or less invasive for the animals, and why this is important for animal welfare research.

10.5 Conclusions

Neuroscience is increasingly applied to animal behaviour and welfare studies but mainly in a multidisciplinary manner where the two disciplines remain separated. As well as expanding the range of methods that are used, it is important to integrate them in an interdisciplinary fashion where the use of both disciplines leads to 'more than the sum of its parts' (see Chapter 1, this volume). An added value that animal welfare science can bring to neuroscience is the expertise on refining animal experiments (e.g. habituation, training, avoidance of social isolation) based on knowledge of animal behaviour and physiology in response to external conditions such as environmental conditions and human handling. Steps in this direction have already been made, such as with studies where animals voluntarily participate. For the future, synergy between neuroscience and animal welfare science may create a breakthrough in the understanding of animals' mental experiences, perception and consciousness, which will undoubtedly contribute to the current pursuits for moving towards positive animal welfare and obtaining an understanding of what it means for animals to have 'a good life' or 'a life worth living'.

Further Reading

This short book chapter cannot cover the full breadth and depth of the potential of neuroscience for animal behaviour and welfare studies. It is meant to give a glimpse of what is possible and what has been done so far. In the references, you will find a wealth of good publications that provide details on the background of the topics and methods. References given in bold are particularly recommended. As is clear from the reference list, most of the studies are not published in the typical animal behaviour and welfare journals (but see *Physiology and Behavior*) but in neuroscience journals or interdisciplinary journals. For this reason, keep an eye on publications from such journals, for example *Nature Reviews Neuroscience, Neuroscience and Biobehavioral Reviews* and *Behavioural Brain Research.*

References

Alexander, R., Aragón, O.R., Bookwala, J., Cherbuin, N., Gatt, J.M. *et al.* (2021) The neuroscience of positive emotions and affect: implications for cultivating happiness and wellbeing. *Neuroscience & Biobehavioral Reviews* 121, 220–249. DOI: 10.1016/j.neubiorev.2020.12.002.

Armstrong, E.A., Voelkl, B., Voegeli, S., Gebhardt-Henrich, S.G., Guy, J.H. *et al.* (2020) Cell proliferation in the adult chicken hippocampus correlates with individual differences in time spent in outdoor areas and tonic immobility. *Frontiers in Veterinary Science* 7, 587. DOI: 10.3389/fvets.2020.00587.

Arroyo, L., Carreras, R., Valent, D., Peña, R., Mainau, E. *et al.* (2016) Effect of handling on neurotransmitter profile in pig brain according to fear related behaviour. *Physiology & Behavior* 167, 374–381. DOI: 10.1016/j.physbeh.2016.10.005.

Blakemore, C., MaCarthur Clark, J., Nevalainen, T., Oberdorfer, M. and Sussman, A. (2012) Implementing the 3Rs in neuroscience research: a reasoned approach. *Neuron* 75(6), 948–950. DOI: 10.1016/j.neuron.2012.09.001.

Broom, D.M. (1987) Applications of neurobiological studies to farm animal welfare. In: Wiepkema, P.R. and van Adrichem, P.W.M. (eds) *Biology of Stress in Farm Animals: An Integrative Approach.* Springer, Dordrecht, The Netherlands, pp. 101–110.

Broom, D.M. and Zanella, A.J. (2004) Brain measures which tell us about animal welfare. *Animal Welfare* 13, S41–S46.

Brown, S.M., Peters, R. and Lawrence, A.B. (2017) Up-regulation of IGF-1 in the frontal cortex of piglets exposed to an environmentally enriched arena. *Physiology & Behavior* 173, 285–292. DOI: 10.1016/j.physbeh.2017.02.030.

Brown, S.M., Bush, S.J., Summers, K.M., Hume, D.A. and Lawrence, A.B. (2018) Environmentally enriched pigs have transcriptional profiles consistent with neuroprotective effects and reduced microglial activity. *Behavioural Brain Research* 350, 6–15. DOI: 10.1016/j.bbr.2018.05.015.

Bullock, T.H. (1990) Goals of neuroethology. *BioScience* 40(4), 244–248. DOI: 10.2307/1311260.

Bunford, N., Andics, A., Kis, A., Miklósi, Á. and Gácsi, M. (2017) *Canis familiaris* as a model for non-invasive comparative neuroscience. *Trends in Neurosciences* 40(7), 438–452. DOI: 10.1016/j.tins.2017.05.003.

Campos, D.P., Abatti, P.J., Bertotti, F.L., De Paula Vieira, A., Hill, J.A.G. *et al.* (2019) Short-term fibre intake estimation in goats using surface electromyography of the masseter muscle. *Biosystems Engineering* 183, 209–220. DOI: 10.1016/j.biosystemseng.2019.04.021.

Carter, C.S. and Keverne, E.B. (2002) The neurobiology of social affiliation and pair bonding. In: Pfaff, D.W., Arnold, A.P., Fahrbach, S.E., Etgen, A.M. and Rubin, R.T. (eds) *Hormones, Brain and Behavior.* Academic Press, pp. 299–337.

Carter, M. and Shieh, J.C. (2015) *Guide to Research Techniques in Neuroscience*, 2nd edn. Academic Press.

Clouard, C., Jouhanneau, M., Meunier-Salaün, M.-C., Malbert, C.-H. and Val-Laillet, D. (2012) Exposures to conditioned flavours with different hedonic values induce contrasted behavioural and brain responses in pigs. *PLoS One* 7(5), e37968. DOI: 10.1371/journal.pone.0037968.

Cook, P., Prichard, A., Spivak, M. and Berns, G.S. (2018) Jealousy in dogs? Evidence from brain imaging. *Animal Sentience* 22, 117.

Coquery, N., Meurice, P., Janvier, R., Bobillier, E., Quellec, S. *et al.* (2018) fMRI-based brain responses to quinine and sucrose gustatory stimulation for nutrition research in the minipig model: a proof-of-concept study. *Frontiers in Behavioral Neuroscience* 12, 151. DOI: 10.3389/fnbeh.2018.00151.

de Haas, E.N. and van der Eijk, J.A.J. (2018) Where in the serotonergic system does it go wrong? Unravelling the route by which the serotonergic system affects feather pecking in chickens. *Neuroscience & Biobehavioral Reviews* 95, 170–188. DOI: 10.1016/j.neubiorev.2018.07.007.

Fang, M., Li, J., Rudd, J.A., Wai, S.M., Yew, J.C.C. *et al.* (2006) fMRI mapping of cortical centers following visual stimulation in postnatal pigs of different ages. *Life Sciences* 78(11), 1197–1201. DOI: 10.1016/j.lfs.2005.06.030.

Gieling, E.T., Schuurman, T., Nordquist, R.E. and van der Staay, F.J. (2011) The pig as a model animal for studying cognition and neurobehavioral disorders. *Current Topics in Behavioral Neurosciences* 7, 359–383. DOI: 10.1007/7854_2010_112.

Gohel, V. and Mehendale, N. (2020) Review on electromyography signal acquisition and processing. *Biophysical Reviews* 12, 1361–1367. DOI: 10.1007/s12551-020-00770-w.

Gualtieri, F., Armstrong, E.A., Longmoor, G.K., D'Eath, R.B., Sandilands, V. *et al.* (2019) Unpredictable chronic mild stress suppresses the incorporation of new neurons at the caudal pole of the chicken hippocampal formation. *Scientific Reports* 9(1), 1–13. DOI: 10.1038/s41598-019-43584-x.

Guzowski, J.F., Timlin, J.A., Roysam, B., McNaughton, B.L., Worley, P.F. *et al.* (2005) Mapping behaviorally relevant neural circuits with immediate-early gene expression. *Current Opinion in Neurobiology* 15(5), 599–606. DOI: 10.1016/j.conb.2005.08.018.

Hale, M.E. (2019) Toward diversification of species models in neuroscience. *Brain, Behavior and Evolution* 93(2–3), 166–168. DOI: 10.1159/000500499.

Huber, L. and Lamm, C. (2017) Understanding dog cognition by functional magnetic resonance imaging. *Learning & Behavior* 45(2), 101–102. DOI: 10.3758/s13420-017-0261-6.

Ji, D. and Wilson, M.A. (2007) Coordinated memory replay in the visual cortex and hippocampus during sleep. *Nature Neuroscience* 10(1), 100–107. DOI: 10.1038/nn1825.

Jørgensen, L.M., Weikop, P., Svarer, C., Feng, L., Keller, S.H. *et al.* (2018) Cerebral serotonin release correlates with [^{11}C]AZ10419369 PET measures of 5-HT$_{1B}$ receptor binding in the pig brain. *Journal of Cerebral Blood Flow & Metabolism* 38(7), 1243–1252. DOI: 10.1177/0271678X17719390.

Kanitz, E., Puppe, B., Tuchscherer, M., Heberer, M., Viergutz, T. *et al.* (2009) A single exposure to social isolation in domestic piglets activates behavioural arousal, neuroendocrine stress hormones, and stress-related gene expression in the brain. *Physiology & Behavior* 98(1–2), 176–185. DOI: 10.1016/j.physbeh.2009.05.007.

Karl, S., Boch, M., Virányi, Z., Lamm, C. and Huber, L. (2019) Training pet dogs for eye-tracking and awake fMRI. *Behavior Research Methods* 52(2), 838–856. DOI: 10.3758/s13428-019-01281-7.

Keifer, J. and Summers, C.H. (2016) Putting the "biology" back into "neurobiology": the strength of diversity in animal model systems for neuroscience research. *Frontiers in Systems Neuroscience* 10, 69. DOI: 10.3389/fnsys.2016.00069.

Kells, N.J., Beausoleil, N.J., Chambers, J.P., Sutherland, M.A., Morrison, R.S. *et al.* (2017) Electroencephalographic responses of anaesthetized pigs (*Sus scrofa*) to tail docking using clippers or cautery iron performed at 2 or 20 days of age. *Veterinary Anaesthesia and Analgesia* 44(5), 1156–1165. DOI: 10.1016/j.vaa.2017.02.003.

Klemm, W.R. (1992) Are there EEG correlates of mental states in animals? *Neuropsychobiology* 26(3), 151–165. DOI: 10.1159/000118911.

Kops, M.S., Kjaer, J.B., Güntürkün, O., Westphal, K.G.C., Korte-Bouws, G.A.H. *et al.* (2017) Brain monoamine levels and behaviour of young and adult chickens genetically selected on feather pecking. *Behavioural Brain Research* 327, 11–20. DOI: 10.1016/j.bbr.2017.03.024.

Lambe, N.R., Young, M.J., McLean, K.A., Conington, J. and Simm, G. (2003) Prediction of total body tissue weights in Scottish Blackface ewes using computed tomography scanning. *Animal Science* 76(2), 191–197. DOI: 10.1017/S1357729800053443.

Lawrence, A.B., Newberry, R.C. and Špinka, M. (2018) Positive welfare: what does it add to the debate over pig welfare? In: Špinka, M. (ed.) *Advances in Pig Welfare*. Woodhead Publishing, Duxford, UK, pp. 415–444.

Lind, N.M., Moustgaard, A., Jelsing, J., Vajta, G., Cumming, P. *et al.* (2007) The use of pigs in neuroscience: modeling brain disorders. *Neuroscience & Biobehavioral Reviews* 31(5), 728–751. DOI: 10.1016/j.neubiorev.2007.02.003.

Malvaut, S., Constantinescu, V.-S., Dehez, H., Doric, S. and Saghatelyan, A. (2020) Deciphering brain function by miniaturized fluorescence microscopy in freely behaving animals. *Frontiers in Neuroscience* 14, 819. DOI: 10.3389/fnins.2020.00819.

Manger, P.R. and Siegel, J.M. (2020) Do all mammals dream? *Journal of Comparative Neurology* 528(17), 3198–3204. DOI: 10.1002/cne.24860.

McBride, S. and Hemmings, A. (2009) A neurologic perspective of equine stereotypy. *Journal of Equine Veterinary Science* 29(1), 10–16. DOI: 10.1016/j.jevs.2008.11.008.

McBride, S.D. and Morton, A.J. (2018) Indices of comparative cognition: assessing animal models of human brain function. *Experimental Brain Research* 236(12), 3379–3390. DOI: 10.1007/s00221-018-5370-8.

Murrell, J.C. and Johnson, C.B. (2006) Neurophysiological techniques to assess pain in animals. *Journal of Veterinary Pharmacology and Therapeutics* 29(5), 325–335. DOI: 10.1111/j.1365-2885.2006.00758.x.

Panksepp, J., Lane, R.D., Solms, M. and Smith, R. (2017) Reconciling cognitive and affective neuroscience perspectives on the brain basis of emotional experience. *Neuroscience & Biobehavioral Reviews* 76(Pt B), 187–215. DOI: 10.1016/j.neubiorev.2016.09.010.

Paul, E.S., Sher, S., Tamietto, M., Winkielman, P. and Mendl, M.T. (2020) Towards a comparative science of emotion: affect and consciousness in humans and animals. *Neuroscience & Biobehavioral Reviews* 108, 749–770. DOI: 10.1016/j.neubiorev.2019.11.014.

Poirier, C., Bateson, M., Gualtieri, F., Armstrong, E.A., Laws, G.C. *et al.* (2019) Validation of hippocampal biomarkers of cumulative affective experience. *Neuroscience & Biobehavioral Reviews* 101, 113–121. DOI: 10.1016/j.neubiorev.2019.03.024.

Rault, J.L., van den Munkhof, M. and Buisman-Pijlman, F.T. (2017) Oxytocin as an indicator of psychological and social well-being in domesticated animals: a critical review. *Frontiers in Psychology* 8, 1521.

Rault, J.-L., Lawrence, A.J. and Ralph, C.R. (2018) Brain-derived neurotrophic factor in serum as an animal welfare indicator of environmental enrichment in pigs. *Domestic Animal Endocrinology* 65, 67–70. DOI: 10.1016/j.domaniend.2018.05.007.

Rault, J.-L., Truong, S., Hemsworth, L., Le Chevoir, M., Bauquier, S. *et al.* (2019) Gentle abdominal stroking ('belly rubbing') of pigs by a human reduces EEG total power and increases EEG frequencies. *Behavioural Brain Research* 374, 111892. DOI: 10.1016/j.bbr.2019.04.006.

Rault, J.-L., Lai, A., Hemsworth, L., Le Chevoir, M., Bauquier, S. *et al.* (2020) Wireless 'under the skull' epidural EEG and behavior in piglets during nitrous oxide or carbon dioxide gas euthanasia. *Physiology & Behavior* 227, 113142. DOI: 10.1016/j.physbeh.2020.113142.

Sauleau, P., Lapouble, E., Val-Laillet, D. and Malbert, C.-H. (2009) The pig model in brain imaging and neurosurgery. *Animal* 3(8), 1138–1151. DOI: 10.1017/S1751731109004649.

Seminowicz, D.A., Laferriere, A.L., Millecamps, M., Yu, J.S.C., Coderre, T.J. *et al.* (2009) MRI structural brain changes associated with sensory and emotional function in a rat model of long-term neuropathic pain. *Neuroimage* 47(3), 1007–1014. DOI: 10.1016/j.neuroimage.2009.05.068.

Stomp, M., d'Ingeo, S., Henry, S., Cousillas, H. and Hausberger, M. (2021) Brain activity reflects (chronic) welfare state: evidence from individual electroencephalography profiles in an animal model. *Applied Animal Behaviour Science* 236(1), 105271. DOI: 10.1016/j.applanim.2021.105271.

Stracke, J., Otten, W., Tuchscherer, A., Puppe, B. and Düpjan, S. (2017) Serotonin depletion induces pessimistic-like behavior in a cognitive bias paradigm in pigs. *Physiology & Behavior* 174, 18–26. DOI: 10.1016/j.physbeh.2017.02.036.

van der Staay, F.J., Arndt, S.S. and Nordquist, R.E. (2009) Evaluation of animal models of neurobehavioral disorders. *Behavioral and Brain Functions* 5, 11. DOI: 10.1186/1744-9081-5-11.

van der Staay, F.J., Schuurman, T., Hulst, M., Smits, M., Prickaerts, J. *et al.* (2010) Effects of chronic stress: a comparison between tethered and loose sows. *Physiology & Behavior* 100(2), 154–164. DOI: 10.1016/j.physbeh.2010.02.020.

Verhoeven, M.T.W., Gerritzen, M.A., Hellebrekers, L.J. and Kemp, B. (2015) Indicators used in livestock to assess unconsciousness after stunning: a review. *Animal* 9(2), 320–330. DOI: 10.1017/S1751731114002596.

Villadsen, J., Hansen, H.D., Jørgensen, L.M., Keller, S.H., Andersen, F.L. *et al.* (2018) Automatic delineation of brain regions on MRI and PET images from the pig. *Journal of Neuroscience Methods* 294, 51–58. DOI: 10.1016/j.jneumeth.2017.11.008.

Wilson, M.A. and McNaughton, B.L. (1994) Reactivation of hippocampal ensemble memories during sleep. *Science* 265(5172), 676–679. DOI: 10.1126/science.8036517.

Yee, J.R., Kenkel, W.M., Kulkarni, P., Moore, K., Perkeybile, A.M. *et al.* (2016) BOLD fMRI in awake prairie voles: a platform for translational social and affective neuroscience. *Neuroimage* 138, 221–232. DOI: 10.1016/j.neuroimage.2016.05.046.

Zanella, A.J., Broom, D.M., Hunter, J.C. and Mendl, M.T. (1996) Brain opioid receptors in relation to stereotypies, inactivity, and housing in sows. *Physiology & Behavior* 59(4–5), 769–775. DOI: 10.1016/0031-9384(95)02118-3.

11

The Microbiota–Gut–Brain Axis in Determining Social Behaviours of Animals

Nienke van Staaveren[1], Paul Forsythe[2], Jerine A.J. van der Eijk[3], Dietmar Fuchs[4], Johanna M. Gostner[4], Claire Mindus[1], T. Bas Rodenburg[5] and Alexandra Harlander[1*]

[1]University of Guelph, Guelph, Canada; [2]University of Alberta, Edmonton, Alberta, Canada; [3]Wageningen University and Research, Wageningen, The Netherlands; [4]Medical University of Innsbruck, Innsbruck, Austria; [5]Utrecht University, Utrecht, The Netherlands

11.1 Introduction

The high prevalence of damaging behaviours resulting in physical harm and distress in farm animals poses important animal welfare, societal and economic concerns. These behavioural issues have long been thought to indicate frustration due to animals' barren environments. For example, the lack or absence of foraging opportunities for chickens and pigs in commercial farms is proposed to be the main driver of damaging behaviours such as severe feather pecking or tail biting. However, this ethological view does not fully explain all harmful behavioural interactions in farmed animals and why these behaviours become repetitive, fluctuate over time and vary among individuals exposed to similar environments. Therefore, animal welfare scientists are increasingly turning to other disciplines in an effort to elucidate the physiological mechanisms contributing to behavioural problems. The microbiota–gut–brain axis has emerged as an important contributor to behavioural modulation.

The physically, nutritionally and sensorially restricted environments that house farm animals can act as a powerful stressor. Research in humans and rodents indicates that stressors can alter the neurobiology and microbial communities of an animal in ways that are associated with

*Corresponding author: aharland@uoguelph.ca

 Bridging Research Disciplines to Advance Animal Welfare Science: A practical guide (ed. I. Camerlink)
DOI: 10.1079/9781789247893.0011

behavioural disorders. Therefore, the role of the microbial community, microbe-mediated changes to the neurobiology and bidirectional communication of different components of the microbiota–gut–brain axis in the development of behavioural disorders is receiving increasing attention. It represents a promising avenue of research that can provide novel ways to tackle damaging behaviour in farmed animals.

As the study of the microbiota–gut–brain axis in farm animals is a relatively new field in animal science and welfare, this chapter is organized into six pillars: (i) introduction to observations linking the microbiota, the brain and behaviour; (ii) an overview of the microbiota–gut–brain axis and its application to animal welfare; (iii) techniques; (iv) pitfalls; (v) a conclusion; and (vi) suggestions for further reading. This chapter is not a comprehensive overview of microorganisms and the gut–brain interaction. Instead, it introduces the general topic of gut microorganisms, their potential role in animal welfare, and how intestinal microbiota management can keep farm animals physically and psychologically healthy.

11.1.1 A shift in focus from pathogens to beneficial microorganisms

Veterinary medicine has traditionally focused on identification of pathogens and the prevention and treatment of disease. Nevertheless, the vast majority of microorganisms inhabiting the body, collectively termed the microbiota, are innocuous or even beneficial. With this understanding, there is increasing enthusiasm to better understand their functions and benefits. The microbiota is found on external surfaces, such as the skin, or internal surfaces, such as the oral cavity's mucosal surface, the gastrointestinal tract and the urogenital tract. Composed of a combination of prokaryotes, eukaryotes and viruses, these microbial communities are proposed to impact the establishment of homeostasis by contributing to physiological processes, such as nutrient metabolism, defence against pathogens, immune system development and activation, and neurotransmitter production. Most recently, their influence has been proposed to extend to modulating behaviour (reviewed by Sekirov *et al.*, 2010; Jandhyala *et al.*, 2015; Trinh *et al.*, 2018).

11.1.2 The role of the environment in shaping animals' microbial communities

While a diverse microbiota minimizes pathogen colonization, an imbalance of the normal community structure (i.e. microbial dysbiosis) contributes to the risk of developing a pathogenic infection. The individual's genetic background, age and diet may impact the ease with which environmental microorganisms are incorporated into the microbiota (Sekirov *et al.*, 2010; Jandhyala *et al.*, 2015).

Maternal transfer of microorganisms is a contributor to the microbiota of an individual. Children born by caesarean section have distinct microbiota compared with those who underwent vaginal birth (Dominguez-Bello *et al.*, 2010). These differences have proven to confer long-lasting health consequences (Neu and Rushing, 2011). Differences in microbiota are also observed in farm animals, whereby pigs born by caesarean section and chicks reared without adult hens have distinct microbiota profiles from piglets that underwent vaginal delivery and newly hatched chicks kept in the presence of an adult hen, respectively (Kraimi *et al.*, 2019).

The built environment or man-made structures can also play an essential role in microbiota establishment. Prolonged exposure to indoor environments, coupled with reduced outdoor exposure, is associated with the development of mental health issues in humans (Hoisington *et al.*, 2019). Given that many farm animals are kept indoors for most or the entirety of their lives, the built environment is presumably a critical consideration in microbiota development (Kers *et al.*, 2018).

11.1.3 The role of nutrition and feeding behaviour in shaping animals' microbial communities

Diet is yet another important contributing factor to the establishment and subsequent shaping of the microbiota (Sekirov *et al.*, 2010; Jandhyala *et al.*, 2015). The dietary intake of the host is the primary energy source for the microbiota and influences microbial fermentation, bacterial numbers and species composition (Flint, 2012).

In the past, antibiotics were added to feed to increase feed efficiency and prevent disease; however, this strategy has led to a surge in antimicrobial resistance, rendering the treatment of infections difficult in humans and animals (Seal *et al.*, 2013). Consequently, treating and preventing disease using prebiotics or probiotics has become a promising alternative to ensure healthy and productive rearing conditions (Hill *et al.*, 2014). Recent scientific studies show that orally administered live or killed microorganisms and bacterial products promote a healthy gastrointestinal tract, thereby providing a viable and substantiated alternative to reducing antibiotic use in farmed animals (Seal *et al.*, 2013).

Foraging is an important behaviour that links the diet, the microbiota and behaviour. When animals are subject to a period of hunger, induced by feed restriction, it causes them to display increased physical activity, engaging in foraging behaviour to find food (Day *et al.*, 1995; Dixon *et al.*, 2014). Food can be a major driver for social interactions, including positive interactions due to animals foraging together or negative interactions due to competition. Microbiota composition is known to trigger aggregation or mate choice, and to modify foraging decisions and feeding preferences, further shaping the composition of the gut microbiota (reviewed by Pasquaretta *et al.*, 2018). The mechanisms controlling the reciprocal influence on health and behaviour, however, are not yet elucidated.

11.2 Overview of the Gut–Brain Axis: Concepts and Definitions

The bridge connecting the environment and behaviour is the central nervous system (CNS). The current body of research points to a significant role of the gut microbiota in the communication pathways between the peripheral nervous system and the CNS. Given this strong modulatory potential on neurobiological function, the gut–brain axis has dominated the field of research investigating the relationship between gut microbiota and behaviour (Morais *et al.*, 2020).

Direct and indirect signalling by the microbiota can shape the nervous system (e.g. neurotransmitter regulation or interaction with the vagus nerve) and the immune system (e.g. cytokine modulation). As such, these pathways present an opportune target for developing microbe-based preventatives or treatments for a range of emotional or behavioural problems (Bravo *et al.*, 2011; Morais *et al.*, 2020). Despite significant research efforts in humans and rodents, the gut–brain axis of farm animals is far less studied. Therefore, further research is needed to determine whether microbiota-based approaches used to treat behavioural issues in rodents and humans can also be employed to improve farm animal welfare (Kraimi *et al.*, 2019).

11.2.1 Gut microbes and the enteric nervous system in gut–brain signalling

Gut microbes are essential for the normal development of the enteric nervous system (McVey Neufeld *et al.*, 2013; Collins *et al.*, 2014; McVey Neufeld *et al.*, 2015). Germ-free mice that lack a microbiome have lower nerve density and a reduced number of neurons per ganglion in the jejunum and ileum compared with conventional mice (Collins *et al.*, 2014). The gut contains two types of sensory nerves that carry signals from the gastrointestinal tract to the brain (i.e. afferents): (i) extrinsic primary afferents with somata outside the gut; and (ii) intrinsic primary afferent neurons (IPANs), characterized by somata located within the gut wall. The latter represent the majority of sensory fibres innervating the intestinal mucosa (Kunze and Furness, 1999). The absence of gut microbes decreases the excitability of myenteric IPANs (Collins *et al.*, 2014), which decreases the frequency and amplitude of muscle contractions in the intestine. In addition to regulating gut motility, the interaction between gut microbes and IPANs constitutes a microbiota–gut–brain axis component. This accounts for the ability of certain bacteria to modulate brain chemistry and behaviour, as it allows the relay of signals from the lumen to vagal sensory ganglia (Perez-Burgos *et al.*, 2014). Gut microbes may also directly activate vagal chemoreceptors by producing substances that can be transported across the epithelial barrier to the portal circulation, namely short-chain fatty acids (Lal *et al.*, 2001). Alternatively, gut microbes may stimulate the release of paracrine mediators such as serotonin (5-hydroxytryptamine (5-HT)), histamine, cholecystokinin (CCK), adenosine triphosphate (ATP) and glucagon-like peptides from mucosal enteroendocrine cells (Li, 2007; Dockray, 2009). Importantly, enteroendocrine cells synapse with vagal neurons and rapidly transduce gut luminal signals using glutamate as a neurotransmitter (Kaelberer *et al.*, 2018).

11.2.2 The immune system in gut–brain signalling

There exists a well-established link between the immune system and behaviour. In humans, both depression and anxiety are associated with dysregulated immunological processes and an increase in markers of inflammation (Irwin and Miller, 2007; Liukkonen *et al.*, 2011). Moreover, pro-inflammatory cytokine-based treatments increase the risk of developing symptoms of depression, and this risk can be reduced by antidepressants, suggesting a causal relationship between inflammation and mood disorders (Capuron *et al.*, 2002). Similarly, in rodents, activation of the innate immune response leads to the development of symptoms associated with anxiety and depressive-like behaviour, and cognitive dysfunction (Wilson *et al.*, 2002; Godbout *et al.*, 2008; Raison and Miller, 2013). While direct afferent vagal signalling is necessary for the central effects of neuroactive mediators produced by microorganisms, recent studies indicate that the immune system also plays a critical role in mediating behavioural effects of gut microbes (Liu *et al.*, 2020). Indeed, specific peripheral immune cells have been linked with the modulation of behaviour. T-regulatory cells (Tregs) have been shown to mediate the anxiolytic and antidepressant-like effects of *Lactobacillus rhamnosus* in mice. The increase in Tregs coincides with a decrease in peripheral activated monocytes, which are also known to induce anxiety-like behaviour when they migrate to the brain (Wohleb *et al.*, 2015).

Biochemical circuits that are relevant in this context are the metabolism of aromatic amino acids and their interplay with tetrahydrobiopterin (BH_4). Besides being building blocks of proteins, tryptophan and phenylalanine are precursor molecules of indoleamine and catecholamine neurotransmitters, respectively. In humans, the strong relationship between the increased breakdown of peripheral tryptophan along the kynurenine axis and the increase in immune activation markers is well established (Geisler *et al.*, 2018). A pro-inflammatory oxidative milieu induces indoleamine 2,3-dioxygenase 1 (IDO-1), the main enzyme responsible for tryptophan catabolism along the kynurenine axis in the blood. Chronic immune stimulation also correlates with reduced phenylalanine turnover (Neurauter *et al.*, 2008). Oxidative conditions support the depletion of tetrahydrobiopterin, a cofactor that is required for the conversion of phenylalanine to tyrosine by phenylalanine hydroxylase, as well as for tyrosine 3-monooxygenase to form the catecholamine precursor levodopa (L-DOPA). Moreover, synthesis of the 5-HT precursor 5-hydroxytryptophan depends on BH_4. Thus, in humans, an inflammation-induced shift of phenylalanine and tryptophan metabolic routes can further dysregulate dopaminergic and 5-HT metabolism, respectively. These, in turn, play an essential role in inflammation-associated behavioural and emotional changes (Geisler *et al.*, 2018).

In addition, dietary tryptophan and probably also microbial tryptophan catabolites coming from the tryptophanase–indole pathway exhibit immunomodulatory properties relevant for neuroinflammation, as was shown in a mouse model for multiple sclerosis (Sonner *et al.*, 2019). Thus, a multitude of interactions has to be considered. Certainly, there are differences, for example in tryptophan metabolism (e.g. IDO-1 is lacking in chickens: Yuasa *et al.*, 2015). Beyond this, more information is needed to fully explore the potential of gut–brain signalling of peripheral amino acids and their endogenous as well as their microbial-derived catabolites.

11.2.3 Neuroimmune interactions in gut–brain signalling

Both Tregs and intact vagal signalling are demonstrably important for bacterial communication with the brain. It is noteworthy that the vagus nerve innervates tissues involved in immune functions, such as the thymus, spleen, lung, liver and gastrointestinal tract. The functional implications of this anatomical relationship between the vagus nerve and the immune system were recognized with the identification of a neural circuit that controls the inflammatory response in a reflex-like manner (Tracey, 2007). In this reflex, vagal sensory afferent nerves detect mediators of the inflammatory response, such as cytokines, which are released by macrophages and other immune cells after stimulation with lipopolysaccharide (LPS) or other microbial-associated molecular patterns (MAMPs) (Pavlov and Tracey, 2012). This leads to vagal cholinergic efferent signalling, which, in turn, suppresses the activity of pro-inflammatory immune cells. Microbe-driven changes to behaviour are therefore hypothesized, depending on peripheral immune changes facilitated by this efferent arm of the anti-inflammatory reflex (Tracey, 2007).

More recently, hepatic vagal sensory afferent nerves were found to be responsible for indirectly sensing the gut microenvironment and relaying the sensory inputs to the nucleus tractus solitarius (NTS) of the brainstem (Teratani *et al.*, 2020). Signalling to the NTS activates vagal efferent nerves and, subsequently, enteric neurons that regulate the intestinal Treg population (Teratani *et al.*, 2020).

Taken together, the evidence suggests that the effect of gut microbes and/or MAMPs on intestinal motility may be linked to behavioural changes through a reflex loop. This involves an IPAN-to-vagus nerve synapse (Perez-Burgos *et al.*, 2014), leading to activation of the NTS. Subsequently, efferent vagal and enteric nervous system modulation of the peripheral immune system (Teratani *et al.*, 2020) regulates central neuroinflammation, stress responses and anxiety (Wohleb *et al.*, 2015; Liu *et al.*, 2020).

11.2.4 Application to farm animal welfare: a focus on feather pecking in chickens

The contribution of the microbiota–gut–brain axis to feather pecking challenges traditional ethological views and provides innovative, corrective solutions. Indeed, the role of the gastrointestinal tract, gut microbes, and the nervous and immune systems in feather pecking development has gained increased attention (Fig. 11.1). Birds that are descendants of a feather-pecking genetic line have distinct gut microbiota and short-chain fatty-acid profiles compared with birds from a non-pecking line (Meyer *et al.*, 2012, 2013). Birkl *et al.* (2018) and van der Eijk *et al.* (2019b) report a lower abundance of *Lactobacillus* species in the caecal excreta of birds genetically selected for feather-pecking behaviour. From a functional perspective, applying *L. rhamnosus* to the caecal gut lumen reduces the frequency and increases the amplitude of caecal contractions. Furthermore, the velocity and amplitude of caecal contractions were positively associated with the number of feather-pecking bouts (van Staaveren *et al.*, 2020). In addition to regulating chicken gut motility, gut microbes may also directly (e.g. via the vagus nerve) or indirectly (e.g. mediated via microbial metabolites or stimulation of enteroendocrine cells to release 5-HT) modulate brain chemistry and behaviour. Consequently, dissecting the gut–microbe network, identifying its contribution to behaviour and using this information to develop nutraceuticals are exciting avenues of research with important implications for preventative therapy development. To this end, Mindus *et al.* (2021) reported that *L. rhamnosus* ingestion prevented feather pecking in stressed chickens and limited caecal dysbiosis. The authors also showed that the probiotic bacteria lead to increased Tregs in the spleen and the caecal tonsils. Further investigations are needed to assess the potential of *Lactobacillus* species to act as a corrective therapy to established feather-pecking behaviour. Finally, a recent report by van der Eijk *et al.* (2020) is the first to investigate the effect of early-life microbiota transplantation on feather pecking, 5-HT levels and immune profiles in chickens. Interestingly, while the impact on the gut microbiota composition was minimal, there were significant age- and genotype-dependent changes to behaviour and the immune system.

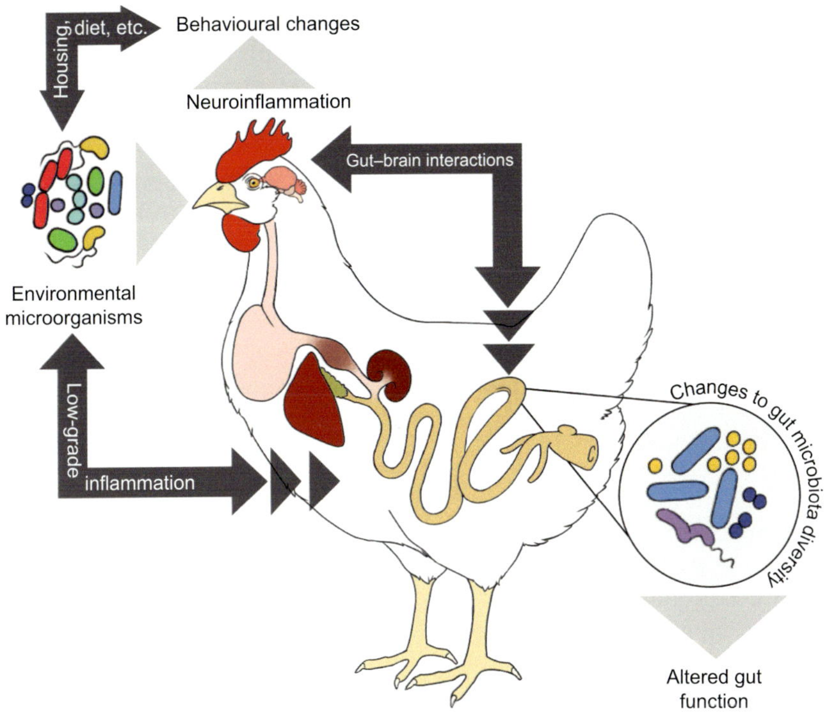

Fig. 11.1. The gut–brain axis in chickens. Environmental factors (e.g. diet, housing, microorganisms) play a key role in the immune status of chickens. Changes to the environment can produce a dysregulated gut microbiota, resulting in the secretion of microbial metabolites that are neuroactive, alter (neuro)immune signalling and can cause changes to aromatic amino acid turnover rates. The accumulation of these effects can result in behavioural changes, which in turn can influence environmental factors and microorganisms that chickens are exposed to in the future, further affecting the cycle. Figure reproduced with permission from Renée Garant.

11.3 Techniques

11.3.1 Techniques to manipulate the gut microbiota

The use of germ-free mice, mice with a defined microbiota, cross-fostering and faecal transplants are some of the tools and techniques used in rodent research to evaluate potential behavioural changes associated with gut microbiota manipulation. These techniques allow a reductionist approach to examine the influence of a single strain or simple, multi-strain treatment. Studies employing these methods generally identify whether the microbiota plays any role in a model phenotype. In farm animal research, a gut microbiota-associated phenotype of interest is modelled by comparison of antibiotic-treated and untreated animals. Following an antibiotic regimen, it is possible to repopulate the gut with a desired microbial population via a faecal microbiota transfer. Alternatively to depletion and re-establishment of the resident microbial population, the existing gut microbiome can be manipulated using live microorganisms (probiotics), substrates utilized by potential beneficial microorganisms (prebiotics) or a combination thereof to study the association between gut microbiota and behaviour. The latter strategy using probiotics and prebiotics has the potential to treat or even prevent behavioural problems. Regardless of the technique used to manipulate the microbiota, sampling and analysis are required to assess the manipulation's effectiveness under investigation.

11.3.2 Sampling techniques to harvest microorganisms

The ideal sampling method for a live animal's gut microbiota should be non-invasive. Faecal samples or faecal swabs are often used as a proxy for the intestinal gut microbiota. It is, nevertheless, important to note that faecal microorganisms are distinct from the luminal and mucosa-associated populations. For example, Yan *et al.* (2019) found that faecal samples reasonably mirrored the phylum composition found in the intestine; however, it did not accurately represent the bacterial genera to the same degree. Direct sampling of the mucosal microbiota in live animals requires an invasive technique, as it needs biopsy/endoscopy to retrieve mucosa-associated microorganisms. While it is undeniably easier to harvest gut microbiota samples from various anatomical sites from euthanized animals, this sampling method is often only feasible within a research setting.

11.3.3 Molecular techniques to identify microorganisms

While the concept of the microbiota–gut–brain axis has been around for centuries, culture-independent microbial detection methods have enabled the identification of microbes that are highly resistant to culture-based assays. This, in turn, has spurred a boom in microbiota–gut–brain axis research. 16S ribosomal RNA (rRNA) gene sequencing for taxonomic identification is the most utilized method in the field of microbiota research. It does, however, provide little information compared with metagenomic sequencing, which randomly sequences DNA fragments from the complete genomes of viruses, prokaryotes and eukaryotes within a sample. The resulting data provides both taxonomic and functional information on the microbiota. Metagenomics can identify rare species and how microbes communicate or compete. Specialized statistical tools can use 16S rRNA sequencing data to predict metagenomics, such as PICRUSt, which uses a reconstruction algorithm to predict gene frequencies. Metabolomics identifies and characterizes the interactions between bacterial metabolites and the host. As such, this type of study prioritizes the identification of functions.

11.3.4 Techniques to measure the impact of microbiota on the immune system

Immune biomarkers can be measured in tissues or in blood; however, it can be challenging to determine the sampling site that will provide the most reliable data. Recent work found that birds' immune responses from divergently selected lines for feather pecking were distinct, thereby linking behaviour to an immune phenotype (Buitenhuis *et al.*, 2006; van der Eijk *et al.*, 2019a). Interestingly, similar work using genetic lines of pigs with varying propensities for tail biting suggests differences in immune marker levels accompanied by changes to neurotransmitter levels (Brunberg *et al.*, 2016; Nordgreen *et al.*, 2020; Veit *et al.*, 2021). Enzyme-linked immunosorbent assays (ELISAs) can quantify specific antibody titres, and fluorescence-activated cell sorting (FACS) can detect changes in immune cell populations (van der Eijk *et al.*, 2019a). In the latter, flow cytometry is used to separate cells based on cell-surface markers. Furthermore, *ex vivo* stimulation of isolated, blood-derived monocytes with MAMPs, such as LPS, can provide insight into (innate) immune competence (van der Eijk *et al.*, 2019a). To study pro-inflammatory responses, cytokine and acute-phase protein concentrations can be measured in the serum or plasma of challenged animals.

11.3.5 Techniques to elucidate gut–brain axis signalling

The response of the axis to microbiota changes can be measured using biophysical, biochemical, electrophysiological and computational modelling techniques. The integrity of vagal afferent signalling is critical to the communication between specific gut microbes and the brain. Vagotomy has been shown to block neuron activation in the brain and, concomitantly, block the anxiolytic effects and changes to social deficits that are normally observed in response to probiotics, such as *L. rhamnosus* or *Bifidobacterium longum* (Bercik *et al.*, 2011; Bravo *et al.*, 2011; Sgritta *et al.*, 2019; Bharwani *et al.*, 2020). Indeed, vagal-dependent gut–brain signalling via IPANs was first demonstrated by exposing the intestine to *L. rhamnosus*. This signalling was attenuated after nicotinic receptor or total synaptic blockade (Sgritta *et al.*, 2019). Further support for the relationship between gut microbes, IPANs and the vagus is provided by the observed absence of mesenteric afferent signalling following IPAN stimulation of germ-free mice. This signalling response is restored following conventionalization (i.e. establishing a microbiota) of the adult animals (McVey Neufeld *et al.*, 2015). Gaining a better understanding of gut–brain axis signalling necessitates investigation of the complex interactions between the vagus nerve, the immune system and the brain. Delineating these interactions requires the ability to differentiate between afferent and efferent vagal pathways involved in facilitating the effects of gut microbes on the CNS. Such techniques include selective deafferentation using specific neurotransmitter receptor agonists conjugated to toxins (Diepenbroek *et al.*, 2017), or transgenic approaches, as have been employed in the ablation of nociceptive nerves (Talbot *et al.*, 2015). Similarly, the role of the immune system in signalling between the gut and the brain can be investigated using antibodies or transgenic animals to deplete specific cell types (Wohleb *et al.*, 2015; Liu *et al.*, 2020), or to block migration of peripheral immune cells to the brain by targeting chemotactic factors or receptors (Wohleb *et al.*, 2015). The role of distinct immune cell types in gut–brain communication can also be investigated through isolation and adoptive transfer of the cells followed by assessment of behaviour and brain function in recipient animals (Liu *et al.*, 2020).

11.4 Pitfalls

While a significant portion of the existing literature addresses interactions of the host with its bacterial inhabitants, a scant number of studies identify how intestinal inhabitants, including viruses, protozoa and fungi, affect the bacterial microbiota and influence model behaviours. While the interplay between numerous microbiota members and the host signalling pathway is admittedly complex, the rapid expansion of biomedical research and the laboratory animal infrastructure in the past decade promises that exciting revelations about health and disease are on the horizon.

The translation of small rodent or human studies to farm animal welfare research is challenging, as the techniques discussed previously are often not feasible on a large scale or in commercial settings, often due to practicality and costs. For instance, rodent studies use small numbers of inbred animals. Because farm animals are outbred, large numbers are required for studies to have meaningful and significant outcomes. Even if a sufficient farm animal population is available and the study is affordable, performing an experiment such as a probiotic trial requires careful consideration of the variables that can significantly impact the outcome. For instance, rodent studies often use germ-free models or antibiotic treatment regimens to eliminate the microbiota to investigate the effect of a single bacterium or a defined mixture of bacteria. Determining the bacterial strain to be administered in farm animals, its formulation (e.g. live or dead bacteria), delivery mode (e.g. water or feed), dosage (e.g. amend or scale-up from the rodent model; administered daily or weekly) and duration of the study can be difficult with little supporting literature. Molecular techniques can predict the metabolic impact of microbes within the host. The multi-factorial nature of behavioural disorders makes the selection of the molecular technique targeting a single pathway challenging. The close interconnectedness of the microbiota–gut–brain axis processes makes it difficult to integrate each component of the axis into a simple set of experiments. In these instances, putative underlying mechanisms, such as gut dysmotility or visceral hypersensitivity associated with behavioural disorders, can be used to test the potential therapeutic value of a specific bacteria. For example, while *ex vivo* investigations of certain probiotics on the enteric nervous system (West *et al.*, 2017; Bharwani *et al.*, 2020; van Staaveren *et al.*, 2020) are essential from a fundamental point of view, whether these observations are replicated *in vivo* is unclear. While this remains an important step, behaviour is determined by complex interactions between the components of the gut–brain axis. Additionally, evolutionary differences in physiology and normal gut microbiota composition may render findings from human or rodent studies poorly translated to farm animal species without additional research (Birkl *et al.*, 2019). To this end, it is not always possible to make inferences about the functionality of specific microbes and their association with behaviour from previous rodent research because of the lag in understanding and characterizing farm animal gut microbiota and other physiological processes (Waite and Taylor, 2015).

The technological advances of high-throughput gene sequencing and bioinformatics analysis have expanded the study of gut microbiota. However, the variability of the methods and approaches (e.g. culture-based studies, quantitative PCR, 16S rRNA studies, use of pooled or individual samples, environmental factors), and results from different sequence centres, as well as differences in the level of detail in results reported (e.g. α- or β-diversity, phylum, class or genus level) make comparisons between studies difficult. Additionally, the number of animals used in these studies is often low, as the analysis is expensive (Stanley *et al.*, 2014; Jandhyala *et al.*, 2015). The decreasing costs of DNA sequencing have fuelled microbiome research. Nevertheless, the cost to process the sample numbers required to get reliable data needs to be assessed. Additionally, it is noteworthy that sequencing does not differentiate between dead and live microbiota, and the metabolome is not taken into account. Another limitation to the prolific and efficient use of sequencing technology is highly qualified personnel with specific expertise that can manipulate the sequencing data to extract microbial profiles, work with reference databases for taxonomic assessments, and explain potential mechanistic/physiological pathways and associate them with behaviour. Furthermore, there is a recurring emphasis on achieving a ‘healthy’, ‘balanced’, ‘normal’ gut microbiota, as ‘dysbiosis’, ‘imbalance’ and ‘abnormal’ gut microbiota are linked to disease or behavioural disorders (Sekirov *et al.*, 2010; Jandhyala *et al.*, 2015). However, a normal or ideal gut microbiota remains undefined and is likely to be variable for each individual animal.

Moving forward, the microbiota research community as a whole faces several challenges in elucidating key players of the microbiota–gut–brain axis. Many intestinal microbes remain uncultivable, which limits their characterization. Novel culture methods would be required to overcome this identification barrier, which consequently impedes elucidating their interaction with the host. Moreover, there is a need for a standardized complex microbiota for use in germ-free models. A dialogue to determine the standardized composition and whether the microbiota of existing commercial colonies can be used for this purpose should be initiated. The maintenance of germ-free animals should also be standardized.

Another major limitation is that biochemical circuits may differ among species due to distinct enzymatic repertoires or activities, e.g. those involved in tryptophan (Yuasa *et al.*, 2015) or nitric oxide biochemistry (Schneemann and Schoedon, 2002). These are just two examples out of many that underline the relevance of the fine-tuned coordination of immunometabolic pathways in different species. Notably, differential responses to inflammatory stimuli have been demonstrated even in a comparison of inbred mouse strains (Hoover-Plow *et al.*, 2008). This implies that investigations of relevant pathways on molecular and biochemical levels should, at best, go hand in hand with genetic analysis. Nevertheless, not all responses will be predictable, and the previous discussion has already emphasized the influence of nutrition and the microbiome on immunological status.

11.5 Conclusions

We are beginning to understand the potential role of the microbiota–gut–brain axis and its ability to impact behaviour; however, the bidirectional communication between the gut, bacteria, viruses, fungi and protozoa in other body parts and the brain is mediated through the neuroimmune system and neurotransmitters, among other molecules. Dissecting this intricate network, identifying its contribution towards modifying behaviour and using this information to develop nutraceuticals to improve the welfare of farm animals are exciting avenues of research.

Further Reading

For further information we recommend the resources that are provided in bold in the reference list. In addition, we recommend the following resources:

Lyte M. and Cryan, J.F. (eds) (2014) *Microbial Endocrinology: The Microbiota–Gut–Brain Axis in Health and Disease.* Springer, New York.

Mens, A.J.W., van Krimpen, M.M. and Kwakkel, R.P. (2020) Nutritional approaches to reduce or prevent feather pecking in laying hens: any potential to intervene during rearing? *World's Poultry Science Journal* 76, 591–610.

van Staaveren and Harlander (2020) Causes and prevention of injurious pecking in chickens. In: Nicol, C. (ed.) *Understanding the Behaviour and Improving the Welfare of Chickens.* Burleigh Dodds Science Publishing, Cambridge, UK.

Villageliũ, D.N. and Lyte, M. (2017) Microbial endocrinology: why the intersection of microbiology and neurobiology matters in poultry health. *Poultry Science* 96, 2501–2508.

References

Bercik, P., Park, A.J., Sinclair, D., Khoshdel, A., Lu, J. *et al.* (2011) The anxiolytic effect of *Bifidobacterium longum* NCC3001 involves vagal pathways for gut-brain communication. *Neurogastroenterology and Motility* 23(12), 1132–1139. DOI: 10.1111/j.1365-2982.2011.01796.x.

Bharwani, A., West, C., Champagne-Jorgensen, K., McVey Neufeld, K.-A., Ruberto, J. *et al.* (2020) The vagus nerve is necessary for the rapid and widespread neuronal activation in the brain following oral administration of psychoactive bacteria. *Neuropharmacology* 170, 108067. DOI: 10.1016/j.neuropharm.2020.108067.

Birkl, P., Bharwani, A., Kjaer, J.B., Kunze, W., McBride, P. *et al.* (2018) Differences in cecal microbiome of selected high and low feather-pecking laying hens. *Poultry Science* 97(9), 3009–3014. DOI: 10.3382/ps/pey167.

Birkl, P., Chow, J., Forsythe, P., Gostner, J.M., Kjaer, J.B. *et al.* (2019) The role of tryptophan–kynurenine in feather pecking in domestic chicken lines. *Frontiers in Veterinary Science* 6, 209. DOI: 10.3389/fvets.2019.00209.

Bravo, J.A., Forsythe, P., Chew, M.V., Escaravage, E. and Savignac, H.M. (2011) Ingestion of *Lactobacillus* strain regulates emotional behavior and central GABA receptor expression in a mouse via the vagus nerve. *Proceedings of the National Academy of Sciences USA* 108, 16050–16055.

Brunberg, E.I., Rodenburg, T.B., Rydhmer, L., Kjaer, J.B., Jensen, P. *et al.* (2016) Omnivores going astray: a review and new synthesis of abnormal behavior in pigs and laying hens. *Frontiers in Veterinary Science* 3, 57. DOI: 10.3389/fvets.2016.00057.

Buitenhuis, A.J., Kjaer, J.B., Labouriau, R. and Juul-Madsen, H.R. (2006) Altered circulating levels of serotonin and immunological changes in laying hens divergently selected for feather pecking behavior. *Poultry Science* 85(10), 1722–1728. DOI: 10.1093/ps/85.10.1722.

Capuron, L., Gumnick, J.F., Musselman, D.L., Lawson, D.H., Reemsnyder, A. *et al.* (2002) Neurobehavioral effects of interferon-alpha in cancer patients: phenomenology and paroxetine responsiveness of symptom dimensions. *Neuropsychopharmacology* 26(5), 643–652. DOI: 10.1016/S0893-133X(01)00407-9.

Collins, J., Borojevic, R., Verdu, E.F., Huizinga, J.D. and Ratcliffe, E.M. (2014) Intestinal microbiota influence the early postnatal development of the enteric nervous system. *Neurogastroenterology and Motility* 26(1), 98–107. DOI: 10.1111/nmo.12236.

Day, J.E.L., Kyriazakis, I. and Lawrence, A.B. (1995) The effect of food deprivation on the expression of foraging and exploratory behaviour in the growing pig. *Applied Animal Behaviour Science* 42, 193–206.

Diepenbroek, C., Quinn, D., Stephens, R., Zollinger, B., Anderson, S. *et al.* (2017) Validation and characterization of a novel method for selective vagal deafferentation of the gut. *American Journal of Physiology – Gastrointestinal and Liver Physiology* 313(4), G342–G352. DOI: 10.1152/ajpgi.00095.2017.

Dixon, L.M., Brocklehurst, S., Sandilands, V., Bateson, M., Tolkamp, B.J. *et al.* (2014) Measuring motivation for appetitive behaviour: food-restricted broiler breeder chickens cross a water barrier to forage in an area of wood shavings without food. *PLoS One* 9(7), e102322. DOI: 10.1371/journal.pone.0102322.

Dockray, G.J. (2009) Cholecystokinin and gut-brain signalling. *Regulatory Peptides* 155(1–3), 6–10. DOI: 10.1016/j.regpep.2009.03.015.

Dominguez-Bello, M.G., Costello, E.K., Contreras, M., Magris, M., Hidalgo, G. *et al.* (2010) Delivery mode shapes the acquisition and structure of the initial microbiota across multiple body habitats in newborns. *Proceedings of the National Academy of Sciences USA* 107(26), 11971–11975. DOI: 10.1073/pnas.1002601107.

Flint, H.J. (2012) The impact of nutrition on the human microbiome. *Nutrition Reviews* 70 (Suppl 1), S10–S13. DOI: 10.1111/j.1753-4887.2012.00499.x.

Geisler, S., Sperner-Unterweger, B., Fuchs, D. and Gostner, J.M. (2018) Immunometabolism in the pathogenesis of depressive disorders – therapeutic considerations. *Current Topics in Medicinal Chemistry* 18(16), 1408–1415. DOI: 10.2174/1568026618666180410141042.

Godbout, J.P., Moreau, M., Lestage, J., Chen, J., Sparkman, N.L. *et al.* (2008) Aging exacerbates depressive-like behavior in mice in response to activation of the peripheral innate immune system. *Neuropsychopharmacology* 33(10), 2341–2351. DOI: 10.1038/sj.npp.1301649.

Hill, C., Guarner, F., Reid, G., Gibson, G.R., Merenstein, D.J. *et al.* (2014) The International scientific association for probiotics and prebiotics consensus statement on the scope and appropriate use of the term probiotic. *Nature Reviews Gastroenterology & Hepatology* 11(8), 506–514. DOI: 10.1038/nrgastro.2014.66.

Hoisington, A.J., Stearns-Yoder, K.A., Schuldt, S.J., Beemer, C.J. and Maestre, J.P. (2019) Ten questions concerning the built environment and mental health. *Building and Environment* 155, 58–69.

Hoover-Plow, J.L., Gong, Y., Shchurin, A., Busuttil, S.J., Schneeman, T.A. *et al.* (2008) Strain and model dependent differences in inflammatory cell recruitment in mice. *Inflammation Research* 57(10), 457–463. DOI: 10.1007/s00011-008-7062-5.

Irwin, M.R. and Miller, A.H. (2007) Depressive disorders and immunity: 20 years of progress and discovery. *Brain, Behavior and Immunity* 21(4), 374–383. DOI: 10.1016/j.bbi.2007.01.010.

Jandhyala, S.M., Talukdar, R., Subramanyam, C., Vuyyuru, H., Sasikala, M. *et al.* (2015) Role of the normal gut microbiota. *World Journal of Gastroenterology* 21(29), 8787–8803. DOI: 10.3748/wjg.v21.i29.8787.

Kaelberer, M.M., Buchanan, K.L., Klein, M.E., Barth, B.B., Montoya, M.M. *et al.* (2018) A gut–brain neural circuit for nutrient sensory transduction. *Science* 361(6408), eaat5236. DOI: 10.1126/science.aat5236.

Kers, J.G., Velkers, F.C., Fischer, E.A.J., Hermes, G.D.A., Stegeman, J.A. *et al.* (2018) Host and environmental factors affecting the intestinal microbiota in chickens. *Frontiers in Microbiology* 9, 235. DOI: 10.3389/fmicb.2018.00235.

Kraimi, N., Dawkins, M., Gebhardt-Henrich, S.G., Velge, P., Rychlik, I. *et al.* (2019) Influence of the microbiota-gut-brain axis on behavior and welfare in farm animals: a review. *Physiology & Behavior* 210, 112658. DOI: 10.1016/j.physbeh.2019.112658.

Kunze, W.A. and Furness, J.B. (1999) The enteric nervous system and regulation of intestinal motility. *Annual Review of Physiology* 61, 117–142. DOI: 10.1146/annurev.physiol.61.1.117.

Lal, S., Kirkup, A.J., Brunsden, A.M., Thompson, D.G. and Grundy, D. (2001) Vagal afferent responses to fatty acids of different chain length in the rat. *American Journal of Physiology – Gastrointestinal and Liver Physiology* 281(4), G907–G915. DOI: 10.1152/ajpgi.2001.281.4.G907.

Li, Y. (2007) Sensory signal transduction in the vagal primary afferent neurons. *Current Medicinal Chemistry* 14(24), 2554–2563. DOI: 10.2174/092986707782023334.

Liu, Y., Mian, M.F., McVey Neufeld, K.-A. and Forsythe, P. (2020) $CD4^+CD25^+$ T cells are essential for behavioral effects of *Lactobacillus rhamnosus* JB-1 in male BALB/c mice. *Brain, Behavior, and Immunity* 88, 451–460. DOI: 10.1016/j.bbi.2020.04.014.

Liukkonen, T., Räsänen, P., Jokelainen, J., Leinonen, M., Järvelin, M.-R. *et al.* (2011) The association between anxiety and C-reactive protein (CRP) levels: results from the Northern Finland 1966 birth cohort study. *European Psychiatry* 26(6), 363–369. DOI: 10.1016/j.eurpsy.2011.02.001.

McVey Neufeld, K.A., Mao, Y.K., Bienenstock, J., Foster, J.A. and Kunze, W.A. (2013) The microbiome is essential for normal gut intrinsic primary afferent neuron excitability in the mouse. *Neurogastroenterology and Motility Society* 25(2), 183–190. DOI: 10.1111/nmo.12049.

McVey Neufeld, K.A., Perez-Burgos, A., Mao, Y.K., Bienenstock, J. and Kunze, W.A. (2015) The gut microbiome restores intrinsic and extrinsic nerve function in germ-free mice accompanied by changes in calbindin. *Neurogastroenterology and Motility Society* 27(5), 627–636. DOI: 10.1111/nmo.12534.

Meyer, B., Bessei, W., Bessei, A.W., Vahjen, W., Zentek, J. and Harlander-Matauschek, A. *et al.* (2012) Dietary inclusion of feathers affects intestinal microbiota and microbial metabolites in growing Leghorn-type chickens. *Poultry Science* 91(7), 1506–1513. DOI: 10.3382/ps.2011-01786.

Meyer, B., Zentek, J. and Harlander-Matauschek, A. (2013) Differences in intestinal microbial metabolites in laying hens with high and low levels of repetitive feather-pecking behavior. *Physiology & Behavior* 110-111, 96–101. DOI: 10.1016/j.physbeh.2012.12.017.

Mindus, C., van Staaveren, N., Bharwani, A., Fuchs, D, Gostner, J., Kjaer J. B., Kunze W., Mian M. F., Shoveller A. K., Forsythe P., and Harlander-Matauschek A. (2021) Ingestion of *Lactobacillus rhamnosus* modulates chronic stress-induced feather pecking in chickens. *Scientific Reports,* in press.

Morais, L.H., Schreiber, H.L. and Mazmanian, S.K. (2020) The gut microbiota-brain axis in behaviour and brain disorders. *Nature Reviews Microbiology* 19(4), 241–255. DOI: 10.1038/s41579-020-00460-0.

Neu, J. and Rushing, J. (2011) Cesarean versus vaginal delivery: long-term infant outcomes and the hygiene hypothesis. *Clinics in Perinatology* 38(2), 321–331. DOI: 10.1016/j.clp.2011.03.008.

Neurauter, G., Schröcksnadel, K., Scholl-Bürgi, S., Sperner-Unterweger, B., Schubert, C. *et al.* (2008) Chronic immune stimulation correlates with reduced phenylalanine turnover. *Current Drug Metabolism* 9(7), 622–627. DOI: 10.2174/138920008785821738.

Nordgreen, J., Edwards, S.A., Boyle, L.A., Bolhuis, J.E., Veit, C. *et al.* (2020) A proposed role for pro-inflammatory cytokines in damaging behavior in pigs. *Frontiers in Veterinary Science* 7, 646. DOI: 10.3389/fvets.2020.00646.

Pasquaretta, C., Gómez-Moracho, T., Heeb, P. and Lihoreau, M. (2018) Exploring interactions between the gut microbiota and social behavior through nutrition. *Genes* 9(11), 534. DOI: 10.3390/genes9110534.

Pavlov, V.A. and Tracey, K.J. (2012) The vagus nerve and the inflammatory reflex–linking immunity and metabolism. *Nature Reviews Endocrinology* 8(12), 743–754. DOI: 10.1038/nrendo.2012.189.

Perez-Burgos, A., Mao, Y.-K., Bienenstock, J. and Kunze, W.A. (2014) The gut-brain axis rewired: adding a functional vagal nicotinic 'sensory synapse'. *FASEB Journal* 28(7), 3064–3074. DOI: 10.1096/fj.13-245282.

Raison, C.L. and Miller, A.H. (2013) Malaise, melancholia and madness: the evolutionary legacy of an inflammatory bias. *Brain, Behavior, and Immunity* 31, 1–8. DOI: 10.1016/j.bbi.2013.04.009.

Schneemann, M. and Schoedon, G. (2002) Species differences in macrophage NO production are important. *Nature Immunology* 3(2), 102. DOI: 10.1038/ni0202-102a.

Seal, B.S., Lillehoj, H.S., Donovan, D.M. and Gay, C.G. (2013) Alternatives to antibiotics: a symposium on the challenges and solutions for animal production. *Animal Health Research Reviews* 14(1), 78–87. DOI: 10.1017/S1466252313000030.

Sekirov, I., Russell, S.L., Antunes, L.C.M. and Finlay, B.B. (2010) Gut microbiota in health and disease. *Physiological Reviews* 90(3), 859–904. DOI: 10.1152/physrev.00045.2009.

Sgritta, M., Dooling, S.W., Buffington, S.A., Momin, E.N., Francis, M.B. *et al.* (2019) Mechanisms underlying microbial-mediated changes in social behavior in mouse models of autism spectrum disorder. *Neuron* 101(2), 246–259. DOI: 10.1016/j.neuron.2018.11.018.

Sonner, J.K., Keil, M., Falk-Paulsen, M., Mishra, N., Rehman, A. *et al.* (2019) Dietary tryptophan links encephalogenicity of autoreactive T cells with gut microbial ecology. *Nature Communications* 10(1), 4877. DOI: 10.1038/s41467-019-12776-4.

Stanley, D., Geier, M.S., Hughes, R.J., Denman, S.E. and Moore, R.J. (2014) Highly variable microbiota development in the chicken gastrointestinal tract. *PLoS One* 8, e84290.

Talbot, S., Abdulnour, R.-E.E., Burkett, P.R., Lee, S., Cronin, S.J.F. *et al.* (2015) Silencing nociceptor neurons reduces allergic airway inflammation. *Neuron* 87(2), 341–354. DOI: 10.1016/j.neuron.2015.06.007.

Teratani, T., Mikami, Y., Nakamoto, N., Suzuki, T., Harada, Y. *et al.* (2020) The liver-brain-gut neural arc maintains the Treg cell niche in the gut. *Nature* 585(7826), 591–596. DOI: 10.1038/s41586-020-2425-3.

Tracey, K.J. (2007) Physiology and immunology of the cholinergic antiinflammatory pathway. *The Journal of Clinical Investigation* 117(2), 289–296. DOI: 10.1172/JCI30555.

Trinh, P., Zaneveld, J.R., Safranek, S. and Rabinowitz, P.M. (2018) One health relationships between human, animal, and environmental microbiomes: a mini-review. *Frontiers in Public Health* 6, 235. DOI: 10.3389/fpubh.2018.00235.

van der Eijk, J.A.J., Verwoolde, M.B., de Vries Reilingh, G., Jansen, C.A., Rodenburg, T.B. *et al.* (2019a) Chicken lines divergently selected on feather pecking differ in immune characteristics. *Physiology & Behavior* 212, 112680. DOI: 10.1016/j.physbeh.2019.112680.

van der Eijk, J.A.J., de Vries, H., Kjaer, J.B., Naguib, M., Kemp, B. *et al.* (2019b) Differences in gut microbiota composition of laying hen lines divergently selected on feather pecking. *Poultry Science* 98(12), 7009–7021. DOI: 10.3382/ps/pez336.

van der Eijk, J.A.J., Rodenburg, T.B., de Vries, H., Kjaer, J.B., Smidt, H. *et al.* (2020) Early-life microbiota transplantation affects behavioural responses, serotonin and immune characteristics in chicken lines divergently selected on feather pecking. *Scientific Reports* 10(1), 2750. DOI: 10.1038/s41598-020-59125-w.

van Staaveren, N., Krumma, J., Forsythe, P., Kjaer, J.B., Kwon, I.Y. *et al.* (2020) Cecal motility and the impact of Lactobacillus in feather pecking laying hens. *Scientific Reports* 10(1), 12978. DOI: 10.1038/s41598-020-69928-6.

Veit, C., Janczak, A.M., Ranheim, B., Vas, J., Valros, A. *et al.* (2021) The effect of LPS and ketoprofen on cytokines, brain monoamines, and social behavior in group-housed pigs. *Frontiers in Veterinary Science* 7, 1096. DOI: 10.3389/fvets.2020.617634.

Waite, D.W. and Taylor, M.W. (2015) Exploring the avian gut microbiota: current trends and future directions. *Frontiers in Microbiology* 6, 673. DOI: 10.3389/fmicb.2015.00673.

West, C., Wu, R.Y., Wong, A., Stanisz, A.M., Yan, R. *et al.* (2017) *Lactobacillus rhamnosus* strain JB-1 reverses restraint stress-induced gut dysmotility. *Neurogastroenterology and Motility Society* 29(1), e12903. DOI: 10.1111/nmo.12903.

Wilson, C.J., Finch, C.E. and Cohen, H.J. (2002) Cytokines and cognition – the case for a head-to-toe inflammatory paradigm. *Journal of the American Geriatrics Society* 50(12), 2041–2056. DOI: 10.1046/j.1532-5415.2002.50619.x.

Wohleb, E.S., McKim, D.B., Sheridan, J.F. and Godbout, J.P. (2015) Monocyte trafficking to the brain with stress and inflammation: a novel axis of immune-to-brain communication that influences mood and behavior. *Frontiers in Neuroscience* 8, 447. DOI: 10.3389/fnins.2014.00447.

Yan, W., Sun, C., Zheng, J., Wen, C., Ji, C. *et al.* (2019) Efficacy of fecal sampling as a gut proxy in the study of chicken gut microbiota. *Frontiers in Microbiology* 10, 2126. DOI: 10.3389/fmicb.2019.02126.

Yuasa, H.J., Mizuno, K. and Ball, H.J. (2015) Low efficiency IDO2 enzymes are conserved in lower vertebrates, whereas higher efficiency IDO1 enzymes are dispensable. *The FEBS Journal* 282(14), 2735–2745. DOI: 10.1111/febs.13316.

12

Improving Animal Welfare Using Genetic and Genomic Tools

Wendy M. Rauw[1]*, Jack C.M. Dekkers[2] and Luis Gomez-Raya[1]

[1]Instituto Nacional de Investigación y Tecnología Agraria y Alimentaria (INIA-CSIC), Madrid, Spain; [2]Iowa State University, Ames, Iowa, USA

12.1 Introduction

12.1.1 Measuring animal welfare

The need to address and improve animal welfare has been well recognized for many decades (see Chapter 1, this volume). In the 1990s, it became clear that selection for narrowly defined breeding goals, based primarily on production traits, may result in metabolic, reproduction, health and behavioural problems in highly selected farm animals (Rauw *et al.*, 1998). Since then, the focus in animal breeding has shifted to implementation of broader breeding goals by including functional traits associated with good welfare. Animal scientists now emphasize that welfare can also be *improved* by genetic selection (e.g. Jensen *et al.*, 2008; Rauw, 2016). However, the first challenge encountered when aiming to select for improved animal welfare is deciding which traits are considered to be representative of welfare. This is a formidable task. As animal welfare has many facets, there have been many attempts to best define it (Stafleu *et al.*, 1996) and no universally endorsed definition of animal welfare has emerged (Mellor, 2016) (covered in more detail in Chapter 1, this volume). According to Broom (1991), animal welfare is a function of an animal's needs. Reduced welfare follows from a deficiency to an animal's mental and bodily stability that arouses its motivational state such that it seeks to remedy its need(s) through behavioural and physiological (i.e. coping) responses. Successful coping results in adaptation, while difficulty or failure to cope results in unpleasant

*Corresponding author: rauw.wendy@inia.es

 Bridging Research Disciplines to Advance Animal Welfare Science: A practical guide (ed. I. Camerlink)
DOI: 10.1079/9781789247893.0012

subjective feelings and possibly suffering and is, therefore, indicative of poor welfare (Dawkins, 1990; Broom, 1991). Following Broom's (1991) definition of welfare, we can identify: (i) the situation or event that results in a deficiency to the animal's stability; (ii) the animal's motivational state that results from its perception of the significance of the situation or event; and (iii) the animal's ability to cope successfully. In controlled livestock production environments, it is clear that (i) results from and can be modified by management decisions, such as provision of feed, water, space, thermal and acoustic comfort, social interactions, the ability to express necessary natural behaviours, and absence of predators, infectious agents and other stressors, while minimizing suffering during human (veterinary) interactions, transport and slaughter. Given that stressful events and situations are not fully evitable (and it can be debated if that would even be desirable) and that stress emotional states and coping ability are heritable animal characteristics, animal welfare within a production system can be improved by genetic selection.

12.1.2 Event appraisal

While research on coping ability in humans benefits from self-reported coping scales and measures, the lack of such assessment methods for non-human animals has resulted in coping responses often being directly equated to a personality concept (reviewed by Rauw *et al.*, 2017). Thorough reviews on the truly multidimensional nature of animal personality are given by Koski (2014) and Finkemeier *et al.* (2018). In humans, a personality trait that modulates event appraisal includes 'dispositional optimism', defined as the 'tendency to expect good outcomes across important life domains' (Scheier and Carver, 2018, p. 1082). According to Nes and Segerstrom (2006), optimistic people are typically less distressed, leading to engagement and increased effort (approach coping), as opposed to disengagement and giving up (avoidance coping). However, once engaged, optimistic people use problem-focused proactive strategies (actively seeking to change or eliminate the stressor itself), as well as emotion-focused reactive strategies (to reduce or manage the emotional consequences of the stressor), suggesting a level of independence between event appraisal (or motivation) and subsequent stress-coping strategies. Operant-conditioning tasks have been developed to investigate (some expression of) optimism and pessimism in poultry, cattle, pigs, horses, dogs, monkeys, mice and bees (Lagisz *et al.*, 2020). Another indicator of event appraisal can be found in the concepts of fear and fearfulness. According to Boissy (1995, p. 176), fearfulness can be considered a mental trait of the individual 'that determines the extent to which an individual becomes frightened in potentially alarming situations', such as when confronted with novelty, a predator or with social stress. Once the event is appraised as being fearful, the subsequent coping response may be characterized by proactive as well as reactive strategies. Boissy *et al.* (2007) emphasized that, while prevention of negative emotions is the general objective of animal welfare science, providing animals with positive experiences that result in positive emotions (or 'happiness') may be the core component of good welfare. To date, positive experiences have been investigated through play behaviour (Oliveira *et al.*, 2010; Held and Špinka, 2011).

Although little is known about the genetic background of event appraisal or emotional states, a genetic background to optimism in humans has been suggested (e.g. Mosing *et al.*, 2009). Fearfulness also has a clear genetic background (e.g. Agnvall *et al.*, 2012), and results by Walker and Byers (1991) suggest a genetic basis for locomotor play in house mice. Therefore, animal welfare could be improved by genetic selection for animals with 'optimistic', 'fearless' or 'happy' emotional states.

12.1.3 Coping responses

Negative event appraisal is followed by a coping response, which refers to the behavioural and physiological responses an individual employs to manage (i.e. eliminate, reduce or tolerate) the internal or external demands of a situation or event that is appraised as stressful. Coping styles have been described extensively for farm animal species, including pigs (Kanitz *et al.*, 2019), dairy cattle (Neave *et al.*, 2020) and chickens (Zidar *et al.*, 2017), as well as for aquaculture fish species (Castanheira *et al.*, 2017). A range of tests have been developed to distinguish individuals that are proactive versus reactive, and some studies have investigated the heritability of these test responses (reviewed by Canario *et al.*, 2013). Coping styles in farm animal species are directly related to animal welfare, in that they modulate an animal's ability to respond and adapt to the semi-natural production environment, which is often barren, preventing the expression of variable and differentiated behaviour, with few features that the animal can manipulate, and without the ability to build stable and natural social relationships (Wechsler, 1995). This implies that animal welfare can be improved by selection for coping styles when (indicators of) styles that are (more) favourable to adequately respond and adapt to farm animal production systems can be identified and are genetically determined. Generally, genetics explains less than 35% of the phenotypic variation in coping styles (Bell and Aubin-Horth, 2010). Behavioural traits associated with coping strategies generally have low to moderate heritabilities (Canario *et al.*, 2013), including aggressive behaviour in pigs (Desire *et al.*, 2016), novel-environment indicator traits and feather pecking in laying hens (Rodenburg *et al.*, 2003; van der Eijk *et al.*, 2018), and restraint resistance behaviour in lambs, sheep, cattle, horses and pigs (Zambra *et al.*, 2015). Therefore, several of these traits could be considered for genetic improvement of behavioural adaptation to a wide range of farming conditions.

The nature of coping styles implies that correlations exist between the different behavioural and physiological responses, and therefore selection for (a few) well-chosen traits may result in correlated responses in other characteristics of the coping style. However, this also implies that selection against one particular undesirable behaviour characteristic could result in unwanted correlated modifications to the individual's underlying adaptive capacity. For example, Fauré *et al.* (2003, p. 228) wrote: 'It is particularly striking that genetic selection for one behavioural reaction, one physiological response or one performance measure affected the way the quail responded to a wide variety of stressful events.' In addition, natural variation in coping styles highlights the effectiveness of different patterns in response to different environmental and management conditions. Therefore, it should not be assumed a priori that selection should favour one coping style over another (Koolhaas and Van Reenen, 2016).

12.1.4 Robustness and resilience

It is generally understood that a reduction in animal welfare will negatively affect farm animal production and profitability. As reviewed by Rauw and Gomez-Raya (2017), this can readily be understood because the processes involved in stress responses incur a cost. Animals prioritize maintenance costs (supporting survival and longevity), and therefore growth and reproduction will be negatively affected when stress is (acute and) severe or chronic (Moberg, 2000; Glazier, 2009). In addition, stress is largely found to be immunosuppressive (Rauw, 2012), and it is generally understood that impaired health is negatively associated with animal welfare (Dawkins, 2008). However, high production does not necessarily imply good welfare, as exemplified by undesirable side effects of selection for high production efficiency (Rauw *et al.*, 1998). As a result, and in response to environmental, social, immunological and dietary challenges, the last few decades have seen an increasing interest in a more holistic perspective for genetic improvement of farm animals, with a focus on animal robustness (Rauw and Gomez-Raya, 2015). Concomitant to robustness is the concept of resilience, which can be defined as the capacity of animals to cope with episodic or sporadic perturbations in their environment (Colditz and Hine, 2016). Resilience is often used in relation to immunological challenges, such as disease resistance (Berghof *et al.*, 2019).

To date, in many breeding programmes, robustness is approached by inclusion of functional traits in the selection index, such as longevity, health and fertility; inclusion of behavioural traits, and estimation of breeding values for environmental sensitivity of production performance through the use of reaction norms has been proposed (reviewed by Rauw and Gomez-Raya, 2015). Resilience can be improved by focusing on the ability of an animal to withstand a (routine management) challenge, either through reduced sensitivity or an enhanced coping response (Colditz and

Hine, 2016). Because of a lack of clearly defined indicator traits, resilience has not yet been included in standard breeding programmes (Berghof *et al.*, 2019). However, deployment of automated on-farm monitoring technologies now enables development of novel phenotypes of resilience and robustness based on adaptive capacity over time (Friggens *et al.*, 2017). Elgersma *et al.* (2018) showed that fluctuations in daily milk yield in dairy cows are heritable and can be used as resilience indicators to select for healthy, resilient and long-lasting dairy cows. Putz *et al.* (2019) used day-to-day variations in feed intake behaviour as novel phenotypes for (disease) resilience, and estimated heritabilities between 0.15 and 0.26. Berghof *et al.* (2019) proposed measuring resilience based on variance, autocorrelation and/or skewness of deviations of expected production levels over time, and showed a clear benefit of including selection for resilience measures in breeding programmes. Sánchez-Molano *et al.* (2019) demonstrated the applicability of reaction norm (i.e. 'response') functions to derive novel animal resilience phenotypes to weather variability, such as in response to the temperature–humidity index (Bohmanova *et al.*, 2007); heritability estimates of resilience phenotypes in goats to changes in weather ranged from 0.09 to 0.11 (Sánchez-Molano *et al.*, 2019). Although caution should be applied when extrapolating results between species, the presence of a genetic basis for these novel traits that are representative of resilience and robustness suggests that these traits can be used to improve farm animal welfare through genetic selection.

12.2 Concepts

Some of the main concepts and their interrelationship are described in Fig. 12.1.

Causal genes are the genes that are directly responsible for the trait phenotype. For only a small number of causal genes has the location on the genome been firmly established; for the majority, their exact location is unknown. The latter situation is depicted in Fig. 12.1:

1. *Microsatellites* are codominant genetic markers containing repeated nucleotide sequences that are present sporadically across the genome. Because their location on the genome has been mapped (solid box), they can be used to identify so-called *quantitative trait loci* (QTLs), which are regions on the genome that contain one or more causal genes (dotted box). In the example, allele A at the causal gene (and thus its phenotypic expression) is in linkage phase with allele MICRO1 at the genetic marker.
2. Because microsatellites can be some distance from the causal gene, crossover can occur in the chromosomal region between the microsatellite and the causal gene during sexual reproduction. This can change the linkage phase such that allele A will become linked with allele MICRO2

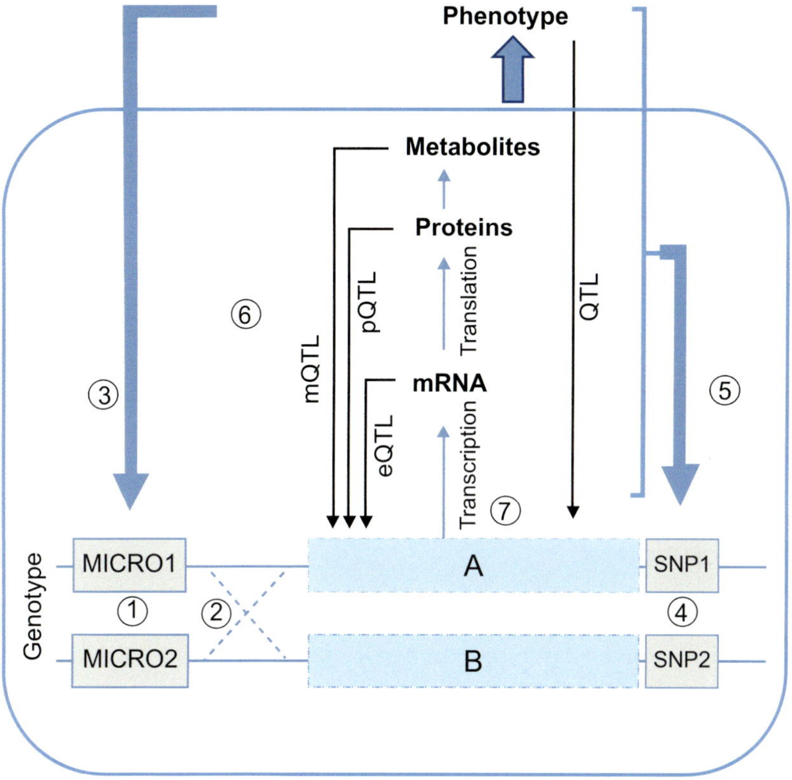

Fig. 12.1. The main concepts and their interrelationships for improving phenotypic traits using genetic and genomic tools. Circled numbers relate to the concepts described in section 12.2. MICRO, microsatellite; SNP, single-nucleotide polymorphism; QTL, quantitative trait locus; eQTL, expression QTL; mQTL, metabolite QTL; pQTL, protein QTL.

in some individuals. As a result, associations between the genetic marker and the causal gene must be established in each family.

3. Because the causal gene is unknown, evidence for a QTL is established based on statistical association between the trait phenotype and the marker genotype. The marker genotype can be used to aid selection of animals, which is called *marker-assisted selection* (MAS). In this example, if allele A produces a favourable phenotype, animals can be selected for the presence of allele MICRO1.
4. Because coverage of the genomes by *single-nucleotide polymorphisms* (SNPs) is much denser, SNP markers can be found that are in such close proximity to the causal gene that crossovers are unlikely to change the linkage phase between the SNP and the causal gene.
5. In *genome-wide association studies* (GWAS), the entire genome is searched for associations between the phenotype and genotypes from SNP arrays

comprised of around 50 thousand to 1 million SNPs distributed across the genome. Because of this dense coverage, one or more SNPs can be found to be in very close proximity for every causal gene, or even within the sequence of the causal gene itself. Conversely, the identified QTL region can contain many genes that could be candidates for the causal gene. With *genomic selection*, animals can now be accurately selected based on the effects associated with their genotypes for all SNPs on a genome-wide array, even before their phenotype has been observed.

6. Quantitative trait phenotypes can be used to map QTLs, expression of gene (messenger) transcripts can be used to map *expression QTLs* (eQTLs), protein abundance can be used to map *protein QTLs* (pQTL) and metabolite abundance can be used to map *metabolite QTLs* (mQTL).
7. Stress can activate *epigenetic mechanisms*, which result in modulation of DNA transcription, and therefore of phenotype (see Chapter 13, this volume).

12.3 Techniques

12.3.1 Traditional genetic selection

For centuries, humans have practised selection in livestock populations. As opposed to discontinuous traits, for which phenotype is classified into discrete classes (e.g. litter size, survival or death), most traits of economic importance in livestock production are continuous (e.g. milk production, growth rate), giving rise to the theory of quantitative genetics (Falconer and Mackay, 1996). Most behavioural traits are also clearly quantitative in nature, such as measures of aggression, mood and intelligence (Flint, 2003). As the key to genetic selection is the presence of genetic variation for the trait, a key development in animal breeding was Fisher's introduction of a novel method that allowed the separation of variability according to different sources, called analysis of variance (ANOVA) (Fisher and Mackenzie, 1923). Fisher initially believed that most of the variation in phenotypes between animals was due to genetics, but later he recognized that environment was also an important source of variation, although this 'nuisance' was considered to be dismissed or at least minimized (Tabery, 2008). The phenotypic (P) variance (V_P) of a quantitative trait can thus be attributed to genetic (G; V_G) and environmental (E; V_E) factors:

$$V_P = V_G + V_E$$

based on P = G + E. Fisher (1918) theorized that variation of continuous, quantitative traits is produced by the collective effects of an individual's genotype at an infinite number of loci, each contributing an infinitesimal effect to phenotype, along with a large number of environmental factors, which theoretically results in a normal distribution of observed phenotypes. The genetic contribution to phenotype and to phenotypic variation,

V_G, can be further partitioned into variation attributed to the sum of the additive effects at all alleles at all loci that affect the trait, V_A, in addition to the sum of dominance effects of alleles over their complementary allele, V_D, and the sum of epistatic effects, V_I, resulting from interactions between genotypes at different loci:

$$V_G = V_A + V_D + V_I$$

Genes (alleles) can have different effects, depending on the environment, so some variation in phenotypes can also be attributed to the interaction between genes and the environment that the animal is exposed to, $V_{G\times E}$, such that:

$$V_P = (V_A + V_D + V_I) + V_E + V_{G\times E}$$

As only the additive genetic effects (breeding values) are predictably heritable from parents to progeny, the (narrow sense) heritability of a trait, h^2, is defined as V_A/V_P, i.e. the proportion of the phenotypic variance that results from variance in the additive effects of all alleles that affect the trait.

An issue with the classical ANOVA design for estimation of variance components is that it assumes balanced data, which are rarely available. Methods such as minimum variance quadratic unbiased estimation (MIVQUE) and restricted maximum likelihood estimation (REML) overcame such issues. Best linear unbiased prediction (BLUP) for estimation of breeding values, using linear mixed models that use genetic relationships between individuals based on pedigree, along with vast advances in statistics and computing power, provided animal breeders with the tools to implement breeding programmes that have resulted in unprecedented genetic improvement of production traits in farm animals over the past 40 years (Henderson, 1990).

If the proportion of phenotypic variance that is contributed by additive genetic variance (i.e. heritability) is greater for a trait, response to selection and genetic progress for that trait is expected to be greater. Traits with high heritability are therefore desired by animal breeders. Heritabilities that exceed 0.40 are considered high, while heritabilities below 0.15 are low and heritabilities between 0.15 and 0.40 are medium (Getabalew *et al.*, 2019). For example, Miglior *et al.* (2017) showed that heritabilities in dairy cattle are low to intermediate for temperament, milking speed, metabolic diseases, β-hydroxybutyrate in milk, mastitis, calving ease, fertility, longevity, feed and leg conformation, and somatic cell score, and medium to high for body conformation, mammary conformation, milk production, and fat and protein production or content. In contrast, Mir *et al.* (2017, p. 2997) reported that reviewed heritability estimates for meat quality traits in broilers are 'amazingly high', ranging from 0.35 to 0.81. In contrast to morphological traits, with average heritabilities of 0.46 (Mousseau and Roff, 1987), behavioural traits and traits that are related to fitness have lower heritabilities. In a meta-analysis, Dochtermann *et al.* (2019) estimated the average heritability for behavioural traits to be 0.24, which is similar to the average heritability of fitness traits of 0.26 calculated by Mousseau and Roff (1987). The low heritability of fitness traits can be explained in two fundamentally different ways. On the one hand, because natural selection acts on fitness, alleles that are favourable to fitness have been fixed over the many generations of natural selection, while unfavourable alleles have been eliminated, thus reducing V_A and therefore heritability. On the other hand, fitness traits are complex traits that are determined by a very large number of loci, which implies a *larger* V_A. However, the non-heritable part of phenotypic variation is also disproportionally large for fitness traits, in particular V_E and $V_{G×E}$, which can be understood by the observation that behavioural and fitness traits are highly plastic and are typically expressed repeatedly and variably in response to environmental conditions. The high V_E and $V_{G×E}$, in addition to high V_A, results in even higher V_P and lower heritability (Stirling *et al.*, 2002). Indeed, Dochtermann *et al.* (2019) demonstrated the presence of substantial genetic variation but also large environmental variation for fitness traits.

Behavioural traits can also depend on the social environment when animals live in groups. In these settings, the phenotype of an individual also depends on the genetics for social interactions of its group mates, also known as social indirect genetic effects, which create some sort of $G_{Animal1} \times G_{Animal2}$ interactions that may have causal effects on V_A (Meffert *et al.*, 2002; Ellen *et al.*, 2014). Novel selection methods have been developed that specifically address the fact that traditional selection methods neglect social indirect genetic effects on the performance of group mates, which can increase the response to selection considerably (Wade *et al.*, 2010; Ellen *et al.*, 2014).

12.3.2 MAS and genomic selection

The infinitesimal model of quantitative genetics that is the basis of traditional selection methods considers that phenotypic traits are controlled by a very large number of unlinked genes, each with an infinitesimal effect on phenotypic expression (Fisher, 1918; Falconer and Mackay, 1996). Traditional selection methods are essentially a black box technique because they are not based on knowledge of the physiological processes and pathways on which genes act. That changed with the advent of molecular genetic methods, resulting in the ability to genotype individuals for genetic markers, which are simple polymorphic DNA sequences, and to identify regions of the genome that control the phenotype, so-called QTLs. The first attempt to map QTLs is attributed to Sax (1923). However, it was not until the early 1980s that the development and utilization of restriction fragment length polymorphism (RFLPs) as genetic markers (Botstein *et al.*, 1980) allowed full coverage of the human and animal genomes with markers for

QTL mapping. The main principle of QTL mapping is to identify genetic markers that are near the causal genes that affect a quantitative trait, i.e. for which the average performance of offspring inheriting alternative alleles from a parent differs. Using RFLPs distributed across the genome and specific experimental crosses, as initially proposed by Thoday (1961), Lander and Botstein (1989) proposed interval mapping to identify regions in the genome that harbour QTLs. The main principle of their method was to use flanking markers to localize QTLs within a chromosomal region by statistical testing of a difference in the performance of progeny inheriting alternate alleles at two marker loci. The testing is done sequentially every 1 cM in a given interval. Later, RFLPs were substituted by microsatellites (codominant markers containing repeated nucleotide sequences with 2–10 nucleotides per repeated unit), which were applied to map QTLs in pigs (Andersson *et al.*, 1994), dairy cattle (Georges *et al.*, 1995; Gomez-Raya *et al.*, 1998) and other animals. Once QTLs or genetic markers associated with QTLs are identified, the genotype at these markers can be used to aid selection of animals, which is called MAS (Fernando and Grossman, 1989; Lander and Botstein, 1989). However, crossovers that can occur in the chromosomal region between the marker and causal gene during sexual reproduction can rearrange the association between the genotype at the genetic marker and the causal gene. As a result, the association between marker genotype and phenotype (or 'linkage phase') found in one population or family (and therefore what is considered to be the desired marker genotype) cannot simply be extrapolated to other families or populations. This problem is greater if the linkage distance between the marker and the causal gene is greater, because this results in a higher chance of crossovers occurring. The low coverage of animal genomes, with distances between microsatellites of up to 10–20 cM, forced early QTL mapping efforts to use linkage disequilibrium within families, which required marker–phenotype associations to determine which marker genotype was favourable to be conducted within each family of interest. These initial QTL linkage analysis designs also resulted in the interval for QTL location spanning hundreds of genes that could be the causal gene for the QTL (Blasco and Toro, 2014). An alternative to the use of random markers across the genome was to use known sequences within or very close to candidate genes that were physiologically related to the trait of interest, such as casein genes as candidate genes for milk production (Lien *et al.*, 1995). However, the causal genes of the observed phenotypic performance have proven to be very difficult to establish.

As proposed by Fernando and Grossman (1989), the application of MAS required the joint evaluation of the effect of the genotype at one (or more) markers and a BLUP evaluation for the rest of the genome using pedigree relationships. Using computer simulations, Gomez-Raya and Klemetsdal (1999) showed that there is a conflict between the short- and long-term response to selection when selecting jointly for a marker and phenotype versus a marker and BLUP. The strategies and challenges for the detection and application of QTLs and genetic markers in MAS were reviewed by Dekkers and Hospital (2002). Despite the challenges, using these strategies, large numbers of QTLs were detected for many traits, including for conformation and functional traits in dairy cattle (Schrooten *et al.*, 2000). QTLs for temperament and locomotion in cattle and for fearfulness, adaptability and feather pecking in poultry were reviewed by Jensen *et al.* (2008). As phenotypic data are, in principle, no longer required after the association between a QTL marker and phenotype has been established, MAS is particularly useful for traits with low heritabilities that are highly influenced by environmental variation, those that are measured in one sex only, for selection at an early age before the phenotype can be measured, and if a trait is difficult to measure or requires sacrificing the animal (van der Werf, 2000). Challenges to the commercial application of MAS in livestock at these early stages were reviewed by Dekkers (2004), who also indicated that the improvement in response to MAS compared with traditional methods is not expected to be large, because the identified QTLs typically only capture a limited proportion of the genetic variance for a trait, and breeding goals typically include many traits. In addition, the typically small sample sizes of QTL detection studies resulted in many false-positive and false-negative results, as well as overestimation of the effects of the detected QTLs (Xu, 2003; Dekkers, 2004). Therefore, although the ability to directly select genetic markers linked to QTLs was expected to lead to faster genetic gains, these limitations hampered the use of MAS (Dekkers, 2004).

Many of the limitations of MAS were overcome by the developments in human genetics, massive and cheap detection of SNPs and of genetic variation at a single nucleotide in the DNA allowed the construction of SNP arrays with thousands or hundreds of thousands of markers for linkage disequilibrium mapping (Sachidanandam *et al.*, 2001). This technology was subsequently extended to domesticated animals (Goddard and Hayes, 2009). In contrast to traditional genetic markers, the dense coverage of the genomes by SNPs and high-density SNP genotyping facilitated the application of GWAS, which allowed SNPs to be found that are in such close proximity to the causal gene (i.e. they are so tightly linked) that the linkage phase does not have to be established in each family (Smith and Smith, 1993). It should be noted that most of the SNPs that are found to be significantly associated with phenotype in a GWAS do not themselves contribute to trait variation, as only a small proportion are found directly within the genes or regulatory sequences (Hayes and Goddard, 2010). Nevertheless, although GWAS allowed more QTLs to be detected that could be used across families, it did not solve the problem of false positives and false negatives and overestimation of significant marker effects. Specifically, because of limited statistical power to detect QTLs of small effects, GWAS only detect QTLs with moderate to large effects. As a result, the QTLs that are detected in GWAS together typically only explain a limited proportion of the genetic variance for the trait, which became known as 'The case of the missing heritability' in human genetics (Maher, 2008).

To mitigate the impact of the missing heritability problem, another paradigm shift in the use of genetic markers for breeding value estimation was needed, which led to the concept of genomic prediction and genomic selection (Meuwissen *et al.*, 2001). In contrast to MAS, in genomic selection, estimates of the effects of all genotyped markers across the genome are used to estimate an individual's breeding value, rather than estimates only of the effects of significant markers. Furthermore, to prevent the issue of overfitting when tens of thousands of markers are fitted simultaneously in the model, marker effects are fitted as random effects, with a variety of possible prior distributions of marker effects (Meuwissen *et al.*, 2001; Schmid and Bennewitz, 2017). Subsequent research showed that these marker-effects breeding value estimation models are equivalent to the standard BLUP mixed linear model that is used for estimation of breeding values based on pedigree, but with pedigree-based relationships among animals replaced by genomic relationships that are computed based on SNP genotypes (Habier *et al.*, 2007; VanRaden, 2008; Gomez-Raya *et al.*, 2021). With these advances, breeding values can now be based on the estimates of effects on phenotype of all genotyped SNPs (i.e. not only those with a significant effect) and by using genetic relationships among animals that are based on marker genotypes instead of pedigree (Hayes and Goddard, 2010; Legarra *et al.*, 2014). By using these methods for GWAS, this information can also be used to investigate the genetic architecture of trait phenotypes, including how many QTLs control the trait, with what effect sizes and where they are located (Schmid and Bennewitz, 2017).

The QTLs that have been identified for the major livestock species are compiled in the Animal QTL Database (www.animalgenome.org/cgi-bin/QTLdb/index; accessed 10 August 2021), which currently contains information on 208,982 QTLs across 1817 traits in cattle, pigs and chickens; for only a small number of these QTLs has the causal gene been firmly established (Blasco and Toro, 2014). Of the reported QTLs for cattle, pigs and chickens, 205, 973 and 1291, respectively, are related to behavioural traits, mostly associated with motivational processes (Canario *et al.*, 2013).

12.3.3 Transcriptomics, epigenomics, proteomics, metabolomics and adaptomics

A wide range of novel omics methods are now available that can be used to further identify genes that are associated with phenotypic variation of polygenic traits. In addition, omics methods can be used to elucidate the molecular mechanisms and pathways resulting from interactions between genes and networks that may explain the complex phenotypic expression of polygenic traits, such as between behaviour, physiological variables and the immune system (Zuk and Balenger, 2014; Calisi and MacManes, 2015). High-throughput transcriptome sequencing, or RNA sequencing, can be used to identify and quantify the expression of gene transcripts (mRNAs) (Bell and Aubin-Horth, 2010). Because gene expression is responsive to the environment, it provides a means to investigate genotype × environment interactions (Bell and Aubin-Horth, 2010). Subsequently, by treating the level of expression of a gene as another phenotype for QTL mapping or GWAS, regions on the genome that control this abundance can be identified, called expression QTLs (eQTLs) (Nettleton and Wang, 2006). In an application of these methods, Gley *et al.* (2018, 2019) linked QTLs (or causal variants) for coping behaviour in pigs in response to a backtest with transcriptomic variation in the adrenal glands, hypothalamus, amygdala and hippocampus. They identified several (candidate) genes involved in various biological pathways that were differentially expressed between pigs that differed in coping behaviour, demonstrating the interplay between genotype and gene-expression profiles. Fogelholm *et al.* (2019) used QTL analysis followed by eQTL analysis of gene-expression data from hypothalamus tissue to identify candidate genes for tonic immobility in chickens. Transcriptomics can also be used to associate variation in trait phenotypes with non-coding RNAs, including microRNAs (miRNAs), long non-coding RNAs (lncRNAs) and small interference RNAs (siRNAs) (Kasper *et al.*, 2020).

It is well known that stress exposure can result in the modulation (reducing or silencing) of DNA transcription and gene expression through epigenetic mechanisms (see Chapter 13, this volume), including DNA methylation and histone and chromatin reconfiguration (Skvortsova *et al.*, 2018). These modifications can occur in different life phases as a result of brief or chronic events of stress and can result in long-term modifications of behaviour and stress susceptibility (Jensen, 2014). Because epigenetic effects can span three generations (i.e. they can affect not only the epigenome of a gestating female but also that of her fetus, and even that of the fetal germ line (see Chapter 13, this volume)), this implies that the welfare of offspring may be shaped by welfare implications suffered by their mother.

Through post-transcriptional mechanisms, mRNAs are translated into proteins. Using protein abundance as a phenotypic trait for QTL mapping or GWAS, regions of the genome that control protein abundance can be identified, called protein QTLs (pQTLs) (Chick *et al.*, 2016). Besides its function in encoding proteins, genomic DNA is also closely involved in the genetic control of metabolite levels through the actions of enzyme and transporter protein mechanisms. Metabolites are the structurally, chemically and functionally heterogeneous ensemble of all small molecules present in cells, tissues and biofluids that are formed by metabolic pathways (see Chapter 11, this volume, on the microbiome), including (but not limited to) lipids, sugars, ions, metabolic intermediates and nucleic acids

(Cacciatore and Loda, 2015). The ensemble of an individual's metabolites, or its biochemical state, constitutes its metabolic phenotype, or metabotype, and reflects the individual's metabolic state at a particular time (Suhre and Gieger, 2012). Because it is situated at the interface between genotype and phenotype, the metabotype is considered an 'intermediate phenotype' that can function as a biomarker or risk factor for a phenotype (e.g. a biomarker for feed efficiency or a risk factor for disease). The genetic factors that control the abundance of metabolites, i.e. metabolomic QTLs (mQTLs), can be mapped by performing metabolome-wide association studies (mGWAS) (Suhre and Gieger, 2012; Suravajhala *et al.*, 2016). To date, the Livestock Metabolome Database (https://lmdb.ca/metabolites; accessed 16 June 2021) contains detailed information on 1202 metabolites in cattle, sheep, goats, horses and pigs. Together with genetics and other omics technologies, metabolomics can be used to develop next-generation phenotyping approaches to investigate the biological bases of complex traits, including those of stress susceptibility, behaviour and other welfare-related traits (Fontanesi, 2016; Kasper *et al.*, 2020).

The systematic application of omics methods to food animals has given rise to the field of 'feedomics'. Feedomics identifies the complex linkage between animal feed input and farm animal end products at a holistic system biological level, integrating omics methods applied to feed science, animal nutrition, animal production, health and welfare (Sun and Guan, 2018). Similarly, the concept of 'adaptomics', which has been proposed to describe the molecular (omics) solutions that plant species develop to adapt to local environmental demands (Koch *et al.*, 2018), could be developed for farm animals in the context of animal stress-coping behaviour and animal welfare.

12.3.4 Phenomics, phenome-wide association studies and big data analytics

The ability of GWAS to identify only a proportion of the genetic variance for a trait emphasized the importance of continued use of phenotyping approaches (Houle *et al.*, 2010). The use of novel, high-throughput phenotyping methods and developments in bioinformatics approaches for the use and analysis of these data gave rise to the discipline of 'phenomics', defined as 'the acquisition of high-dimensional phenotypic data on an organism-wide scale' (Houle *et al.*, 2010, p. 855). This is facilitated by rapid developments in automated technology for frequent trait recording (Wurtz *et al.*, 2019). The resulting high-density longitudinal data may enable the transition from GWAS to phenome-wide studies (PheWAS) for the analysis of complex traits. Whereas GWAS test tens of thousands of SNPs for associations with a single or a small number of phenotypes, PheWAS flip the direction of inference, with the simultaneous study of a broad range of phenotypic outcomes, as they are brought about by distinct genetic candidate gene(s) through the process of pleiotropy (Bush *et al.*,

2016). This method, which is currently applied in human research, may be particularly relevant to traits associated with animal welfare that are dependent on environmental exposure or that are defined as environmental (stress) response traits.

A holistic characterization of phenotypic and genetic (omics) variation, at the level of genes, cells, organs and the organism, will advance the thorough investigation of the links between genotype and phenotype in response to the environment of complex traits (Houle *et al.*, 2010). It is clear that these advances have resulted in datasets that are ever-increasing in size (Perakakis *et al.*, 2018). It is generally considered more correct in scientific research to state clearly defined hypotheses in advance and design experiments that clearly define what to test and how often to measure it, while 'data dredging' (and multiple testing) after data has been collected is considered to lack statistical rigour (Banerjee *et al.*, 2009). However, the vast developments in phenomics and bioinformatics, including data integration algorithms and machine learning tools, advocate for 'why not measure it all?' (Houle *et al.*, 2010). Development of computational tools for big data analytics is the next challenge to utilize this in-depth information for improved and more sustainable animal production and to improve animal welfare (Koltes *et al.*, 2019).

12.4 Pitfalls

Genetic mapping efforts are typically complicated by the effects of genotype × environment interactions on phenotypic trait expression. This is particularly true for welfare traits, as welfare is related to the animal's response to environmental situations or events. Although QTL mapping efforts can be successful, it is fair to acknowledge that most QTLs that have been and will be identified with a significant effect on welfare traits are not expected to explain much more than 1–2% of the genetic variance (Hayes and Goddard, 2010; Barton *et al.*, 2017). For example, most of the 14 SNPs that were significantly associated with resilience phenotypes to weather variability in dairy goats explained less than 1% of the phenotypic variance in a study by Sánchez-Molano *et al.* (2019). Only in some cases have QTLs with major effects on behaviour been identified. For example, Cohen *et al.* (2001) identified significant QTLs for performance measures from an elevated plus maze in mice that together explained more than 30% of the phenotypic variation in behaviour. For these traits, genomic prediction is particularly relevant, as it allows the effects of all QTLs that affect a polygenic complex trait to be included. However, although genomic prediction is based on estimates of the effects of genotyped markers instead of phenotypic measurements to estimate an individual's breeding value, it is important to note that the relationship between phenotype and genetic markers has to be established first. Therefore, a critical first step towards the implementation of genomic prediction is the establishment of a 'training data set' of animals that are both phenotyped for target traits and genotyped with genome-wide markers to train the prediction model (Akdemir and Isidro-Sánchez, 2019). The benefit:cost ratio of acquiring advanced individual phenotype recording systems and the time required to manage, analyse and interpret the data generated should be taken into account when deciding whether welfare traits should be improved with genetic and genomic tools. Although an increase in production levels that results from selection of QTLs with even small effects is evident because of the direct economic benefits, the cost-effectiveness of characterizing QTLs for behavioural traits with a small effect may not be immediately evident (Flint, 2003).

12.5 Conclusions

With the exception of selection against fearfulness or for tameness (e.g. Fauré *et al.*, 2003; Haskell *et al.*, 2014), improvement of animal welfare through genetic selection for positive event appraisal in farm animals is unlikely in the near future because of limited knowledge. However, several coping response traits, as well as indicator traits of overall resilience and robustness to environmental stress and variability, can be considered for genetic improvement of animal welfare. Genomic selection is particularly relevant for these traits. Transcriptomic, epigenomic, proteomic, metabolomic and phenomic technologies and phenome-wide association methods may further disentangle the genetic architecture of genes and networks associated with animal welfare traits in response to the environment. Although farm animal welfare can be improved by genetic selection, this should not be undertaken without, first and foremost, keeping farm animals in production systems that provide them with acceptable welfare conditions through management decisions.

Further Reading

Literature that is particularly recommended for further reading is indicated in bold in the reference list.

Acknowledgements

This chapter was made possible by funding from ERA-Net SusAn project 'Sustainability of pig production through improved feed efficiency' (SusPig, https://suspig-eranet.jimdosite.com; accessed 10 August 2021) and project AGL2016-75942-R (IBERFIRE).

References

Agnvall, B., Jöngren, M., Strandberg, E. and Jensen, P. (2012) Heritability and genetic correlations of fear-related behaviour in Red Junglefowl–possible implications for early domestication. *PLoS One* 7(4), e35162. DOI: 10.1371/journal.pone.0035162.

Akdemir, D. and Isidro-Sánchez, J. (2019) Design of training populations for selective phenotyping in genomic prediction. *Scientific Reports* 9(1), 1446. DOI: 10.1038/s41598-018-38081-6.

Andersson, L., Haley, C.S., Ellegren, H., Knott, S.A., Johansson, M. *et al.* (1994) Genetic mapping of quantitative trait loci for growth and fatness in pigs. *Science* 263(5154), 1771–1774. DOI: 10.1126/science.8134840.

Banerjee, A., Chitnis, U.B., Jadhav, S.L., Bhawalkar, J.S. and Chaudhury, S. (2009) Hypothesis testing, type I and type II errors. *Industrial Psychiatry Journal* 18(2), 127–131. DOI: 10.4103/0972-6748.62274.

Barton, N.H., Etheridge, A.M. and Véber, A. (2017) The infinitesimal model: definition, derivation, and implications. *Theoretical Population Biology* 118, 50–73. DOI: 10.1016/j.tpb.2017.06.001.

Bell, A.M. and Aubin-Horth, N. (2010) What can whole genome expression data tell us about the ecology and evolution of personality? *Philosophical Transactions of the Royal Society of London B: Biological Sciences* 365(1560), 4001–4012. DOI: 10.1098/rstb.2010.0185.

Berghof, T.V.L., Poppe, M. and Mulder, H.A. (2019) Opportunities to improve resilience in animal breeding programs. *Frontiers in Genetics* 9, 692. DOI: 10.3389/fgene.2018.00692.

Blasco, A. and Toro, M.A. (2014) A short critical history of the application of genomics to animal breeding. *Livestock Science* 166, 4–9.

Bohmanova, J., Misztal, I. and Cole, J.B. (2007) Temperature-humidity indices as indicators of milk production losses due to heat stress. *Journal of Dairy Science* 90(4), 1947–1956. DOI: 10.3168/jds.2006-513.

Boissy, A. (1995) Fear and fearfulness in animals. *Quarterly Review of Biology* 70, 165–191.

Boissy, A., Manteuffel, G., Jensen, M.B., Moe, R.O., Spruijt, B. *et al.* (2007) Assessment of positive emotions in animals to improve their welfare. *Physiology & Behavior* 92(3), 375–397. DOI: 10.1016/j.physbeh.2007.02.003.

Botstein, D., White, R.L., Skolnick, M. and Davis, R.W. (1980) Construction of a genetic linkage map in man using restriction fragment length polymorphisms. *American Journal of Human Genetics* 32(3), 314–331.

Broom, D.M. (1991) Animal welfare: concepts and measurement. *Journal of Animal Science* 69(10), 4167–4175. DOI: 10.2527/1991.69104167x.

Bush, W.S., Oetjens, M.T. and Crawford, D.C. (2016) Unravelling the human genome–phenome relationship using phenome-wide association studies. *Nature Reviews* 17(3), 129–145. DOI: 10.1038/nrg.2015.36.

Cacciatore, S. and Loda, M. (2015) Innovation in metabolomics to improve personalized healthcare. *Annals of the New York Academy of Sciences* 1346(1), 57–62. DOI: 10.1111/nyas.12775.

Calisi, R.M. and MacManes, M.D. (2015) RNAseq-ing a more integrative understanding of animal behavior. *Behavioral Sciences* 6, 65–68.

Canario, L., Mignon-Grasteau, S., Dupont-Nivet, M. and Phocas, F. (2013) Genetics of behavioural adaptation of livestock to farming conditions. *Animal* 7(3), 357–377. DOI: 10.1017/S1751731112001978.

Castanheira, M.F., Conceição, L.E.C., Millot, S., Rey, S. and Bégout, M.L. (2017) Coping styles in fish: consequences for aquaculture. *Reviews in Aquaculture* 9, 23–41.

Chick, J.M., Munger, S.C., Simecek, P., Huttlin, E.L., Choi, K. *et al.* (2016) Defining the consequences of genetic variation on a proteome-wide scale. *Nature* 534(7608), 500–520. DOI: 10.1038/nature18270.

Cohen, R.M., Kang, A. and Gulick, C. (2001) Quantitative trait loci affecting the behavior of A/J and CBA/J intercross mice in the elevated plus maze. *Mammalian Genome* 12(7), 501–507. DOI: 10.1007/s00335-001-2047-5.

Colditz, I.G. and Hine, B.C. (2016) Resilience in farm animals: biology, management, breeding and implications for animal welfare. *Animal Production Science* 56, 1961–1983.

Dawkins, M.S. (1990) From an animal's point of view: motivation, fitness and animal welfare. *Behavioral and Brain Sciences* 13, 1–9.

Dawkins, M.S. (2008) The science of animal suffering. *Ethology* 114, 937–945.

Dekkers, J.C.M. (2004) Commercial application of marker- and gene-assisted selection in livestock: strategies and lessons. *Journal of Animal Science* 82 (E-Suppl), E313–E328. DOI: 10.2527/2004.8213_supplE313x.

Dekkers, J.C.M. and Hospital, F. (2002) The use of molecular genetics in the improvement of agricultural populations. *Nature Reviews Genetics* 3(1), 22–32. DOI: 10.1038/nrg701.

Desire, S., Turner, S.P., D'Eath, R.B., Doeschl-Wilson, A.B., Lewis, C.R.G. *et al.* (2016) Prediction of reduction in aggressive behaviour of growing pigs using skin lesion traits as selection criteria. *Animal* 10(8), 1243–1253. DOI: 10.1017/S1751731116000112.

Dochtermann, N.A., Schwab, T., Anderson Berdal, M., Dalos, J. and Rayouté, R. (2019) The heritability of behavior: a meta-analysis. *The Journal of Heredity* 110, 403–410.

Elgersma, G.G., de Jong, G., van der Linde, R. and Mulder, H.A. (2018) Fluctuations in milk yield are heritable and can be used as a resilience indicator to breed healthy cows. *Journal of Dairy Science* 101(2), 1240–1250. DOI: 10.3168/jds.2017-13270.

Ellen, E.D., Rodenburg, T.B., Albers, G.A.A., Bolhuis, J.E., Camerlink, I. *et al.* (2014) The prospects of selection for social genetic effects to improve welfare and productivity in livestock. *Frontiers in Genetics* 5, 377. DOI: 10.3389/fgene.2014.00377.

Falconer, D.S. and Mackay, T.F.C. (1996) *Introduction to Quantitative Genetics.* Longmans Green, Harlow, UK.

Fauré, J.M., Bessei, W. and Jones, R.B. (2003) Direct selection for improvement of animal well-being. In: Muir, W.M. and Aggrey, S.E. (eds) *Poultry Genetics, Breeding, and Biotechnology.* CAB International, Wallingford, UK, pp. 221–246.

Fernando, R. and Grossman, M. (1989) Marker assisted selection using best linear unbiased prediction. *Genetics Selection Evolution* 21, 467–477.

Finkemeier, M.-A., Langbein, J. and Puppe, B. (2018) Personality research in mammalian farm animals: concepts, measures, and relationship to welfare. *Frontiers in Veterinary Science* 5, 131. DOI: 10.3389/fvets.2018.00131.

Fisher, R.A. (1918) The correlation between relatives on the supposition of Mendelian inheritance. *Transactions of the Royal Society of Edinburgh* 52, 399–433.
Fisher, R.A. and Mackenzie, W.A. (1923) Studies in crop variation. II. The manorial response of different potato varieties. *Journal of Agricultural Science* 13, 311–320.
Flint, J. (2003) Analysis of quantitative trait loci that influence animal behavior. *Journal of Neurobiology* 54(1), 46–77. DOI: 10.1002/neu.10161.
Fogelholm, J., Inkabi, S., Höglund, A., Abbey-Lee, R., Johnsson, M. *et al.* (2019) Genetical genomics of tonic immobility in the chicken. *Genes* 10(5), 341. DOI: 10.3390/genes10050341.
Fontanesi, L. (2016) Merging metabolomics, genetics, and genomics in livestock to dissect complex production traits. In: Kadarmideen, H.N. (ed.) *Systems Biology in Animal Production and Health,* Vol. 1. Springer, Cham, Switzerland, pp. 43–62.
Friggens, N.C., Blanc, F., Berry, D.P. and Puillet, L. (2017) Review: deciphering animal robustness. A synthesis to facilitate its use in livestock breeding and management. *Animal* 11(12), 2237–2251. DOI: 10.1017/S175173111700088X.
Georges, M., Nielsen, D., Mackinnon, M., Mishra, A., Okimoto, R. *et al.* (1995) Mapping quantitative trait loci controlling milk production in dairy cattle by exploiting progeny testing. *Genetics* 139(2), 907–920.
Getabalew, M., Alemneh, T. and Akeberegn, D. (2019) Heritability and its use in animal breeding. *International Journal of Veterinary Science and Technology* 4, 1–5.
Glazier, D.S. (2009) Resource allocation patterns. In: Rauw, W.M. (ed.) *Resource Allocation Theory Applied to Farm Animal Production.* CAB International, Wallingford, UK, pp. 22–43.
Gley, K., Murani, E., Trakooljul, N., Zebunke, M., Puppe, B. *et al.* (2018) Transcriptome profiles of hypothalamus and adrenal gland linked to haplotype related to coping behavior in pigs. *Scientific Reports* 9, 13038.
Gley, K., Murani, E., Haack, F., Trakooljul, N., Zebunke, M. *et al.* (2019) Haplotypes of coping behavior associated QTL regions reveal distinct transcript profiles in amygdala and hippocampus. *Behavioural Brain Research* 372, 112038. DOI: 10.1016/j.bbr.2019.112038.
Goddard, M.E. and Hayes, B.J. (2009) Mapping genes for complex traits in domestic animals and their use in breeding programmes. *Nature Reviews Genetics* 10(6), 381–391. DOI: 10.1038/nrg2575.
Gomez-Raya, L. and Klemetsdal, G. (1999) Two-stage selection strategies utilizing marker-quantitative trait locus information and individual performance. *Journal of Animal Science* 77(8), 2008–2018. DOI: 10.2527/1999.7782008x.
Gomez-Raya, L., Klungland, H., Vage, D.I., Olsaker, I. and Fimland, E. (1998) Mapping QTL for milk production traits in Norwegian cattle. *Proceedings of the 6th World Congress on Genetics Applied to Livestock Production* 26, 429–432.
Gomez-Raya, L., Rauw, W.M. and Dekkers, J.C.M. (2021) Vector space algebra for scaling and centering relationship matrices under non-Hardy-Weinberg equilibrium conditions. *Genetics, Selection Evolution* 53(1), 7. DOI: 10.1186/s12711-020-00589-9.
Habier, D., Fernando, R.L. and Dekkers, J.C.M. (2007) The impact of genetic relationship information on genome-assisted breeding values. *Genetics* 177(4), 2389–2397. DOI: 10.1534/genetics.107.081190.
Haskell, M.J., Simm, G. and Turner, S.P. (2014) Genetic selection for temperament traits in dairy and beef cattle. *Frontiers in Genetics* 5, 368. DOI: 10.3389/fgene.2014.00368.

Hayes, B. and Goddard, M. (2010) Genome-wide association and genomic selection in animal breeding. *Genome* 53(11), 876–883. DOI: 10.1139/G10-076.

Held, S.D.E. and Špinka, M. (2011) Animal play and animal welfare. *Animal Behaviour* 81, 891–899.

Henderson, C.R. (1990) Statistical methods in animal improvement: historical overview. In: Gianola, D. and Hammond, K. (eds) *Advances in Statistical Methods for Genetic Improvement of Livestock. Advanced Series in Agricultural Sciences*, Vol. 18. Springer, Berlin and Heidelberg, pp. 2–14.

Houle, D., Govindaraju, D.R. and Omholt, S. (2010) Phenomics: the next challenge. *Nature Reviews* 11(12), 855–866. DOI: 10.1038/nrg2897.

Jensen, P. (2014) Behaviour epigenetics – the connection between environment, stress and welfare. *Applied Animal Behaviour Science* 157, 1–7.

Jensen, P., Buitenhuis, B., Kjaer, J., Zanella, A., Mormède, P. *et al.* (2008) Genetics and genomics of animal behaviour and welfare – challenges and possibilities. *Applied Animal Behaviour Science* 113, 383–403.

Kanitz, E., Tuchscherer, M., Otten, W., Tuchscherer, A., Zebunke, M. *et al.* (2019) Coping style of pigs is associated with different behavioral, neurobiological and immune responses to stressful challenges. *Frontiers in Behavioral Neuroscience* 13, 173. DOI: 10.3389/fnbeh.2019.00173.

Kasper, C., Ribeiro, D., de Almeida, A.M., Larzul, C. and Liaubet, L. (2020) Omics applications in animal science – a special emphasis on stress response and damaging behaviour in pigs. *Genes* 11, 920.

Koch, M.A., German, D.A., Kiefer, M. and Franzke, A. (2018) Database taxonomics as key to modern plant biology. *Trends in Plant Science* 23(1), 4–6. DOI: 10.1016/j.tplants.2017.10.005.

Koltes, J.E., Cole, J.B., Clemmens, R., Dilger, R.N., Kramer, L.M. *et al.* (2019) A vision for development and utilization of high-throughput phenotyping and big data analytics in livestock. *Frontiers in Genetics* 10, 1197. DOI: 10.3389/fgene.2019.01197.

Koolhaas, J.M. and Van Reenen, C.G. (2016) Interaction between coping style/personality, stress, and welfare: relevance for domestic farm animals. *Journal of Animal Science* 94(6), 2284–2296. DOI: 10.2527/jas.2015-0125.

Koski, S.E. (2014) Broader horizons for animal personality research. *Frontiers in Ecology and Evolution* 2, 70.

Lagisz, M., Zidar, J., Nakagawa, S., Neville, V., Sorato, E. *et al.* (2020) Optimism, pessimism and judgement bias in animals: a systematic review and meta-analysis. *Neuroscience and Biobehavioral Reviews* 118, 3–17. DOI: 10.1016/j.neubiorev.2020.07.012.

Lander, E.S. and Botstein, D. (1989) Mapping Mendelian factors underlying quantitative traits using RFLP linkage maps. *Genetics* 121(1), 185–199.

Legarra, A., Christensen, O.F., Aguilar, I. and Misztal, I. (2014) Single step, a general approach for genomic selection. *Livestock Science* 166, 54–65.

Lien, S., Gomez-Raya, L., Steine, T., Fimland, E. and Rogne, S. (1995) Associations between casein haplotypes and milk yield traits. *Journal of Dairy Science* 78(9), 2047–2056. DOI: 10.3168/jds.S0022-0302(95)76830-8.

Maher, B. (2008) Personal genomes: the case of the missing heritability. *Nature* 456(7218), 18–21. DOI: 10.1038/456018a.

Meffert, L.M., Hicks, S.K. and Regan, J.L. (2002) Nonadditive genetic effects in animal behavior. *The American Naturalist* 160 (Suppl 6), S198–S213. DOI: 10.1086/342896.

Mellor, D.J. (2016) Updating animal welfare thinking: moving beyond the "Five Freedoms" towards "a life worth living". *Animals* 6, 21.

Meuwissen, T.H., Hayes, B.J. and Goddard, M.E. (2001) Prediction of total genetic value using genome-wide dense marker maps. *Genetics* 157(4), 1819–1829.

Miglior, F., Fleming, A., Malchiodi, F., Brito, L.F., Martin, P. *et al.* (2017) A 100-year review: identification and genetic selection of economically important traits in dairy cattle. *Journal of Dairy Science* 100(12), 10251–10271. DOI: 10.3168/jds.2017-12968.

Mir, N.A., Rafiq, A., Kumar, F., Singh, V. and Shukla, V. (2017) Determinants of broiler chicken meat quality and factors affecting them: a review. *Journal of Food Science and Technology* 54(10), 2997–3009. DOI: 10.1007/s13197-017-2789-z.

Moberg, G.P. (2000) Biological response to stress: implications for animal welfare. In: Moberg, G.P. and Mench, J.A. (eds) *The Biology of Animal Stress.* CAB International, Wallingford, UK, pp. 1–22.

Mosing, M.A., Zietsch, B.P., Shekar, S.N., Wright, M.J. and Martin, N.G. (2009) Genetic and environmental influences on optimism and its relationship to mental and self-rated health: a study of aging twins. *Behavior Genetics* 39(6), 597–604. DOI: 10.1007/s10519-009-9287-7.

Mousseau, T.A. and Roff, D.A. (1987) Natural selection and the heritability of fitness components. *Heredity* 59, 181–197. DOI: 10.1038/hdy.1987.113.

Neave, H.W., Costa, J.H.C., Weary, D.M. and von Keyserlingk, M.A.G. (2020) Long-term consistency of personality traits of cattle. *Royal Society Open Science* 7(2), 191840. DOI: 10.1098/rsos.191849.

Nes, L.S. and Segerstrom, S.C. (2006) Dispositional optimism and coping: a meta-analytic review. *Personality and Social Psychology Review* 10(3), 235–251. DOI: 10.1207/s15327957pspr1003_3.

Nettleton, D. and Wang, D. (2006) Selective transcriptional profiling for trait-based eQTL mapping. *Animal Genetics* 37 (Suppl 1), 13–17. DOI: 10.1111/j.1365-2052.2006.01478.x.

Oliveira, A.F.S., Rossi, A.O., Silva, L.F.R., Lau, M.C. and Barreto, R.E. (2010) Play behaviour in nonhuman animals and the animal welfare issue. *Journal of Ethology* 28, 1.

Perakakis, N., Yazdani, A., Karniadakis, G.E. and Mantzoros, C. (2018) Omics, big data and machine learning as tools to propel understanding of biological mechanisms and to discover novel diagnostics and therapeutics. *Metabolism* 87, A1–A9. DOI: 10.1016/j.metabol.2018.08.002.

Putz, A.M., Harding, J.C.S., Dyck, M.K., Fortin, F., Plastow, G.S. *et al.* (2019) Novel resilience phenotypes using feed intake data from a natural disease challenge model in wean-to-finish pigs. *Frontiers in Genetics* 9, 660. DOI: 10.3389/fgene.2018.00660.

Rauw, W.M. (2012) Immune response from a resource allocation perspective. *Frontiers in Genetics* 3, 267. DOI: 10.3389/fgene.2012.00267.

Rauw, W.M. (2016) *Improving Welfare through Genetic Selection.* Frontiers Media, Lausanne, Switzerland.

Rauw, W.M. and Gomez-Raya, L. (2015) Genotype by environment interaction and breeding for robustness in livestock. *Frontiers in Genetics* 6, 310. DOI: 10.3389/fgene.2015.00310.

Rauw, W.M. and Gomez-Raya, L. (2017) Effects of stress on growth and development. From domestication to factory farming. In: Scanes, C.G. and Hill, R.A. (eds) *Biology of Domestic Animals*. Taylor & Francis, Abingdon, UK, pp. 241–265.

Rauw, W.M., Kanis, E., Noordhuizen-Stassen, E.N. and Grommers, F.J. (1998) Undesirable side effects of selection for high production efficiency in farm animals: a review. *Livestock Production Science* 56, 15–33.

Rauw, W.M., Johnson, A.K., Gomez-Raya, L. and Dekkers, J.C.M. (2017) A hypothesis and review of the relationship between selection for improved production efficiency, coping behavior, and domestication. *Frontiers in Genetics* 8, 134. DOI: 10.3389/fgene.2017.00134.

Rodenburg, T.B., Buitenhuis, A.J., Ask, B., Uitdehaag, K.A., Koene, P. *et al.* (2003) Heritability of feather pecking and open-field response of laying hens at two different ages. *Poultry Science* 82(6), 861–867. DOI: 10.1093/ps/82.6.861.

Sachidanandam, R., Weissman, D., Schmidt, S.C., Kakol, J.M., Stein, L.D. *et al.* (2001) A map of human genome sequence variation containing 1.42 million single nucleotide polymorphisms. *Nature* 409(6822), 928–933. DOI: 10.1038/35057149.

Sánchez-Molano, E., Kapsona, V.V., Ilska, J.J., Desire, S., Conington, J. *et al.* (2019) Genetic analysis of novel phenotypes for farm animal resilience to weather variability. *BMC Genetics* 20(1), 84. DOI: 10.1186/s12863-019-0787-z.

Sax, K. (1923) The association of size differences with seed-coat pattern and pigmentation in *Phaseolus vulgaris*. *Genetics* 8(6), 552–560.

Scheier, M.F. and Carver, C.S. (2018) Dispositional optimism and physical health: a long look back, a quick look forward. *The American Psychologist* 73(9), 1082–1094. DOI: 10.1037/amp0000384.

Schmid, M. and Bennewitz, J. (2017) Genome-wide association analysis for quantitative traits in livestock – a selective review of statistical models and experimental designs. *Archives Animal Breeding* 60, 335–346.

Schrooten, C., Bovenhuis, H., Coppieters, W. and Van Arendonk, J.A. (2000) Whole genome scan to detect quantitative trait loci for conformation and functional traits in dairy cattle. *Journal of Dairy Science* 83(4), 795–806. DOI: 10.3168/jds.S0022-0302(00)74942-3.

Skvortsova, K., Iovino, N. and Bogdanović, O. (2018) Functions and mechanisms of epigenetic inheritance in animals. *Nature Reviews Molecular Cell Biology* 19(12), 774–790. DOI: 10.1038/s41580-018-0074-2.

Smith, C. and Smith, D.B. (1993) The need for close linkages in marker-assisted selection for economic merit in livestock. *Animal Breeding Abstracts* 61, 197–204.

Stafleu, F.R., Grommers, F.J. and Vorstenbosch, J. (1996) Animal welfare: evolution and erosion of a moral concept. *Animal Welfare* 5, 225–234.

Stirling, D.G., Réale, D. and Roff, D.A. (2002) Selection, structure and the heritability of behaviour. *Journal of Evolutionary Biology* 15, 277–289.

Suhre, K. and Gieger, C. (2012) Genetic variation in metabolic phenotypes: study designs and applications. *Nature Reviews Genetics* 13(11), 759–769. DOI: 10.1038/nrg3314.

Sun, H.Z. and Guan, L.L. (2018) Feedomics: promises for food security with sustainable food animal production. *Trends in Analytical Chemistry* 107, 130–141.

Suravajhala, P., Kogelman, L.J.A. and Kadarmideen, H.M. (2016) Multi-omic data integration and analysis using systems genomics approaches: methods and applications in animal production, health and welfare. *Genetics Selection Evolution* 48, 38.

Tabery, J. (2008) R. A. Fisher, Lancelot Hogben, and the origin(s) of genotype-environment interaction. *Journal of the History of Biology* 41(4), 717–761. DOI: 10.1007/s10739-008-9155-y.

Thoday, J.M. (1961) Location of polygenes. *Nature* 191(4786), 368–370. DOI: 10.1038/191368a0.

van der Eijk, J.A.J., Lammers, A., Li, P., Kjaer, J.B. and Rodenburg, T.B. (2018) Feather pecking genotype and phenotype affect behavioural responses of laying hens. *Applied Animal Behaviour Science* 205(5), 141–150. DOI: 10.1016/j.applanim.2018.05.027.

van der Werf, J. (2000) Basics of marker assisted selection. In: Kinghorn, B. and van der Werf, J. (eds) *Identifying and Incorporating Genetic Markers and Major Genes in Animal Breeding Programs.* QTL Course, Belo Horizonte, Brazil, pp. 119–128. Available at: https://jvanderw.une.edu.au/brazilcourse.html (accessed 17 June 2021).

VanRaden, P.M. (2008) Efficient methods to compute genomic predictions. *Journal of Dairy Science* 91(11), 4414–4423. DOI: 10.3168/jds.2007-0980.

Wade, M.J., Bijma, P., Ellen, E.D. and Muir, W. (2010) Group selection and social evolution in domesticated animals. *Evolutionary Applications* 3(5–6), 453–465. DOI: 10.1111/j.1752-4571.2010.00147.x.

Walker, C. and Byers, J.A. (1991) Heritability of locomotor play in house mice, *Mus domesticus. Animal Behaviour* 42(6), 891–897. DOI: 10.1016/S0003-3472(05)80141-1.

Wechsler, B. (1995) Coping and coping strategies: a behavioural view. *Applied Animal Behaviour Science* 43(2), 123–134. DOI: 10.1016/0168-1591(95)00557-9.

Wurtz, K., Camerlink, I., D'Eath, R.B., Fernández, A.P., Norton, T. *et al.* (2019) Recording behaviour of indoor-housed farm animals automatically using machine vision technology: a systematic review. *PLoS One* 14(12), e0226669. DOI: 10.1371/journal.pone.0226669.

Xu, S. (2003) Theoretical basis of the Beavis effect. *Genetics* 165(4), 2259–2268. DOI: 10.1093/genetics/165.4.2259.

Zambra, N., Gimeno, D., Blache, D. and van Lier, E. (2015) Temperament and its heritability in Corriedale and Merino lambs. *Animal* 9(3), 373–379. DOI: 10.1017/S1751731114002833.

Zidar, J., Balogh, A., Favati, A., Jensen, P., Leimar, O. *et al.* (2017) A comparison of animal personality and coping styles in the red junglefowl. *Animal Behaviour* 130(1), 209–220. DOI: 10.1016/j.anbehav.2017.06.024.

Zuk, M. and Balenger, S.L. (2014) Behavioral ecology and genomics: new directions, or just a more detailed map? *Behavioral Ecology* 25(6), 1277–1282. DOI: 10.1093/beheco/aru172.

13 Epigenetic Tools to Identify Stress and Exposure to Detrimental Conditions in Production Animals

Carlos Guerrero-Bosagna*

Department of Integrative Biology, Uppsala University, Uppsala, Sweden; IFM Biology, Linköping University, Linköping, Sweden

13.1 Introduction

Epigenetic mechanisms have been studied extensively in many organisms in relation to diverse biological questions. In production animals, however, despite the enormous implications that epigenetics could have for welfare and production aspects, its application is incipient. One of the concerns regarding production practices is the stress that animals experience while in captivity. Despite regulations around the globe, stressful practices are very common in some regions. In animals undergoing stress, a plethora of hormonal responses are produced. Because of this, short-term stress in animals is usually determined by cortisol and epinephrine levels. However, stress hormones can show an acute but not a sustained exposure to stressful conditions. An important challenge is to determine the history of the exposure of an organism to stress in live specimens. Epigenetic marks in peripheral cells (e.g. blood cells, buccal swabs) can serve as epigenetic biomarkers of a history of stress with enormous potential to be converted to application in real situations within the production environment, particularly in hens and pigs.

13.1.1 Stress in farm animals

While meat production generates food and a livelihood for billions of people around the planet, it is also associated with environmental and

*carlos.guerrero.bosagna@ebc.uu.se

 Bridging Research Disciplines to Advance Animal Welfare Science: A practical guide (ed. I. Camerlink)
DOI: 10.1079/9781789247893.0013

health consequences (OECD, 2016). Given the cultural and economic importance of livestock production and the growing concern about climate change and animal welfare, it is extremely important to create tools that foment sustainable production practices. One of the concerns regarding production practices is the stress that animals experience while in captivity, brought about by the social and physical environment and management practices. Stress is a broad concept that encompasses both physiological and psychological responses. Although different levels of regulation exist around the globe, some detrimental practices are common in some regions, such as beak trimming of laying hens to avoid pecking, exposure to random illumination patterns to promote accretion, and exposure to arousing odours, loud noises, or temperatures that are too low or too high, among others (Morgan and Tromborg, 2006).

While the stress response is fundamental for the survival of animals and they can usually recover quickly from one-time stressful events without lasting damage, frequent or prolonged stress can have detrimental consequences for health. Long-term stress is shown to produce high blood pressure and consistently higher levels of corticosterone (Henry, 1992). In humans, post-traumatic stress leads to compromised immune systems, with reduced numbers of lymphocytes and T-cells, natural killer cell activity, and production of interferon-γ and interleukin-4 (Kawamura *et al.*, 2001). Stress also affects farm animals and is becoming increasingly associated with lower production parameters. In chickens, studies have shown clear associations between welfare and production parameters (Alm, 2015). For example, increased fearfulness relates to lower egg production (Barnett *et al.*, 1994) and weight (de Haas *et al.*, 2013), while poor feather coverage relates to higher basal plasma corticosterone levels (de Haas *et al.*, 2013). Additionally, stress and poor housing conditions affect the immune system. For example, dairy calves housed in smaller stalls present reduced lymphocyte proliferation in comparison with calves in larger stalls (Ferrante *et al.*, 1998), and poor environments have shown to affect immunity in pigs (Luo *et al.*, 2020). Thus, besides stress in farm animals being ethically relevant for animal welfare, it can also negatively affect the production parameters and quality of food available to consumers.

The hypothalamus–pituitary–adrenal (HPA) axis is central for the triggering of the stress response. After the hypothalamus collects and processes signals coming from the senses, it stimulates the pituitary gland to release adrenocorticotropic hormone (ACTH). ACTH, in turn, stimulates the production of the well-known stress hormones cortisol, epinephrine and norepinephrine (McClennen *et al.*, 1998). The action of these hormones rapidly increases blood rate and pressure, as well as mental and physical performance to cope with situations of threat, challenge or opportunity (Dhabhar, 2018). Because of this, short-term stress in animals is usually determined by cortisol and epinephrine levels (Ishibashi *et al.*, 2013; Müller *et al.*, 2013). Although the rapid increase in these stress hormones can show an acute exposure to stressful conditions, measuring long-term stress is quite different (Henry, 1992). To obtain a measure of long-term stress, one should wait until the basal levels of these hormones are re-established after the initial peak observed soon after stress exposure (McClennen *et al.*, 1998). Because capturing and handling the animal to obtain a sample will produce a short-term stress response, the levels of stress hormones observed immediately after capture will overshadow the hormonal effects of long-term stress. While basal stress hormone levels can then be measured after a prolonged time in which the animal is restrained, this method certainly cannot be feasibly applied on a large production scale.

Alternative measurements of stress hormones have been proposed, which include glucocorticoid levels in hair, feathers, urine and faeces (Davis and Maney, 2018). A method that determines long-term stress without relying on hormone measurements involves the analysis of the heterophil:lymphocyte (H:L) ratio. The H:L ratio changes a while after exposure to a stressor (e.g. handling), which allows it to be used to measure basal levels of H:L ratio values without the noise produced by the short-term stress response (Davis and Maney, 2018). However, this method is also unfeasible for application at the scale of animal production systems, as it requires cell microscopy and counting.

13.1.2 Measuring long-term stress using epigenetic measurements

Epigenetics could help to study the effects of long-term stress. Epigenetic modifications are small molecules that attach to the nucleotides that comprise DNA. One type of these molecules is methyl groups: one carbon with three hydrogens. When methyl groups attach to DNA nucleotides (mainly cytosines), they can regulate whether a gene is turned on or off. This epigenetic process is called DNA methylation. Epigenetic modifications can regulate gene expression and may be maintained after cell divisions (Skinner *et al.*, 2010). Each cell type in the body has a different pattern of methyl groups bound to the DNA. The pattern of binding of methyl groups to the DNA depends greatly on lifestyle and diet. Methyl groups, for example, are obtained from the vitamins and proteins we eat (Fig. 13.1).

Epigenetic mechanisms are fundamental players in the development of phenotypes and are sensitive to environmental influences (Jirtle and Skinner, 2007). Thus, external influences that affect early-life stages (pre- and post-birth) can have important consequences on epigenetic processes that ultimately shape the adult phenotype (Guerrero-Bosagna and Skinner, 2012). Several factors are among the environmental influences reported to interfere with epigenetic mechanisms during early development, such as endocrine disruption (Susiarjo *et al.*, 2013), inorganic chemicals (Kile *et al.*, 2014), nutritional compounds (Dolinoy *et al.*, 2007; Guerrero-Bosagna *et al.*, 2008) and stressful conditions (Fagiolini *et al.*, 2009).

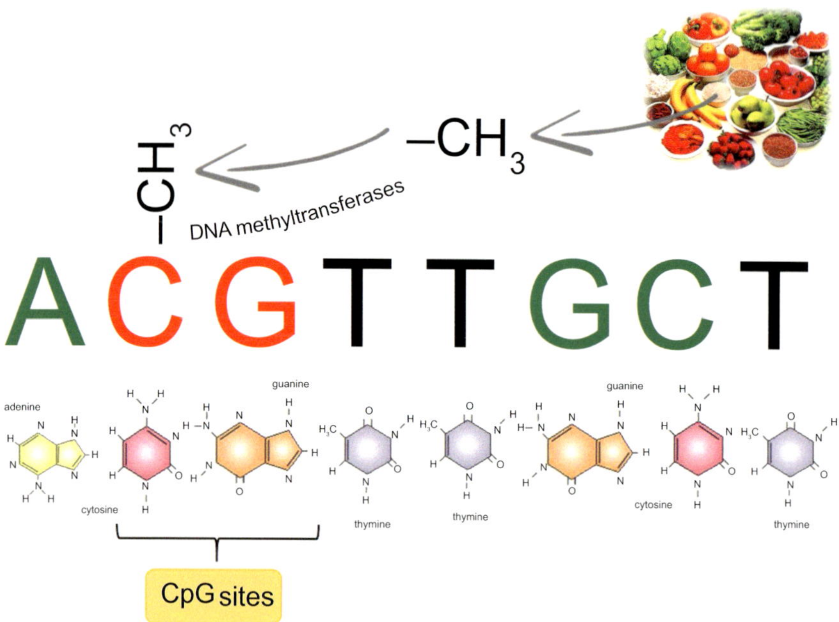

Fig. 13.1. Schematic representation of the epigenetic mechanism of DNA methylation. © C. Guerrero-Bosagna.

Epigenetic mechanisms are reported to be important along the HPA axis to regulate stress response and resilience (Zannas and West, 2014). Epigenetic marks in peripheral cells (e.g. blood cells, buccal swabs) could serve as epigenetic biomarkers of a history of stress (Provençal *et al.*, 2012; Wang *et al.*, 2012). For example, macaques that experienced maternal separation exhibit differential methylation in both the prefrontal cortex and T-cells compared with those that maintained the maternal bond (Provençal *et al.*, 2012). In children subjected to physical aggression, reduced brain serotonin levels correlate with higher DNA methylation levels in specific CpG dinucleotides of the gene *SLC6A4* in T-cells and monocytes (Wang *et al.*, 2012). The ability to detect long-term stress in the epigenome of peripheral cells has enormous potential to be converted for application in real situations within the production environment.

Although research on epigenetics has permeated many fields of biological research, from molecular biology to evolution (Stein and Davis, 2012), and has employed a variety of organism models, including laboratory rodents (Dolinoy *et al.*, 2007; Guerrero-Bosagna *et al.*, 2008; Susiarjo *et al.*, 2013), invertebrates (Lyko *et al.*, 2010), plants (Manning *et al.*, 2006) and yeast (Zhang *et al.*, 2013), the understanding of epigenetic mechanisms in farm animals is just beginning to develop. The consideration of epigenetics is essential to fully understand molecular mechanisms related to the phenotypes of production animals, and thus can lead to uncovering the molecular basis of behavioural traits not well predicted by genotype, such as those related to detrimental behaviours in farm animals (e.g. tail biting in pigs, feather pecking in chickens).

13.1.3 Effects of stressors across generations

Epigenetics can also provide mechanistic cues in processes of transgenerational epigenetic inheritance (TEI), which involves effects of environmental stimuli on exposed individuals and their unexposed descendants (Guerrero-Bosagna and Skinner, 2012). TEI is thought to be mediated by environmentally altered epigenetic marks in the gametes, which are transmitted across generations. Transgenerational effects have been reported in a variety of organisms, including laboratory rodents (Anway *et al.*, 2005; Guerrero-Bosagna *et al.*, 2010, 2012), fish (Baker *et al.*, 2014a, b; Bhandari *et al.*, 2015), quails (Leroux *et al.*, 2017), ducks (Brun *et al.*, 2015) and chickens (Goerlich *et al.*, 2012; Liu *et al.*, 2018). Although initial evidence of TEI was generated by environmental exposures acting in early developmental stages, recent reports in rodents have shown that juvenile or adult exposures could also affect the germ line epigenome with consequences for future generations (Carone *et al.*, 2010; Rodgers *et al.*, 2013; Dias and Ressler, 2014). In chickens, early social isolation has been shown to affect the HPA axis and gene expression in the thalamus/hypothalamus of offspring (Goerlich *et al.*, 2012). Transcriptomic changes in the hypothalamus of chickens raised with unpredictable light patterns are also observed in their offspring (Nätt *et al.*, 2009). In a study that subjected hens to a viral or bacterial infection (F_0), early development and body weight at 1 day of age were affected in the second-generation (F_2) offspring (Liu *et al.*, 2018). These effects observed in parents and their offspring suggest transmission of stress-induced germ-line epigenetic alterations to future generations.

13.2 Concepts

As discussed above, the main concepts involved in epigenetics are DNA modifications, DNA methylation and TEI.

- *Epigenetic modifications* comprise accessory chemical modifications of the DNA itself or of proteins that pack the DNA (i.e. histones), which are able to regulate gene expression and may be maintained after cell divisions. The epigenome is the set of all epigenetic modifications found in a given genome.

- *DNA methylation* is an enzymatic chemical reaction in which CH_3 (methyl) groups are covalently added to nucleotides, preferentially cytosines neighbouring guanines (in the 5′ to 3′ direction); these dinucleotides are known as CpG sites (Fig. 13.1).
- *Transgenerational epigenetic inheritance* is the process of inheritance in which environmental exposures modify the epigenome of the gametes, which will consequently influence the phenotype formation in the descendants. Heritable changes are observed in an unexposed offspring as a consequence of parental exposure to environmental insults.

13.3 Applications and Techniques

The application of epigenetics to measure long-term stress related to animal welfare issues has so far been investigated mainly in poultry and pigs. We initially showed that epigenetic marks in the red blood cells (which are nucleated in birds) of chickens reflect their previous rearing condition, such as in cages or open aviaries (Pértille *et al.*, 2017). These conditions are associated with differential fearfulness and cognitive abilities (Tahamtani *et al.*, 2015; Brantsæter *et al.*, 2016). In this initial experiment, however, epigenetic changes could not be correlated to a particular stress. In follow-up experiments, 4-day-old male chickens were exposed to a known stress, namely social isolation stress, incrementally for 3 weeks. This stress has previously been shown to have long-term and transgenerational effects (Goerlich *et al.*, 2012). Red blood cells were isolated from blood samples taken immediately after the stress treatment ended, and DNA methylation analysis was performed in a reduced fraction of the genome to interrogate for changes between the stress and non-stress conditions.

In order to investigate the repeatability of potential epigenetic marks across different environments, the experiment was performed in parallel at Linköping University in Sweden (SW population) and the Brazilian Agricultural Research Corporation (Embrapa Swine and Poultry National Research Center; BR population). The aim was to identify stress associated with DNA methylation profiles in red blood cells across these populations, despite the variable conditions to which birds are exposed in each facility, and of chickens coming from independent lineages. While the vast majority of the stress-related DNA methylation changes were specific to each geographical location, three significant ($P < 0.05$) differentially methylated regions overlapped between the BR and SW lineages. These sites are putative transversal epigenetic biomarkers of stress in chickens from different lineages, breeding programmes and biomes (Pértille *et al.*, 2020).

In order to simultaneously address genetic (single-nucleotide polymorphisms; see Chapter 12, this volume) and epigenetic (DNA methylation) variations in reduced genomes of a large number of animals in a cost-effective manner, we developed a technique that integrated two well-established methods: genotyping by sequencing (GBS) to investigate genetic variation, and methylated DNA immunoprecipitation (MeDIP) to investigate methylomic variation. The GBS method uses ligation steps in which barcode adapters identify individual samples, enabling the creation of sequencing libraries with DNA pooled from several individuals. Subsequently, MeDIP captures the methylated fraction of this pool of samples. Sequencing of these two fractions is then performed. This combination of methods, called GBS-MeDIP, was initially described in chickens (Pértille *et al.*, 2017, 2019, 2020; Guerrero-Bosagna *et al.*, 2020) and has recently been employed in pigs within the context of epigenetic and genetic modifications in sperm related to fertility (Pértille *et al.*, 2021). This technique allows the interrogation of DNA methylation in a large number of individuals at reasonable sequencing costs and will thereby contribute to the application of epigenetic tools in agriculture.

Besides using epigenetic changes as markers of long-term stress in peripheral cells, they can also help to understand the emergence of behavioural phenotypes in production animals, which are generally highly genetically homogeneous. In production chickens, we have recently shown that epigenetic patterns relate to behavioural patterns. Behavioural activity patterns were characterized in chickens through three classification schemes: (i) daily specific features of behavioural routines (entropy); (ii) daily spatiotemporal activity patterns (dynamic time warping); and (iii) social leading behaviour (leading index) (Guerrero-Bosagna *et al.*, 2020). DNA methylation in the nidopallium (a region in the avian brain) was analysed in connection to these behavioural activity patterns. The nidopallium was chosen for being functionally equivalent in birds to the prefrontal cortex in mammals. Interestingly, unique differentially methylated regions (DMRs) were identified between behavioural patterns emerging within these different classification schemes. Entropy had the higher number of DMRs. Functionally, promoters were shown to be mainly affected by DMRs found in the dynamic time warping. Pathway enrichment analysis of DMR-associated genes revealed that entropy related mainly to cell-cycle checkpoints, leading index to mitochondrial function, and dynamic time warping to gene expression regulation (Guerrero-Bosagna *et al.*, 2020). Based on our results in the nidopallium, this study suggests that different biological functions within neurons could be responsible for the emergence of distinct behaviour patterns, and that epigenetic variation within brain tissues could be an important factor to explain behavioural variation in genetically similar individuals.

13.4 Pitfalls

The use of epigenetic marks to identify physiological conditions such as stress has pitfalls, which need to be pointed out in order to be appropriately addressed. One of them is that epigenetic patterns vary with each cell type in the body. Thus, if DNA samples contain cells from different sources, it is hard to disentangle whether a detected epigenetic variation could rather be produced by variation in the cell types within a population of cells. Because of this, it is very important that, where possible, clean and purified cell types are employed for epigenetic analyses. In chickens, it is advantageous that red blood cells are nucleated, peripheral and easy to purify. However, in mammals, red blood cells are not nucleated, prompting the need for the use of alternative cell types. Future research should focus on what is the most suitable cell type to be investigated for epigenetic biomarkers of stress in production mammals.

Different methods exist for the assessment of reduced fractions of the genome. To the best of our knowledge, our GBS-MeDIP method is the cheapest per sample. However, it requires good knowledge of both laboratory work and bioinformatic analysis. The method also involves sequencing the samples using the Illumina platform, which is available at many sequencing facilities around the world. Another method that is frequently used to analyse the methylome of a reduced fraction of the genome is reduced representation bisulphite sequencing (RRBS). This method could also be a good alternative to use for preliminary investigations of putative epigenetic variants related to stress, particularly as it has been around for many years and is provided as a service by many facilities around the world.

However, despite the method used for initial screening, the development of epigenetic tools that are feasible to use in real production contexts is still ongoing. The complete development of these tools will require consistent confirmation, on a local genomic scale, of the methylomic changes observed under preliminary genome-wide interrogations.

13.5 Conclusion

Our current studies are giving epigenetics a central role as a tool to evaluate detrimental exposures and to understand the emergence of behavioural patterns in farm animals. We expect the results obtained in the near future to represent a tipping point in relation to the use of epigenetic tools within the production context. The results emerging from our ongoing experiments should pave the way for the development of toolboxes based on epigenetic marks that will result in improved animal welfare. The aim is that epigenetic toolboxes will identify whether animals have been exposed to detrimental conditions (e.g. stress, exposure to chemicals) in the production environment. This will allow rapid evaluation of previous exposures or health/welfare status using peripheral cells from live animals.

The potential incorporation of these tools in the production environment could help the transition towards animal production practices that fulfil the current and future demands in terms of animal welfare and sustainability.

Further Reading

For more background on the methods available to evaluate stress in farm animals, both in general and based on epigenetic measurements, see the references in bold.

Acknowledgements

C.G.-B. greatly appreciated funding by the Swedish Research Agency for Sustainable Development (FORMAS), grants 2017-00946 and 2018-01074.

References

Alm, M. (2015) Welfare indicators in chickens. PhD thesis, Swedish University of Agricultural Sciences, Uppsala, Sweden.

Anway, M.D., Cupp, A.S., Uzumcu, M. and Skinner, M.K. (2005) Epigenetic transgenerational actions of endocrine disruptors and male fertility. *Science* 308(5727), 1466–1469. DOI: 10.1126/science.1108190.

Baker, T.R., King-Heiden, T.C., Peterson, R.E. and Heideman, W. (2014a) Dioxin induction of transgenerational inheritance of disease in zebrafish. *Molecular and Cellular Endocrinology* 398(1–2), 36–41. DOI: 10.1016/j.mce.2014.08.011.

Baker, T.R., Peterson, R.E. and Heideman, W. (2014b) Using zebrafish as a model system for studying the transgenerational effects of dioxin. *Toxicological Sciences* 138(2), 403–411. DOI: 10.1093/toxsci/kfu006.

Barnett, J.L., Hemsworth, P.H., Hennessy, D.P., McCallum, T.H. and Newman, E.A. (1994) The effects of modifying the amount of human contact on behavioural, physiological and production responses of laying hens. *Applied Animal Behaviour Science* 41, 87–100.

Bhandari, R.K., vom Saal, F.S. and Tillitt, D.E. (2015) Transgenerational effects from early developmental exposures to bisphenol A or 17α-ethinylestradiol in medaka, *Oryzias latipes*. *Scientific Reports* 5, 9303. DOI: 10.1038/srep09303.

Brantsæter, M., Nordgreen, J., Rodenburg, T.B., Tahamtani, F.M., Popova, A. *et al.* (2016) Exposure to increased environmental complexity during rearing reduces fearfulness and increases use of three-dimensional space in laying hens (*Gallus gallus domesticus*). *Frontiers in Veterinary Science* 3, 14.

Brun, J.-M., Bernadet, M.-D., Cornuez, A., Leroux, S., Bodin, L. *et al.* (2015) Influence of grand-mother diet on offspring performances through the male line in Muscovy duck. *BMC Genetics* 16, 145. DOI: 10.1186/s12863-015-0303-z.

Carone, B.R., Fauquier, L., Habib, N., Shea, J.M., Hart, C.E. *et al.* (2010) Paternally induced transgenerational environmental reprogramming of metabolic gene expression in mammals. *Cell* 143(7), 1084–1096. DOI: 10.1016/j.cell.2010.12.008.

Davis, A. and Maney, D. (2018) The use of glucocorticoid hormones or leucocyte profiles to measure stress in vertebrates: what's the difference? *Methods in Ecology and Evolution* 9, 1556–1568.

de Haas, E.N., Kemp, B., Bolhuis, J.E., Groothuis, T. and Rodenburg, T.B. (2013) Fear, stress, and feather pecking in commercial white and brown laying hen parent-stock flocks and their relationships with production parameters. *Poultry Science* 92(9), 2259–2269. DOI: 10.3382/ps.2012-02996.

Dhabhar, F.S. (2018) The short-term stress response – mother nature's mechanism for enhancing protection and performance under conditions of threat, challenge, and opportunity. *Frontiers in Neuroendocrinology* 49, 175–192. DOI: 10.1016/j.yfrne.2018.03.004.

Dias, B.G. and Ressler, K.J. (2014) Parental olfactory experience influences behavior and neural structure in subsequent generations. *Nature Neuroscience* 17(1), 89–96. DOI: 10.1038/nn.3594.

Dolinoy, D.C., Huang, D. and Jirtle, R.L. (2007) Maternal nutrient supplementation counteracts bisphenol A-induced DNA hypomethylation in early development. *Proceedings of the National Academy of Sciences USA* 104(32), 13056–13061. DOI: 10.1073/pnas.0703739104.

Fagiolini, M., Jensen, C.L. and Champagne, F.A. (2009) Epigenetic influences on brain development and plasticity. *Current Opinion in Neurobiology* 19(2), 207–212. DOI: 10.1016/j.conb.2009.05.009.

Ferrante, V., Canali, E., Mattiello, S., Verga, M. and Sacerdote, P. (1998) Preliminary study on the effect of size of individual stall on the behavioural and immune reactions of dairy calves. *Journal of Animal and Feed Sciences* 7, 29–36.

Goerlich, V.C., Nätt, D., Elfwing, M., Macdonald, B. and Jensen, P. (2012) Transgenerational effects of early experience on behavioral, hormonal and gene expression responses to acute stress in the precocial chicken. *Hormones and Behavior* 61(5), 711–718. DOI: 10.1016/j.yhbeh.2012.03.006.

Guerrero-Bosagna, C. and Skinner, M.K. (2012) Environmentally induced epigenetic transgenerational inheritance of phenotype and disease. *Molecular and Cellular Endocrinol* 354(1–2), 3–8. DOI: 10.1016/j.mce.2011.10.004.

Guerrero-Bosagna, C.M., Sabat, P., Valdovinos, F.S., Valladares, L.E. and Clark, S.J. (2008) Epigenetic and phenotypic changes result from a continuous pre and post natal dietary exposure to phytoestrogens in an experimental population of mice. *BMC Physiology* 8, 17. DOI: 10.1186/1472-6793-8-17.

Guerrero-Bosagna, C., Settles, M., Lucker, B. and Skinner, M.K. (2010) Epigenetic transgenerational actions of vinclozolin on promoter regions of the sperm epigenome. *PLoS One* 5(9), e13100. DOI: 10.1371/journal.pone.0013100.

Guerrero-Bosagna, C., Covert, T.R., Haque, M.M., Settles, M., Nilsson, E.E. *et al.* (2012) Epigenetic transgenerational inheritance of vinclozolin induced mouse adult onset disease and associated sperm epigenome biomarkers. *Reproductive Toxicology* 34(4), 694–707. DOI: 10.1016/j.reprotox.2012.09.005.

Guerrero-Bosagna, C., Pértille, F., Gomez, Y., Rezaei, S., Gebhardt-Henrich, S.G. *et al.* (2020) DNA methylation variation in the brain of laying hens in relation to differential behavioral patterns. *Comparative Biochemistry and Physiology Part D, Genomics & Proteomics* 35, 100700. DOI: 10.1016/j.cbd.2020.100700.

Henry, J.P. (1992) Biological basis of the stress response. *Integrative Physiological and Behavioral Science* 27(1), 66–83. DOI: 10.1007/BF02691093.

Ishibashi, M., Akiyoshi, H., Iseri, T. and Ohashi, F. (2013) Skin conductance reflects drug-induced changes in blood levels of cortisol, adrenaline and noradrenaline in dogs. *The Journal of Veterinary Medical Science* 75(6), 809–813. DOI: 10.1292/jvms.12-0478.

Jirtle, R.L. and Skinner, M.K. (2007) Environmental epigenomics and disease susceptibility. *Nature Reviews Genetics* 8(4), 253–262. DOI: 10.1038/nrg2045.

Kawamura, N., Kim, Y. and Asukai, N. (2001) Suppression of cellular immunity in men with a past history of posttraumatic stress disorder. *The American Journal of Psychiatry* 158(3), 484–486. DOI: 10.1176/appi.ajp.158.3.484.

Kile, M.L., Houseman, E.A., Baccarelli, A.A., Quamruzzaman, Q., Rahman, M. *et al.* (2014) Effect of prenatal arsenic exposure on DNA methylation and leukocyte subpopulations in cord blood. *Epigenetics* 9(5), 774–782. DOI: 10.4161/epi.28153.

Leroux, S., Gourichon, D., Leterrier, C., Labrune, Y., Coustham, V. *et al.* (2017) Embryonic environment and transgenerational effects in quail. *Genetics, Selection Evolution* 49(1), 14. DOI: 10.1186/s12711-017-0292-7.

Liu, L., Yang, N., Xu, G., Liu, S., Wang, D. *et al.* (2018) Transgenerational transmission of maternal stimulatory experience in domesticated birds. *FASEB Journal* 32, 7002–7017. DOI: 10.1096/fj.201800762RR.

Luo, L., Jansen, C.A., Bolhuis, J.E., Arts, J.A.J., Kemp, B. *et al.* (2020) Early and later life environmental enrichment affect specific antibody responses and blood leukocyte subpopulations in pigs. *Physiology & Behavior* 217, 112799. DOI: 10.1016/j.physbeh.2020.112799.

Lyko, F., Foret, S., Kucharski, R., Wolf, S., Falckenhayn, C. *et al.* (2010) The honey bee epigenomes: differential methylation of brain DNA in queens and workers. *PLoS Biology* 8(11), e1000506. DOI: 10.1371/journal.pbio.1000506.

Manning, K., Tör, M., Poole, M., Hong, Y., Thompson, A.J. *et al.* (2006) A naturally occurring epigenetic mutation in a gene encoding an SBP-box transcription factor inhibits tomato fruit ripening. *Nature Genetics* 38(8), 948–952. DOI: 10.1038/ng1841.

McClennen, S.J., Cortright, D.N. and Seasholtz, A.F. (1998) Regulation of pituitary corticotropin-releasing hormone-binding protein messenger ribonucleic acid levels by restraint stress and adrenalectomy. *Endocrinology* 139(11), 4435–4441. DOI: 10.1210/endo.139.11.6311.

Morgan, K.N. and Tromborg, C.T. (2006) Sources of stress in captivity. *Applied Animal Behaviour Science* 102, 262–302.

Müller, S., Lehnert, K., Seibel, H., Driver, J., Ronnenberg, K. *et al.* (2013) Evaluation of immune and stress status in harbour porpoises (*Phocoena phocoena*): can hormones and mRNA expression levels serve as indicators to assess stress? *BMC Veterinary Research* 9, 145. DOI: 10.1186/1746-6148-9-145.

Nätt, D., Lindqvist, N., Stranneheim, H., Lundeberg, J., Torjesen, P.A. *et al.* (2009) Inheritance of acquired behaviour adaptations and brain gene expression in chickens. *PLoS One* 4(7), e6405. DOI: 10.1371/journal.pone.0006405.

OECD (Organisation for Economic Co-operation and Development) (2016) Meat consumption. Available at: https://data.oecd.org/agroutput/meat-consumption.htm (accessed 17 June 2021).

Pértille, F., Brantsæter, M., Nordgreen, J., Coutinho, L.L., Janczak, A.M. *et al.* (2017) DNA methylation profiles in red blood cells of adult hens correlate with their rearing conditions. *The Journal of Experimental Biology* 220(Pt 19), 3579–3587. DOI: 10.1242/jeb.157891.

Pértille, F., da Silva, V.H., Johansson, A.M., Lindström, T., Wright, D. *et al.* (2019) Mutation dynamics of CpG dinucleotides during a recent event of vertebrate diversification. *Epigenetics* 14(7), 685–707. DOI: 10.1080/15592294.2019.1609868.

Pértille, F., Ibelli, A.M.G., Sharif, M.E., Poleti, M.D., Fröhlich, A.S. *et al.* (2020) Putative epigenetic biomarkers of stress in red blood cells of chickens reared across different biomes. *Frontiers in Genetics* 11, 508809. DOI: 10.3389/fgene.2020.508809.

Pértille, F., Alvarez-Rodriguez, M., da Silva, A.N., Barranco, I., Roca, J. *et al.* (2021) Sperm methylome profiling can discern fertility levels in the porcine biomedical model. *International Journal of Molecular Sciences* 22(5), 2679. DOI: 10.3390/ijms22052679.

Provençal, N., Suderman, M.J., Guillemin, C., Massart, R., Ruggiero, A. *et al.* (2012) The signature of maternal rearing in the methylome in rhesus macaque prefrontal cortex and T cells. *The Journal of Neuroscience* 32(44), 15626–15642. DOI: 10.1523/JNEUROSCI.1470-12.2012.

Rodgers, A.B., Morgan, C.P., Bronson, S.L., Revello, S. and Bale, T.L. (2013) Paternal stress exposure alters sperm microRNA content and reprograms offspring HPA stress axis regulation. *The Journal of Neuroscience* 33(21), 9003–9012. DOI: 10.1523/JNEUROSCI.0914-13.2013.

Skinner, M.K., Manikkam, M. and Guerrero-Bosagna, C. (2010) Epigenetic transgenerational actions of environmental factors in disease etiology. *Trends in Endocrinology and Metabolism* 21(4), 214–222. DOI: 10.1016/j.tem.2009.12.007.

Stein, R. and Davis, D. (2012) Epigenetics: a fascinating field with profound research, clinical and public health implications. *The American Biology Teacher* 74, 213–223.

Susiarjo, M., Sasson, I., Mesaros, C. and Bartolomei, M.S. (2013) Bisphenol a exposure disrupts genomic imprinting in the mouse. *PLoS Genetics* 9(4), e1003401. DOI: 10.1371/journal.pgen.1003401.

Tahamtani, F.M., Nordgreen, J., Nordquist, R.E. and Janczak, A.M. (2015) Early life in a barren environment adversely affects spatial cognition in laying hens (*Gallus gallus domesticus*). *Frontiers in Veterinary Science* 2, 3. DOI: 10.3389/fvets.2015.00003.

Wang, D., Szyf, M., Benkelfat, C., Provençal, N., Turecki, G. *et al.* (2012) Peripheral *SLC6A4* DNA methylation is associated with *in vivo* measures of human brain serotonin synthesis and childhood physical aggression. *PLoS One* 7(6), e39501. DOI: 10.1371/journal.pone.0039501.

Zannas, A.S. and West, A.E. (2014) Epigenetics and the regulation of stress vulnerability and resilience. *Neuroscience* 264, 157–170. DOI: 10.1016/j.neuroscience.2013.12.003.

Zhang, Q., Yoon, Y., Yu, Y., Parnell, E.J. and Garay, J.A. (2013) Stochastic expression and epigenetic memory at the yeast *HO* promoter. *Proceedings of the National Academy of Sciences USA* 110, 14012–14017.

14

Using Principles from Behavioural Ecology to Address Animal Welfare Issues

Gareth Arnott[1]*, Victoria E. Lee[1,2] and Simon P. Turner[2]

[1]Queen's University Belfast, Belfast UK; [2]Scotland's Rural College (SRUC), Edinburgh, UK

14.1 Introduction

Behavioural ecology aims to understand how behaviour evolves and is adapted to particular ecological conditions, including both the physical and the social environment. The term 'adapted' refers to changes brought about during evolution by the process of natural selection. How well an animal is adapted to its environment can be revealed by studying fitness, which comprises an individual's ability to survive and reproduce to pass on genes to future generations. Given that behavioural ecology focuses on examining behaviour related to an animal's environment, there is considerable synergy with animal welfare science where the environment, both physical and social, is a key determinant of welfare. A key difference is that 'environment' in behavioural ecology relates to wild and natural habitats, whereas for animal welfare this often involves examining managed captive environments. One of the founders of the study of animal behaviour, Niko Tinbergen, provided a framework for the objective and scientific investigation of behaviour, now known as Tinbergen's four questions (Tinbergen, 1963):

1. Causation: What stimulus elicits the behaviour, and what is the physiological causation of behaviour?
2. Development or ontogeny: How does the animal's experience during growth and development influence behaviour?

*Corresponding author: g.arnott@qub.ac.uk

DOI: 10.1079/9781789247893.0014

3. Function: How does the behaviour aid survival and reproduction?
4. Evolution: How did the behaviour evolve?

These questions provide a framework for studying behaviour at different complementary levels. For example, we can pose the question, 'Why do male red deer stags fight?' In terms of causation, we could examine physiological factors influencing fighting such as levels of testosterone. In terms of development, we could examine how play fighting in early life influences later-life aggression, or how individuals' fighting experience influences their future behaviour. At the function level, we would determine whether winning fights allows males to secure access to breeding females and enhances their reproductive success. To answer the evolution question would involve phylogeny by examining the evolutionary history of the behaviour in related species.

The causal and developmental questions are referred to as *proximate* because they explain how an individual comes to behave in a given way during its lifetime. The function and evolution questions are known as *ultimate* because they explain why and how the individual has evolved the behaviour. From this framework, we can immediately see the value of studying the proximate questions for applied ethology and animal welfare. In addition, if we consider farm animals, although their human-managed environments release them from natural selection pressures, the ultimate questions remain relevant as these animals carry an 'evolutionary legacy' that influences behaviour (Andersen *et al.*, 2006; Dawkins, 2006). In other words, their behavioural repertoire is qualitatively similar to that of their wild ancestors, with the main differences tending to be quantitative in levels of behaviour shown (Price, 1999). For example, studies of domestic pigs under free-range conditions demonstrate that their behavioural repertoire is the same as their wild boar ancestors (Stolba and Wood-Gush, 1989). Understanding the evolutionary origins and development of animal behaviour is central to improving welfare. Here, principles from behavioural ecology have a useful role to play, particularly as this discipline has developed a strong theoretical basis. In this chapter, we provide an example of how concepts and theories from behavioural ecology can be used to further our understanding of an applied welfare issue. Specifically, using research from our group, we will apply game theory, originally developed in behavioural ecology, illustrating how this can be used to inform and address the welfare issue of regrouping aggression in pigs.

14.2 Concepts

Game theory

Game theory is a branch of applied mathematics, developed initially by economists to model human strategic decision making, with well-known examples such as the 'prisoner's dilemma'. Biologists were quick to spot the utility of game theory for behavioural ecology, with the game modelling the potential interactions of two or more individuals whose interests do not entirely coincide (Andersen *et al.*, 2006). The individuals participating in the game undertake strategies, with the choice made by one player being dependent not only on the direct consequences this may have but also on the behavioural feedback from other players. The first behavioural ecology application of game theory was to animal contests, with John Maynard Smith producing the now classic 'hawk/dove' game (Maynard Smith and Price, 1973). Related to animal contests, game theory examines how different strategies might be used by each contestant and how the winner is determined. The approach models contests in terms of costs and benefits to individuals, illustrating how individuals try to maximize fitness in contests. In this modelling approach, natural selection is expected to act in such a way that a population of animals will come to adopt an evolutionarily stable strategy (ESS). The ESS is the strategy that, if used by most members of the population, confers a higher fitness than any other strategy. For example, in the hawk/dove game (Smith and Price, 1973), contestants differ only in their choice of initial behaviour between 'aggressive' hawks, which use escalated injurious fighting, and 'non-aggressive' doves, which only use displays with no risk of injury. In this game, the ESS is a frequency-dependent mixture of hawk and dove, with the proportions of each depending on the costs and benefits associated with each strategy. In this example, costs relate to injury for hawks and time spent displaying for doves, while benefits relate to the value of the contested resource. By modelling changes in these parameters, it is possible to examine how the proportions of each strategy change at the ESS. For example, if the resource becomes more valuable, then there will be more hawks at the new ESS, while if the cost of fighting increases due to injury, then the new ESS will favour a greater proportion of doves. The hawk/dove model set the scene for the development of more realistic game theory models of animal contests that incorporated opponent asymmetries in fighting ability and rules about information gathering concerning the opponent and the contested resource. A major utility of the game theory approach is that the models yield testable predictions regarding contest outcomes and fighting durations for winners and losers, which have greatly assisted empiricists in their studies.

Resource holding potential

Resource holding potential (RHP) is a term that is used to describe the fighting ability of an animal. Victory frequently goes to the larger or heavier contestant, so body size or weight is often used as a proxy for RHP. This proxy measure of RHP is one that is accessible to researchers to measure during their studies, with size or weight typically correlated with strength and the ability to inflict injury. However, the larger or heavier animal is not always the winner, particularly when the relative difference between opponents is small. This is because there are a number of other aspects influencing RHP including energetic state, general physiological state, development of weaponry, experience, position and skill. In this regard, it is possible to generate a multivariate measure of RHP, reflecting the fact that RHP will be influenced by a number of factors.

Assessment strategies

In recent years, a major focus has been on investigating the information-gathering strategies used by animals during aggressive encounters. In a behavioural ecology context, selection is expected to favour contestants who achieve the best balance between the quality of information obtained about the costs and benefits of engaging in a contest and the efficiency of obtaining this information. In this context, information is required on RHP and resource value. As will be outlined in section 14.3 below, the assessment abilities of animals have practical relevance for our discussion regarding aggression as a welfare issue. Therefore, here we outline the main forms of assessment that are central to various game theory models and have subsequently been identified in empirical studies. *Assessment* refers to the gathering and use of information in order to make decisions. Existing contest game theory models can be grouped into three main types that differ in the information about RHP that contestants are able to gather.

Pure self-assessment

In this strategy, each contestant only has information about its own abilities or state, and fails to gather information about its opponent. The actions of the opponent do not inflict costs, although both opponents incur costs from their own actions. Thus, rivals persist purely in line with their own RHP such that weaker rivals (with lower RHP) will reach their limits first and give up. This strategy is a feature of the 'war of attrition without assessment' (Mesterton-Gibbons *et al.*, 1996) and 'energetic war of attrition' (Payne and Pagel, 1996, 1997) models.

Cumulative assessment

In the cumulative assessment model (CAM; Payne, 1998), both opponents also have information only about their own RHP, fighting until one reaches the threshold of costs that it is prepared to pay, with no direct information gathered about the opponent. However, unlike the other self-assessment models, with CAM, costs also accrue from the opponent's actions, with superior opponents (higher RHP) being better at inflicting costs. Thus, in the CAM, the decision to withdraw is influenced by both an individual's own RHP, with weaker individuals able to bear fewer costs before withdrawing, and the opponent's RHP, with high-quality individuals inflicting costs at a higher rate. As pure self-assessment and CAM involve no direct gathering of information about the opponent, a low and similar level of cognitive complexity has been suggested (Elwood and Arnott, 2012). This is in contrast to the mutual assessment strategy.

Mutual assessment

This involves an assessment of relative RHP difference, generally interpreted as individuals gathering information concerning the RHP of the opponent and comparing this against their own ability. This strategy is central to the 'sequential assessment model' (Enquist and Leimar, 1983) and the 'asymmetric war of attrition' (Parker and Rubenstein, 1981; Hammerstein and Parker, 1982). This strategy has intuitive appeal, as there is good evidence that humans are capable of making such rapid judgements (e.g. Sell *et al.*, 2009) and it reflects the role of aggressive encounters in our evolutionary history. This may partly explain why researchers frequently assume such capabilities in other species. Indeed, it has also been popularized in wildlife documentaries of competing animals with phrases such as 'they are sizing each other up'. With mutual assessment, contestants start with little or no information about the opponent, with early low-cost displays providing limited information, while later escalated phases of the contest provide increasingly accurate information on RHP. Therefore, in situations where there is a large difference in RHP between opponents, the decision to abandon the contest can be made quickly by the weaker individual, and at lower cost with less escalation, than if the difference in RHP is small. Mutual assessment therefore appears to offer a clear selective advantage that the contestant with the lower RHP can end the contest as soon as it perceives its inferiority, assuming that the benefits of gathering this information outweigh any costs involved in acquiring it. Thus, the individual can minimize costs such as time, energy and risk of injury from engaging in a contest that it would inevitably lose. In contrast, pure self-assessment and CAM appear inferior, as the contestant will always incur costs up to a threshold related to its RHP. This raises an important applied angle relating to animal welfare aspects of aggression. If we can equip managed animals with the ability for mutual assessment, this would offer a mechanism to establish outcomes and dominance relationships with less escalated aggression and lower associated contest costs. Evidence discussing this will be outlined in section 14.3 below. Mutual assessment has been suggested to involve a higher level of cognitive ability (Elwood and Arnott, 2012), and at present we have only limited understanding regarding which species have evolved the ability for mutual assessment. Indeed, a recent meta-analysis of contest studies found that the majority of species fight according to self-assessment models, although some show evidence of mutual assessment (Pinto *et al.*, 2019).

14.3 Techniques

14.3.1 Competitive interactions as a welfare issue

In a behavioural ecology context, agonistic interactions serve an important function by determining access to fitness-relevant resources. Indeed, since the outcome of contests determines the unequal distribution of resources, they are major drivers of natural and sexual selection. In many animal societies, the outcome of these interactions results in stable dominance hierarchies that reduce the need for further costly aggressive interactions. While contest behaviour is clearly adaptive in the wild, in captive settings, adequate levels of resources are usually provided to ensure biological functioning (e.g. sufficient provision of food), even if other needs are not being entirely met (e.g. the need to execute foraging behaviour). Importantly, despite the disconnect related to fitness, agonistic behaviour and dominance relationships remain major features of the social life of farm animals. For example, domestic pigs will fight to establish dominance relationships when mixed with unfamiliar individuals, a common occurrence during pig production. They will do so both in the absence of resources to compete over and in the presence of adequate resources to support survival, growth and reproduction (Camerlink *et al.*, 2015).

In pig farming, the aggression associated with regrouping is a major welfare issue (Greenwood *et al.*, 2014; Peden *et al.*, 2018). Aggression is substantially more severe in housed domestic pigs than in free-living domestic pigs or in the wild boar from which they originated, although it is qualitatively the same behaviour in form. This results in a physiological stress response (Coutellier *et al.*, 2007), risk of injury (Turner *et al.*, 2006) and reduced immune function (Morrow-Tesch *et al.*, 1994). These welfare impacts are detrimental to animal performance, compromising growth rates (Stookey and Gonyou, 1994; Coutellier *et al.*, 2007) and reproductive performance (Greenwood *et al.*, 2014). Given these issues, there is considerable research interest in managing aggression in pigs. To date, empirical studies have focused on quantifying variation in aggressive behaviour within and between groups (e.g. Foister *et al.*, 2018), with applied research considering the role of housing conditions and group size (e.g. Turner *et al.*, 2001). However, comparatively little is known about the fundamental mechanisms, information-gathering and decision-making processes that occur during aggressive encounters. Below, we outline the utility of using principles from behavioural ecology, including a game theory approach to provide fundamental insights into the causes of aggression in pigs, with the potential to inform and improve welfare. An important aspect of this is understanding the assessment strategy that pigs are capable of, the factors influencing this and how it can be optimized to enable contests to be resolved at lower levels of escalation. To achieve this, it is necessary to know how to discriminate between the alternative forms of assessment.

14.3.2 How to distinguish between assessment strategies

The assessment strategies outlined above do not differ in predicting contest winners but in when and how the decision to give up is made. Furthermore, a particular problem related to contests is that analyses used in earlier studies failed to use the correct approach to accurately distinguish between them (Arnott and Elwood, 2009). For example, with mutual assessment, when the opponents differ greatly in RHP, the loser should quickly determine that it is the weaker rival and likely to lose, and therefore give up early. In contrast, longer, more escalated contests should occur when the opponents are more evenly matched. Thus, a negative correlation between RHP difference and contest duration was used for many years to support the existence of a mutual assessment strategy. However, Taylor and Elwood (2003) demonstrated that the same relationship can occur with pure self-assessment and CAM. To discriminate between the assessment strategies, it is necessary to examine the effects of individual contestant RHP (as discussed above, this will typically be size or weight) on contest duration (Fig. 14.1).

With pure self-assessment, there will be a positive correlation between loser RHP and contest duration, while the winner's RHP will be weakly positively or not related to duration (Fig. 14.1b, c). In contrast, for both CAM and mutual assessment, while there is still a positive correlation between loser RHP and contest duration, this time the winner's RHP is negatively related to contest duration (Fig. 14.1d,e). Thus, we have a way to distinguish pure self-assessment from CAM and mutual assessment. To discriminate between the latter two alternatives, it is necessary to examine contests in which opponents within a contest are matched for RHP, with absolute RHP varying between contest pairs. With mutual assessment, the key factor that determines contest duration is relative ability; therefore, when examining RHP-matched contests, there will be no relationship between absolute RHP of the pairs and contest duration. However, with CAM, a key determinant of duration is the RHP of the loser, resulting in a positive correlation between pair RHP and duration. In terms of experimental design, it is therefore important to stage contests between randomly paired opponents that differ in RHP, as well as between pairs that are matched for RHP.

14.3.3 Applying a game theory perspective to improve farm animal welfare

In our research, we used the principles outlined above to investigate the assessment strategy being used by pigs during aggressive encounters. This has applied relevance, because if pigs are capable of mutual assessment, this would offer the potential for them to resolve contests at lower cost than if they are using a self-assessment strategy. In addition, the question arises

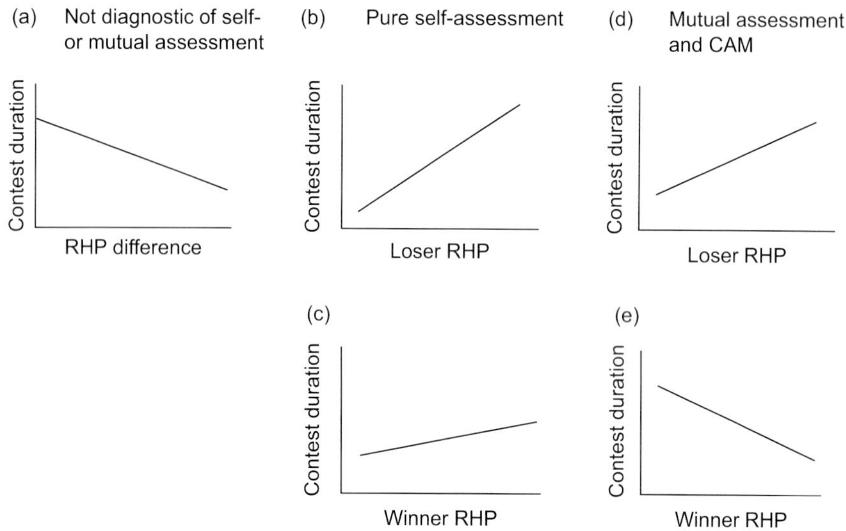

Fig. 14.1. Relationships between RHP and contest duration for alternative forms of assessment. (a) Not diagnostic of self- or mutual assessment; (b, c) pure self-assessment; (d, e) mutual assessment and cumulative assessment model (CAM). © G. Arnott.

regarding how animals develop their assessment abilities, and we therefore also explored the role of experience in shaping the assessment strategy that is used. Dyadic contests were staged between pigs at 10 weeks of age with body weight used as a measure of RHP and pigs randomly matched with an opponent of either similar or varying body weight (Camerlink *et al.*, 2017). To examine the influence of experience, at 12 weeks of age, 55% of the study population was subjected to group mixing with unfamiliar individuals in a manner similar to commercial practice and which results in repeated fights, while controls remained unmixed with familiar individuals. A week after this experience, a second dyadic contest was staged between unfamiliar individuals that were matched for experience. Results from the first contest, when all individuals had minimal prior experience of aggression, revealed evidence of pure self-assessment comprising a positive relationship between loser RHP and contest duration, with no effect of winner RHP. A similar pattern was observed in the second contest for the animals that had experienced regrouping, with contest costs increasing with loser RHP. In contrast, the control dyads in the second contest, which only had experience of one prior contest, showed evidence of mutual assessment, with the duration of the pre-escalation phase increasing with loser RHP and decreasing with increasing winner RHP. In a further complexity for these dyads, there was also evidence of a switch in assessment strategy from mutual assessment during the pre-escalation phase to no evidence of any form of assessment during the phase of escalated mutual fighting (Camerlink *et al.*, 2017). Thus, these results revealed that a brief and relatively mild contest experience can be beneficial in the development of mutual assessment, whereas a more intense experience may result in continued use of a self-assessment strategy.

Following on from the above, we investigated the potential of using an early-life experience to enhance assessment ability. Here, we used a manipulation, termed socialization, whereby two adjacent litters were allowed to mix in the pre-weaning period (Camerlink *et al.*, 2019), which mimics the age at which wild piglets first interact with non-littermates. This period of socialization is designed to facilitate the development of social skills that will be useful during encounters with other individuals in later life (Weller *et al.*, 2020). Compared with non-socialized controls, socialized pigs had shorter contests that were resolved with fewer injuries and they showed more ritualized display behaviour. They were also more decisive in a resident–intruder scenario, more quickly attacking an intruder in their home pen than controls (Weller *et al.*, 2019a). The benefits shown in the dyadic contests were also demonstrated following group mixing and highlight the applied welfare value of early-life socialization for reducing the impact of aggression in later life (Camerlink *et al.*, 2018). Regarding assessment strategy, there was evidence of a novel form of opponent assessment in the socialized group, with a positive relationship between winner RHP and fight duration. This is opposite to the prediction outlined above for mutual assessment. However, the original prediction assumes that fight motivation decreases with the assessment of increased opponent ability. In other words, the bigger the winner, the more quickly the loser perceives this and retreats. The finding of a positive relationship is consistent with increased fight motivation by the smaller contestant, as winner ability increases and is therefore still consistent with a form of mutual assessment, while being in the opposite direction to initial predictions. This form of assessment has now been documented by others (Kareklas *et al.*, 2019) and may be explained using other contest game theory models from a behavioural ecology context. For example, the Napoleon strategy illustrates that if RHP is not completely predictive of fight outcome, then individuals can be as aggressive as their larger opponents (Just and Morris, 2003). In addition, the 'desperado effect' illustrates that small individuals may attack because they have few alternative opportunities to obtain resources (Grafen, 1987). In our dyadic contests, pigs met in an empty arena lacking resources to compete over. However, the desperado effect may still be relevant if small animals had a high motivation to fight for other reasons, for example to avoid a large discrepancy in perceived dominance.

These studies examining the role of early-life socialization raise questions regarding the mechanisms for the beneficial effects on later-life aggression. We predicted that experience gained from play fighting during the socialization period could be a key aspect of this. While the frequency of play fighting did not differ between control and socialized pigs (Weller *et al.*, 2019a), it is likely that benefits accrued in the socialized group from being able to experience a greater variety of play partners including non-littermates. However, there were interesting sex differences in play fighting experience and effects on later-life contest outcomes. While males engaged in more play fighting than females, it was only females that showed a link to later-life contest success, with those that had engaged in more play fighting being more likely to win (Weller *et al.*, 2019b). It is thought that behavioural ecology can help explain this finding related to the evolutionary legacy of behaviour in wild boar. Under free-ranging conditions, female wild boar live in groups with a hierarchy enforced using aggression that is structurally similar to early-life play fighting. In contrast, male wild boar are more solitary and will engage in escalated contests with other males for access to breeding females. These contests can be very damaging, involving the use of their sharp tusks. Thus, there is a difference between male play fighting and the aggressive behaviour performed in later life. However, given the evolutionary persistence of social play in males, this indicates that males gain an alternative benefit from the performance of this behaviour. These benefits may include muscle development, cognition and gaining experience of responding to unexpected situations (Weller *et al.*, 2019b).

We also found that play fighting experience influenced later-life contest assessment strategy (Weller *et al.*, 2019b, 2020). By examining contests between dyads that had either experienced high or low levels of play fighting during the pre-weaning period, this revealed evidence of pure self-assessment in high-play dyads. This was evident in a positive relationship between loser weight (RHP) and contest cost, with a weaker positive association between winner weight and contest cost (Weller *et al.*, 2020). The low-play dyads did not meet the predictions of any assessment strategy. The results suggest that the increased early-life play frequency in the high-play group may have provided competitors with an enhanced prior understanding of their own RHP. A useful technical advance of this study was that it created a composite, multivariate measure of contest cost comprising seven variables spanning behavioural (contest duration), physical (skin lesions on winners and losers) and physiological (changes in blood glucose and lactate for winners and losers) costs. Such a multivariate measure more accurately reflects the intensity of the interaction. For example, relying only on contest duration as a measure of cost can be misleading, as some interactions can be relatively long but low in intensity, whereas others can be brief but highly escalated (Camerlink *et al.*, 2015).

Also drawing on principles from behavioural ecology, as part of our research we investigated how aggressiveness, a component of personality, influences contest behaviour (Briffa *et al.*, 2015). In pigs, aggressiveness, as assessed by a previously validated resident–intruder paradigm (Erhard and Mendl, 1997; D'Eath and Pickup, 2002), did not predict contest outcome. However, more aggressive opponents were more likely to initiate the first attack, with initiators being more likely to win contests that did not escalate to a fight (Camerlink *et al.*, 2015). These results were consistent with aggressiveness acting as an honest 'signal of intent' rather than a component of RHP. Aggressiveness also affects other aspects of contest behaviour in pigs (Camerlink *et al.*, 2016). Among the winners of contests, less aggressive winners spent more time in the pre-escalation phase of a contest compared with more aggressive winners, and showed fewer agonistic behaviours following their opponent's retreat. A lack of investment in pre-escalation assessment likely leads to unnecessary escalation, while excessive aggression towards a defeated opponent contributes to the welfare issues associated with contests. Management or breeding approaches that reduce aggressiveness would be expected to benefit welfare without influencing the effective establishment of dominance relationships.

14.4 Pitfalls

One of the major challenges of applying the game theory approach outlined above to better understanding aggression within an applied welfare context is that of correctly discriminating between the alternative forms of assessment. For this to be achieved, it is important to ensure that the study has an adequate sample size and uses an experimental design that stages agonistic interactions between both RHP-matched and unmatched dyads. Related to this, it is also important to ensure that a validated measure of RHP is used. For example, in pigs, body weight has been validated as a measure of RHP (Camerlink *et al.*, 2015). Furthermore, it is important to control for other relevant variables (e.g. sex, personality and stage of development of animals). If investigating the role of personality on contest behaviour, then it is important to determine that the personality measures chosen are repeatable.

It is then important to ensure that the data are analysed within the framework outlined above, using regression analyses to examine associations between winner and loser RHP and contest costs. Studies using this approach also frequently report inconclusive results, yet still favour a mutual assessment interpretation (Elwood and Arnott, 2012). This should be avoided, particularly as evidence is emerging that across species self-assessment strategies may be more common (Pinto *et al.*, 2019). In addition, if individuals are being examined across multiple contests, then the researchers need to be aware that there are important carryover experience effects, termed winner and loser effects (Hsu *et al.*, 2006). This refers to the fact that previous winners are more likely to win their next contest and previous losers to lose their next encounter. These effects diminish over time, and one way to avoid this potential confound is to ensure that the experimental design involves sufficient time gaps between staged encounters. In addition, researchers need to consider the ethical implications of studying aggression and staging contests. Appropriate measures need to be put in place to minimize the risk of harm. For example, in our studies, there were strict end points and time limits for encounters, with criteria to end an interaction if adverse effects occurred. It can also be possible to study aggression without staging direct physical encounters. For example, in fish, agonistic encounters can be staged with individuals displaying from separate tanks. Similarly, aggressive displays to a mirror have often been used in studies with fish. However, there is a note of caution with this approach, as recent studies indicate important differences when displaying to a real opponent compared with a mirror image.

More broadly, researchers keen to use principles and theories from behavioural ecology to address applied animal welfare issues need to carefully consider how the former are used. For example, the outcome measures in behavioural ecology studies typically relate directly or indirectly to fitness (survival and reproduction) measures. In contrast, within an applied setting, the animal performance and welfare measures will be different, and careful consideration needs to be given to the proxy measures used and their relevance to behavioural ecology theories.

14.5 Conclusions

The studies outlined above are generating knowledge that is providing a better understanding of the information-gathering and decision-making processes used during aggressive encounters. Additional themes derived from behavioural ecology that remain to be explored and that are likely to influence aggression and welfare in farm animals and other captive species include the role of cognitive ability (Reichert and Quinn, 2017), skill (Briffa and Lane, 2017), individual differences in assessment ability (Chapin *et al.*, 2019), affective state (Crump *et al.*, 2020) and social competence (Taborsky and Oliveira, 2012). The knowledge generated so far from using a behavioural ecology approach, together with that from future studies, can be applied to improve welfare by minimizing contest costs and facilitating the development of social skills. To date, the majority of game theory studies in farm animals have focused on pigs. There is good scope to translate this research to other farmed species, as well as more generally to species managed in captivity. For example, there are welfare concerns associated with aggression in farmed fish (Adams *et al.*, 2000), and cattle form dominance hierarchies, with regrouping and aggression influencing welfare (Boyle *et al.*, 2012). In addition, studies using farmed species enable a degree of control and longitudinal follow-up that is typically not possible when studying wild animals. On farms, the social environments of individual animals can be quantified and experimentally manipulated from early life, providing opportunities to address fundamental questions regarding the development of social behaviour. In this context, ideas from behavioural ecology can also inform our understanding of positive social interactions that could enhance welfare. For example, social network analyses have been widely used in behavioural ecology studies and are now beginning to be translated to a welfare context. In addition, play behaviour is beneficial for welfare (Boissy *et al.*, 2007), and an understanding of its evolutionary role and how it can be promoted would be useful within an applied context. Other potentially useful topics from behavioural ecology that could be translated to an applied welfare context include communication, foraging decisions and mate choice. Finally, welfare researchers should keep in mind that how captive and domestic animals respond within human-managed systems has been shaped by their evolutionary past. This has equipped them with a behavioural repertoire (e.g. vigilance, predator avoidance, aggression) to respond to fitness-relevant threats. Thus, behavioural ecology can be used to inform how we manage and understand the behavioural responses that occur within captive settings to ensure better welfare.

Further Reading

For further information on behavioural ecology and animal welfare, we recommend the references in bold in the reference list.

References

Adams, C., Huntingford, F., Turnbull, J., Arnott, S. and Bell, A. (2000) Size heterogeneity can reduce aggression and promote growth in Atlantic salmon parr. *Aquaculture International* 8(6), 543–549. DOI: 10.1023/A:1009255612529.

Andersen, I.L., Nævdal, E., Bøe, K.E. and Bakken, M. (2006) The significance of theories in behavioural ecology for solving problems in applied ethology – possibilities and limitations. *Applied Animal Behaviour Science* 97(1), 85–104. DOI: 10.1016/j.applanim.2005.11.020.

Arnott, G. and Elwood, R.W. (2009) Assessment of fighting ability in animal contests. *Animal Behaviour* 77(5), 991–1004. DOI: 10.1016/j.anbehav.2009.02.010.

Boissy, A., Manteuffel, G., Jensen, M.B., Moe, R.O., Spruijt, B. *et al.* (2007) Assessment of positive emotions in animals to improve their welfare. *Physiology & Behavior* 92(3), 375–397. DOI: 10.1016/j.physbeh.2007.02.003.

Boyle, A.R., Ferris, C.P. O'Connell, N.E. and O'Connell, N. (2012) Are there benefits in introducing dairy heifers to the main dairy herd in the evening rather than the morning? *Journal of Dairy Science* 95(7), 3650–3661. DOI: 10.3168/jds.2011-4362.

Briffa, M. and Lane, S.M. (2017) The role of skill in animal contests: a neglected component of fighting ability. *Proceedings of the Royal Society B: Biological Sciences* 284(1863), 20171596. DOI: 10.1098/rspb.2017.1596.

Briffa, M., Sneddon, L.U. and Wilson, A.J. (2015) Animal personality as a cause and consequence of contest behaviour. *Biology Letters* 11(3), 20141007. DOI: 10.1098/rsbl.2014.1007.

Camerlink, I., Turner, S.P., Farish, M. and Arnott, G. (2015) Aggressiveness as a component of fighting ability in pigs using a game-theoretical framework. *Animal Behaviour* 108(4), 183–191. DOI: 10.1016/j.anbehav.2015.07.032.

Camerlink, I., Arnott, G., Farish, M. and Turner, S.P. (2016) Complex contests and the influence of aggressiveness in pigs. *Animal Behaviour* 121(2), 71–78. DOI: 10.1016/j.anbehav.2016.08.021.

Camerlink, I., Turner, S.P., Farish, M. and Arnott, G. (2017) The influence of experience on contest assessment strategies. *Scientific Reports* 7(1), 14492. DOI: 10.1038/s41598-017-15144-8.

Camerlink, I., Farish, M., D'Eath, R.B., Arnott, G. and Turner, S.P. (2018) Long term benefits on social behaviour after early life socialization of piglets. *Animals* 8(11), 192. DOI: 10.3390/ani8110192.

Camerlink, I., Turner, S.P., Farish, M. and Arnott, G. (2019) Advantages of social skills for contest resolution. *Royal Society Open Science* 6(5), 181456. DOI: 10.1098/rsos.181456.

Chapin, K.J., Peixoto, P.E.C. and Briffa, M. (2019) Further mismeasures of animal contests: a new framework for assessment strategies. *Behavioral Ecology* 30(5), 1177–1185. DOI: 10.1093/beheco/arz081.

Coutellier, L., Arnould, C., Boissy, A., Orgeur, P., Prunier, A. *et al.* (2007) Pig's responses to repeated social regrouping and relocation during the growing-finishing period. *Applied Animal Behaviour Science* 105(1–3), 102–114. DOI: 10.1016/j.applanim.2006.05.007.

Crump, A., Bethell, E.J., Earley, R., Lee, V.E., Mendl, M. *et al.* (2020) Emotion in animal contests. *Proceedings of the Royal Society B: Biological Sciences* 287(1939), 20201715. DOI: 10.1098/rspb.2020.1715.

Dawkins, M.S. (2006) A user's guide to animal welfare science. *Trends in Ecology & Evolution* 21(2), 77–82. DOI: 10.1016/j.tree.2005.10.017.

d'Eath, R.B. and Pickup, H.E. (2002) Behaviour of young growing pigs in a resident–intruder test designed to measure aggressiveness. *Aggressive Behavior* 28(5), 401–415. DOI: 10.1002/ab.80010.

Elwood, R.W. and Arnott, G. (2012) Understanding how animals fight with Lloyd Morgan's canon. *Animal Behaviour* 84(5), 1095–1102. DOI: 10.1016/j.anbehav.2012.08.035.

Enquist, M. and Leimar, O. (1983) Evolution of fighting behaviour: decision rules and assessment of relative strength. *Journal of Theoretical Biology* 102(3), 387–410. DOI: 10.1016/0022-5193(83)90376-4.

Erhard, H.W. and Mendl, M. (1997) Measuring aggressiveness in growing pigs in a resident-intruder situation. *Applied Animal Behaviour Science* 54(2–3), 123–136. DOI: 10.1016/S0168-1591(97)00069-5.

Foister, S., Doeschl-Wilson, A., Roehe, R., Arnott, G., Boyle, L. *et al.* (2018) Social network properties predict chronic aggression in commercial pig systems. *PLoS One* 13(10), e0205122. DOI: 10.1371/journal.pone.0205122.

Grafen, A. (1987) The logic of divisively asymmetric contests: respect for ownership and the desperado effect. *Animal Behaviour* 35(2), 462–467. DOI: 10.1016/S0003-3472(87)80271-3.

Greenwood, E.C., Plush, K.J., van Wettere, W.H.E.J. and Hughes, P.E. (2014) Hierarchy formation in newly mixed, group housed sows and management strategies aimed at reducing its impact. *Applied Animal Behaviour Science* 160(9), 1–11. DOI: 10.1016/j.applanim.2014.09.011.

Hammerstein, P. and Parker, G.A. (1982) The asymmetric war of attrition. *Journal of Theoretical Biology* 96(4), 647–682. DOI: 10.1016/0022-5193(82)90235-1.

Hsu, Y., Earley, R.L. and Wolf, L.L. (2006) Modulation of aggressive behaviour by fighting experience: mechanisms and contest outcomes. *Biological Reviews* 81(1), 33–74. DOI: 10.1017/S146479310500686X.

Just, W. and Morris, M.R. (2003) The Napoleon complex: why smaller males pick fights. *Evolutionary Ecology* 17(5–6), 509–522. DOI: 10.1023/B:EVEC.0000005629.54152.83.

Kareklas, K., McMurray, R. and Arnott, G. (2019) Increased aggressive motivation towards formidable opponents: evidence of a novel form of mutual assessment. *Animal Behaviour* 153(1), 33–40. DOI: 10.1016/j.anbehav.2019.04.013.

Maynard Smith, J. and Price, G.R. (1973) The logic of animal conflict. *Nature* 246(5427), 15–18. DOI: 10.1038/246015a0.

Mesterton-Gibbons, M., Marden, J.H. and Dugatkin, L.A. (1996) On wars of attrition without assessment. *Journal of Theoretical Biology* 181(1), 65–83. DOI: 10.1006/jtbi.1996.0115.

Morrow-Tesch, J.L., McGlone, J.J. and Salak-Johnson, J.L. (1994) Heat and social stress effects on pig immune measures. *Journal of Animal Science* 72(10), 2599–2609. DOI: 10.2527/1994.72102599x.

Parker, G.A. and Rubenstein, D.I. (1981) Role assessment, reserve strategy, and acquisition of information in asymmetric animal conflicts. *Animal Behaviour* 29(1), 221–240. DOI: 10.1016/S0003-3472(81)80170-4.

Payne, R.J.H. (1998) Gradually escalating fights and displays: the cumulative assessment model. *Animal Behaviour* 56(3), 651–662. DOI: 10.1006/anbe.1998.0835.

Payne, R.J.H. and Pagel, M. (1996) Escalation and time costs in displays of endurance. *Journal of Theoretical Biology* 183(2), 185–193. DOI: 10.1006/jtbi.1996.0212.

Payne, R.J.H. and Pagel, M. (1997) Why do animals repeat displays? *Animal Behaviour* 54(1), 109–119. DOI: 10.1006/anbe.1996.0391.

Peden, R.S.E., Turner, S.P., Boyle, L.A. and Camerlink, I. (2018) The translation of animal welfare research into practice: the case of mixing aggression between pigs. *Applied Animal Behaviour Science* 204(3), 1–9. DOI: 10.1016/j.applanim.2018.03.003.

Pinto, N.S., Palaoro, A.V. and Peixoto, P.E.C. (2019) All by myself? Meta-analysis of animal contests shows stronger support for self than for mutual assessment models. *Biological Reviews* 94(4), 1430–1442. DOI: 10.1111/brv.12509.

Price, E.O. (1999) Behavioral development in animals undergoing domestication. *Applied Animal Behaviour Science* 65(3), 245–271. DOI: 10.1016/S0168-1591(99)00087-8.

Reichert, M.S. and Quinn, J.L. (2017) Cognition in contests: mechanisms, ecology, and evolution. *Trends in Ecology & Evolution* 32(10), 773–785. DOI: 10.1016/j.tree.2017.07.003.

Sell, A., Cosmides, L., Tooby, J., Sznycer, D., von Rueden, C. *et al.* (2009) Human adaptations for the visual assessment of strength and fighting ability from the body and face. *Proceedings of the Royal Society B: Biological Sciences* 276(1656), 575–584. DOI: 10.1098/rspb.2008.1177.

Stolba, A. and Wood-Gush, D.G.M. (1989) The behaviour of pigs in a semi-natural environment. *Animal Production* 48, 419–425.

Stookey, J.M. and Gonyou, H.W. (1994) The effects of regrouping on behavioral and production parameters in finishing swine. *Journal of Animal Science* 72(11), 2804–2811. DOI: 10.2527/1994.72112804x.

Taborsky, B. and Oliveira, R.F. (2012) Social competence: an evolutionary approach. *Trends in Ecology & Evolution* 27(12), 679–688. DOI: 10.1016/j.tree.2012.09.003.

Taylor, P.W. and Elwood, R.W. (2003) The mismeasure of animal contests. *Animal Behaviour* 65(6), 1195–1202. DOI: 10.1006/anbe.2003.2169.

Tinbergen, N. (1963) On aims and methods of ethology. *Zeitschrift fur Tierpsychologie* 20(4), 410–433. DOI: 10.1111/j.1439-0310.1963.tb01161.x.

Turner, S.P., Horgan, G.W. and Edwards, S.A. (2001) Effect of social group size on aggressive behaviour between unacquainted domestic pigs. *Applied Animal Behaviour Science* 74(3), 203–215. DOI: 10.1016/S0168-1591(01)00168-X.

Turner, S.P., Farnworth, M.J., White, I.M.S., Brotherstone, S., Mendl, M. *et al.* (2006) The accumulation of skin lesions and their use as a predictor of individual aggressiveness in pigs. *Applied Animal Behaviour Science* 96(3–4), 245–259. DOI: 10.1016/j.applanim.2005.06.009.

Weller, J.E., Camerlink, I., Turner, S.P., Farish, M. and Arnott, G. (2019a) Socialisation and its effect on play behaviour and aggression in the domestic pig (*Sus scrofa*). *Scientific Reports* 9(1), 4180. DOI: 10.1038/s41598-019-40980-1.

Weller, J.E., Camerlink, I., Turner, S.P., Farish, M. and Arnott, G. (2019b) Playful pigs: early life play-fighting experience influences later life contest dynamics. *Animal Behaviour* 158(3), 269–279. DOI: 10.1016/j.anbehav.2019.09.019.

Weller, J.E., Turner, S.P., Farish, M., Camerlink, I. and Arnott, G. (2020) The association between play fighting and information gathering during subsequent contests. *Scientific Reports* 10(1), 1133. DOI: 10.1038/s41598-020-58063-x.

15

Practical Considerations for the Use of Precision Livestock Farming to Improve Animal Welfare

Janice M. Siegford* and Kaitlin Wurtz

Michigan State University, East Lansing, Michigan, USA

15.1 Introduction

15.1.1 The application of precision livestock farming technology in animal sciences

Raising agricultural animals has become increasingly intensive as farmers work to efficiently produce meat, milk, eggs and fibre such as wool. Commercial farms typically house large numbers of animals together in relatively confined conditions, managed by a small number of stock people. This combination makes it hard for farm workers to identify, monitor and collect information from individual animals, particularly when they all look alike, as is the case, for example, with white laying hens or black beef cattle. Yet with growing societal concern over agricultural animals' quality of life, there is clearly a need to do so. One possible solution is to use technology to help human caretakers measure and manage the animals' behaviour, health, productivity and welfare.

Modern livestock and poultry farms routinely use technology to feed animals, remove manure, ventilate barns and monitor environmental conditions on farms. Precision livestock farming (PLF) is the specific use of technology to measure and manage at the level of the animal in real time to optimize the contributions of individual animals to production (Berckmans, 2017). Sensors collecting PLF data are focused on gathering information from the animals themselves, or from their immediate microenvironments, not just from the barn, pen or field at large. Thus, PLF moves beyond common barn-level technologies that measure temperature, humidity, airflow

*Corresponding author: siegford@msu.edu

 Bridging Research Disciplines to Advance Animal Welfare Science: A practical guide (ed. I. Camerlink)
DOI: 10.1079/9781789247893.0015

or water usage to capture animals' experiences of environmental conditions and management. The premise behind the focus on individuals is the recognition that even among animals of similar genetic stock, gender, age, etc. there is variation in how animals respond to situations. Using technology to craft management based on variation in individual animal responses rather than on pen averages has the potential to improve efficiency as well as to make interventions in a more timely and impactful fashion for animal welfare.

15.1.2 Potential of precision livestock farming to improve animal welfare

Individual monitoring and management

Animal welfare is an individual experience, as the subjective perception of quality of life belongs to an individual animal and not to a collective group of animals in a pen, herd or flock (Broom, 2008; Richter and Hintze, 2019). Monitoring farm animals with PLF may be able to supplement attention by stock people, directing their attention to animals in need of additional attention or specific care. In some cases, PLF systems may be capable of both monitoring the responses of the individual and then managing that animal as an individual in real time (Norton and Berckmans, 2018). To use PLF to improve welfare, however, we need systems that integrate data related to production, health and behaviour measures into comprehensible, actionable information that can be acted on either by the farmers or the PLF system (Guarino *et al.*, 2017).

Continuous assessment

There are a growing number of animal welfare certification programmes or assurance schemes designed to reassure consumers that animal products are produced under humane conditions. Current on-farm audits or assessments that examine compliance with standards are performed by human observers who typically visit each farm for only a few hours once a year. For reasons of practicality, standards often include only a few animal-based measures (e.g. body condition score or cleanliness), and these may be collected from a subset of all animals present on the farm (e.g. 100 cows from a selection of pens are scored for lameness). In contrast, PLF offers the potential for continuous, real-time monitoring of animal welfare (or at least of specific measures relevant to animal welfare), in some cases at the level of the individual animal (Berckmans, 2017). Essentially, it might be possible to 'observe' each animal from its arrival at a farm until it leaves and to safeguard its welfare on farm by making timely interventions based on PLF-generated data describing that animal's specific responses to management and environmental conditions (Norton and Berckmans, 2018).

15.1.3 Precision livestock farming in animal welfare research

Conditions on modern commercial farms that make it difficult for stockpersons to keep track of and manage individual animals also make it difficult for scientists to conduct research at commercial facilities. The task of collecting data from large numbers of homogeneous-looking animals requires financial and labour resources beyond the scope of most scientists' laboratories or research grants. In some cases, it is virtually impossible to track or observe focal animals using traditional methods of visual observation or marking animals for individual identification. However, various types of PLF sensing technology have been developed or used in research settings to answer questions ranging from individual animal time budgets, space use and activity levels to social behaviours such as aggression (see Siegford *et al.*, 2016, and Ellen *et al.*, 2019, for reviews of types of PLF used with various species or to answer particular animal behaviour and welfare-related questions).

Using PLF to capture data from animals also offers the possibility of conducting more welfare research in commercial farm settings. Using automated methods for collecting data provides researchers with the ability to collect data from more farms, from larger numbers of individual animals on each farm and over longer continuous periods. This would be an improvement over collecting data from a very small sample population, which often leads to issues related to how animals were chosen for sampling (Hampton *et al.*, 2019), or collecting limited amounts of data (i.e. over a very short time frame or only at certain time points), which may miss important events or changes (Daigle and Siegford, 2014). As will be discussed later, there are, however, constraints to using PLF for animal welfare research ranging from issues of battery life to having the coding skills needed to collect and process data in ways useful for answering research questions.

15.1.4 Precision livestock farming in the real world

As described earlier, use of PLF on commercial farms offers the potential to continuously monitor many (if not all) animals in real time, facilitating producers' ability to precisely tailor daily management to optimally meet the animals' needs and to detect problems earlier, resulting in more efficient use of resources and more effective interventions. At present, there are commercial PLF products available for species ranging from cattle to poultry. In particular, a wide range of PLF technologies are available for dairy cattle, including activity and rumination monitors and temperature sensors. Some are part of larger milking systems that also record information about a cow's milk production and possibly her weight or consumption of feed. There are also an increasing number of commercially available PLF products for pigs, most often tags coupled with feeding systems, such as electronic sow feeders, that capture information about amount of feed eaten, meal frequency and duration, and weight. Remote sensing systems are also available that use sound and vision to detect pigs, coughing or estimate their body weight, respectively. At present, there are relatively few commercially available PLF technologies for poultry, and most that have been developed are focused at the flock level, rather than on individual monitoring. For example, there are products that have been developed for broilers that use cameras to monitor the distribution of birds in the pen, coupled with scales for automatically capturing body weight to provide insight into how birds are behaving and whether there are problems related to food, water or temperature.

In most cases, these PLF systems generate data for farmers that require human action to manage or resolve. However, data processing by PLF systems in some cases will allow artificial intelligence (AI) to use information to make adjustments without the need for human intervention. For example, if data from cameras reveal that broilers are huddled rather than evenly distributed in the barn, the system could automatically increase the temperature rather than alerting the farmer.

As PLF becomes more available on farm, the potential to use animal outcomes data to assess compliance with animal welfare legislation, standards or assurance schemes will grow. Additionally, PLF can offer consumers and the public a closer look at the lives of animals and provide additional transparency and traceability in the food chain. While such transparency is beneficial for building trust between farmers and consumers, it must be balanced with protection of farmers' privacy and proprietary information related to the technology itself (Adams-Progar *et al.*, 2017).

15.2 Concepts

General concepts and definitions important for an understanding of PLF include the following.

Precision livestock farming

PLF is a multidisciplinary approach with the aim of automatically and continuously monitoring individual animals in real time. This can include monitoring production, health, welfare and environmental factors (Berckmans, 2017).

Body-worn sensors

These are sensors that are affixed to the animal. Examples include leg bands, ear tags, neck collars, halters or harnesses, adhesive patches, boluses in the gastrointestinal or reproductive tracts, and sensors implanted in tissues or organs.

Remote sensors

Remote sensors are placed within an animal's environment and capture visual, auditory or chemical data from the animal or group of animals, or information regarding animals' feed and water consumption or body weight, or environmental data such as ventilation, air temperature, humidity, dust or ammonia levels.

Radio-frequency identification

Radio-frequency identification (RFID) systems use a uniquely coded identifying transponder (i.e. a tag), which, when in proximity to an antenna, is registered by a reading unit and the tag's identity (along with a time stamp and the location of the antenna) is sent to a central computer. Tags in passive RFID systems must be activated and receive energy from powered antennae, while tags in active RFID systems have their own internal power source and broadcast information.

Machine learning

Machine learning is a field of study focused on developing computer programs that automatically improve themselves through experience (i.e. analysis of data). Machine learning is an application of AI.

Machine vision

This is a system designed to automate tasks related to understanding visual data present in images or video. The typical pipeline involves image capture and processing of images to decode salient features or classify data, and it concludes with decision making based on the information that was gained.

15.3 Techniques

A variety of sensing technologies can be used to identify and track individual animals, to monitor activity and behavioural patterns, or to measure physiological or health parameters in an automated way. Types of sensors range from devices that are fairly commonly used on farm, such as scales or temperature and humidity sensors, to novel biosensors that offer promise for measuring physiological variables such as cortisol or glucose levels or for detecting bacteria or viruses linked to disease. Thus, rapid technological developments in recent decades have enabled the creation of biosensors capable of non-invasively detecting almost any possible parameter of interest in an animal in an automated fashion. We have the ability to measure the body weight of a chicken using a simple scale (Turner *et al.*, 1983; Lokhorst, 1996), recognize the face of a pig using computer vision (Hansen *et al.*, 2018; Marsot *et al.*, 2020), detect jaw movements of a cow using pressure sensors, accelerometers or microphones (Andriamasinoro *et al.*, 2016), or sense the presence of disease organisms in the breath of an animal (Christensen *et al.*, 2011). Table 15.1 summarizes the technologies currently being used or being developed for use in PLF applications, along with some of their capabilities and drawbacks.

There are two general ways in which information is collected from the animals of interest using PLF. Sensors may be attached directly to an animal (e.g. accelerometers) or they may be mounted elsewhere in the environment and collect data remotely (e.g. cameras used for computer vision technologies). Body-worn sensors make it simpler to track and monitor specific individual animals, even when they are housed in large groups or are hard to distinguish visually from one another.

Because of the wide array of sensors available and the variety of ways in which they can be deployed, it is critical to first formulate a research objective and a hypothesis to narrow down which technologies and approaches will best meet those aims. In addition to considering the species of interest, their social and physical environment, and their management routines, it is also critical to consider the level of technical expertise among project personnel. Table 15.2 presents a list of questions that can be answered to help guide selection of PLF to meet research objectives.

It is also important to remember that, regardless of whether collecting data using PLF technology or more traditional observational methods (i.e. human observers directly recording live animals or manually decoding videos to extract information), the experiment must be designed in a way that ensures that the resulting data can be statistically analysed. Three critical factors that must be determined for research studies, including those using PLF, are: (i) the unit of analysis (i.e. animal, pen, barn or farm); (ii) the control condition or modelling for factors that could introduce variation beyond treatments of interest; and (iii) a large enough sample size to provide the power to detect a true difference if it exists.

15.3.1 Body-worn sensors

Wearable devices, including biosensors, can be used to monitor individual behaviour patterns, locations, or proximity to and interactions with other members of their group. Identification tags, particularly RFID, are becoming more common in commercial animal agriculture, but wearable sensor systems have not yet been widely developed for commercial use, with the

Table 15.1. General types of PLF technologies and considerations for their use.

Technology	Body worn or remote	Pros	Cons	Readiness	Additional reading
Two-dimensional camera	Remote	Readily available, affordable, robust	Small field of view; sensitive to extreme environments; storage requirements	Cameras: + Programs: 0/–	Ellen *et al*. (2019) on pros and cons of cameras; Wurtz *et al*. (2019a) on machine vision
Depth-sensing camera	Remote	Can overcome poor lighting or contrast	Limited field of view; sensitive to natural light; some use lasers (can compromise infrared camera footage if used concurrently)	Cameras: + Programs: 0/–	Wurtz *et al*. (2019a) for a review of how depth-sensing technology works and examples
Infrared camera	Remote	Low-light conditions; wide range of quality/price points	Very sensitive to environment; requires training and calibration; expensive to capture detail	Cameras: + Programs: +/0	Nääs *et al*. (2014) for a review; Boileau *et al*. (2019) for an example in pigs
Weighing scales	Remote	Many for farm use (affordable options); some have built-in tag readers	May only capture subset of population, depending on placement	Pigs, cattle, poultry: +	Xin and Liu (2017) for a review of weighing scales with PLF functionality in poultry
Microphone	Either	Group- or individual-level information	Requires species-specific knowledge; sometimes advanced data cleaning/ filtering	Group: + Individual: +/0 Programs: +/0/–	Norton and Berckmans (2018) and Andriamasinoro *et al*. (2016)

Continued

Table 15.1. Continued.

Technology	Body worn or remote	Pros	Cons	Readiness	Additional reading
Pressure sensor	Either	Non-invasive detection	Requires calibration and validation; can be challenging to interpret data	0/–	Chien and Chen (2018) on egg weights; Rifkin *et al*. (2019) on gait in goats; Ungar and Rutter (2006) on jaw movements of cattle
Accelerometer	Body worn	Small size; can be used in conjunction with other systems	Attachment to animal; battery life; requires additional coding to detect specific behaviours	+	Halachmi *et al*. (2019) for overview of accelerometers and review of applications
RFID tag	Body worn	Small size; wide variety of options	Range; number of tags read at a time; weight, expense and set-up requirements vary depending on tag type	+	Ellen *et al*. (2019) for description of pros and cons of various types of RFID
GPS tag	Body worn	Work outdoors	Cannot be used indoors	Cattle, sheep, goats: + Pigs, poultry: 0	Bailey *et al*. (2018) for a review of using GPS to track range use of livestock

Continued

Table 15.1. Continued.

Technology	Body worn or remote	Pros	Cons	Readiness	Additional reading
Infrared tag	Body worn	Small size; variety of attachment options; reduced false detection of animals	Requires line of sight between tag and reader	0/+	Siegford *et al*. (2016) for review of use of infrared tags in poultry
Chemical sensors	Either	Can detect wide range of parameters	Samples may become contaminated; may not sample continuously; may require off-board analysis	0/-	Neethirajan *et al*. (2017) and Halachmi *et al*. (2019) for reviews of chemical sensors and 'electronic noses'

+, Ready to use; 0, requires some modifications; –, needs further development; RFID, radio-frequency identification; GPS, global positioning system.

Table 15.2. Questions to consider when determining what type of precision livestock farming (PLF) technology to use. By answering the questions in this table, you will explicitly lay out the information you need to make a decision about what commercial products to buy (e.g. helping you ask the right questions of sales representatives) or what you will need to share with the computer scientists and engineers you work with if you develop technology for your use.

Question	Consider
What species/breed are you working with?	Age, stage, gender
What type of environment will your animals be monitored in?	Test/home environment, indoor/outdoor, group size
What information do you need?	Individual identity, location in environment, behaviours, general activity, proximity to others
Do you need data at the level of the individual or group?	Number of individuals in a group, similarity of appearance, environmental features that could result in occlusion
Is there a commercial PLF product that fits your needs?	Cost, availability of tech support, whether you can extract data as needed
What time frame is required?	Duration of recording, battery life, ability to recharge/replace batteries
How quickly do you want data or resulting information?	In real time/no rush
What personnel are needed on the project?	Expertise in ethology, computer science, engineering, animal science and veterinary medicine as needed to meet goals
What is your project budget?	PLF supplies and equipment – whether commercial or being developed in house, personnel needed/time to train/validate/test PLF
What is the purpose of the PLF?	Collect data, monitor animals, product development

exception of dairy cattle. We describe the use of body-worn sensors by their environmental context and provide equipment considerations.

Equipment considerations

The typical process for body-worn PLF begins with acquisition of data by the sensor (and possibly some on-board processing of data), transmission of the data from the sensor to a repository or more highly powered computational unit, and further processing and decision making. The following are critical pieces of equipment needed when using a sensor-based system:

- *Body-worn sensor.* When choosing a sensor, it is key to ask whether the sensing platform is sensitive enough to detect the variable of interest at the right rate but not so sensitive that data will be obscured by background noise. For example, if using an accelerometer to detect motion in a laying hen in an aviary, the accelerometer will need to detect motion in three axes, have a fast sampling rate and be able to detect gravitational forces in a biologically appropriate range to detect the behaviours, movements and even falls of a laying hen. The size of the sensor must be appropriate to the animal of interest – both in terms of overall weight and size relative to the animal, and with consideration to how and where it will be attached (or inserted). For example, in general a much larger sensor can be carried by a cow than on a hen, but if the cow is to wear the sensor on her ear rather than around her neck, it will need to be lighter in weight. Additionally, the sensor must be attached to the animal of interest in a way that does not interfere with the animal but at the same time ensures the necessary stability or orientation to support the working of the sensor. For example, accelerometers must maintain their orientation on the animal in order to interpret what acceleration on the *x*-, *y*- and *z*-axes mean in terms of animal movement or behaviour.
- *Energy source.* Passive RFID tags, infrared tags, or quick response (QR) or bar codes worn by animals do not require on-board energy sources; the tag or code is read or powered externally to move packets of information containing the identification, a time stamp and the reader location along to the host system. However, other wearable sensors, such as accelerometers, active RFID tags or ultra-wideband (UWB) tags that actively record, process or transmit data require an on-board power source. The sensor power source must last long enough to collect data over the desired time frame, ideally without the need to recapture the animal and recharge the sensor or change batteries. Battery life can be extended by using protocols that conserve energy (Halachmi *et al.*, 2019); for example, recording at high frequency only when animals are active, pre-processing data to reduce how much data is transmitted, or minimizing how often sensors send data or check in with base stations (Quwaider *et al.*, 2010; Liu *et al.*, 2018). Technological advances will ultimately enable use of renewable or regenerative sources for wearable sensors, such as body heat or kinetic movement, and such technologies are becoming increasingly robust and inexpensive (Song *et al.*, 2020) but are not yet mainstream.
- *Transmission technology.* Most sensors are limited in terms of how much data they can store or how much on-board processing of data is possible. Therefore, data need to be transmitted or downloaded on a regular and recurring basis from the sensor to free up storage space, to ensure data are not lost if the sensor is damaged or for additional processing of data. In some cases, sensors such as accelerometers may need to physically connect to a base station or computer to retrieve data; however, many

sensors transmit data wirelessly either to a local computer (processor) or to the cloud. The most common types of transmission methods used in PLF are radio (RFID and UWB tags) and light (infrared tags). The variation in these types of tags and their strengths and limitations have been further explored by Ruiz-Garcia and Lunadei (2011), Adrion *et al.* (2017) and Ellen *et al.* (2019).

- *Software.* Computer code, programmes or software are needed both to set up sensors for data collection and to process and extract data. Consideration must be given to how data will be collected (e.g. what pieces of information, at what frequency, for how long), what kinds of on-board processing of data will be done, what outputs will be generated, and how often the sensor will transmit data. For researchers without coding expertise, it will be important to ascertain how easy the sensor is for a novice to use; for example, whether the sensing platform has a good user interface and supporting documentation to help with setting up, data management or troubleshooting. It will be necessary to know if data collection or processing protocols can be tailored to collect the data of interest. Depending on the sensor's original purpose, it will also be important to explore whether the settings can be adjusted for the species/breed of interest (e.g. laying hen will have a different size and activity levels compared with a broiler) or life stages (e.g. calf versus cow). If using open-source software or a code made publically available as part of a publication or grant, someone on the research team will need to have the coding skills to adapt and validate the code for use in the current project.

Accelerometers

Accelerometers are a specific example of a body-worn sensor used to collect research data. At Michigan State University, it took 6 years to develop a wearable sensor for laying hens. The core team consisted of three applied ethologists and three electrical and computer engineers in addition to a number of students. Although the basis of the system was an off-the-shelf sensor card containing a two-axis accelerometer, a harness with plastic case had to be developed to attach the sensor to the hen, and many adjustments were needed (Fig. 15.1a). A substantial amount of time and a large number of conversations and visits to the farm were also needed to develop a shared language for talking about the project and to develop a trusting, collaborative team that understood the possibilities and constraints of the various disciplines (see Chapters 2 and 3, this volume, on cross-disciplinary research management). In the end, the final product could detect location and specific behavioural activities including dust bathing, feeding, walking and jumping down (Quwaider *et al.*, 2010; Banerjee *et al.*, 2012, 2014; Daigle *et al.*, 2012). Nevertheless, the system remained largely a prototype, limited to collecting data from relatively few birds over several days and

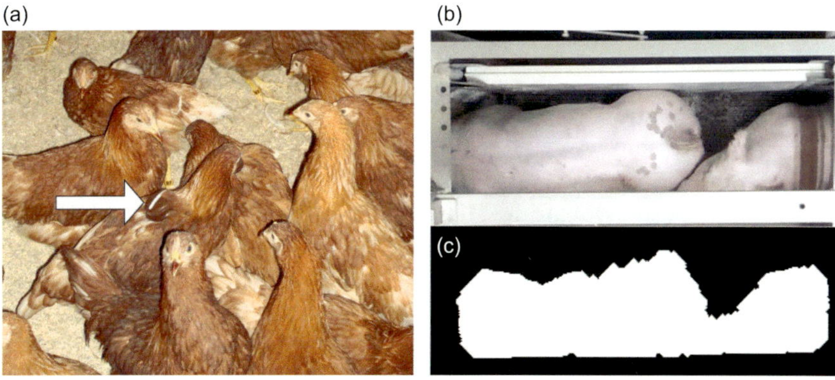

Fig. 15.1. (a) A hen wearing a camouflaged sensor (indicated by the arrow) equipped with an accelerometer in a group within a cage-free housing system. (b) An RGB (red, green and blue) image of two pigs in a feeder. (c) The same image after applying a mask creating a binary image in the process of automating detection of how many pigs are in the feeder, done as a precursor to understanding whether the presence of a second pig affects the first pig's feeding behaviour.

requiring substantial calibration and validation for each new task or housing system.

15.3.2 Machine vision

Machine vision is among the most popular approaches for capturing animal data remotely and relies on the use of multiple algorithms to extract useful information from images. Image processing can aid decision making on farm by automating routine welfare checks and environmental controls. However, the environment and the equipment used to capture these images need to be considered carefully to ensure compatibility with the proposed methods and desired outcome.

Environmental context

For digital camera use, the environmental lighting should be even and diffuse. Lights should be placed to avoid creating shadows. Natural light is uncontrollable and can interfere with the functioning of depth-sensing (i.e. three-dimensional) cameras and typically needs to be kept from entering indoor barn environments. The environment should also be free of reflective surfaces, which can create artificial bright spots in the image. Considering the surface behind the animal(s) (i.e. background) is critical when using digital cameras. In the case of ceiling-mounted cameras providing a top-down field of view, this would be the floor. Most algorithms perform optimally when the contrast between the animals and the background is maximized (e.g. filming pink Yorkshire pigs on a dark, matte background or Black Angus cattle against a light background) and background noise is minimized (e.g. there are no bits of hay or bedding, clumps of faeces or puddles of urine). The pen or area being observed should either not include objects that can block the view of the animals, such as feeders, pen dividers, nests, waterers or feed lines, or camera placement must be carefully considered to avoid occlusion. Some algorithms can track animals after they leave the camera view, but many cannot. Depth cameras, in particular, struggle to accurately capture an animal when it is partly or fully occluded. Many algorithms are limited in the number of animals they can detect or track. Confirm that the chosen method can account for the group size. Keep in mind that animals can occlude each other (e.g. when piglets lie on top of each other). Clustered animals may be difficult for algorithms to distinguish from one another. If animals cross paths or come together, the system may swap their identities when they separate.

15.3.2.2 Equipment considerations

The typical pipeline for machine vision begins with image acquisition, followed by extraction of information, and concludes with some sort of decision making. The following are critical pieces of equipment that need to be considered when designing a machine vision system:

- *Cameras.* The imager needs to provide a high enough quality image to work with the desired algorithms. Considerations need to be made regarding the number of cameras, where they will be located and the resulting view (e.g. top down, side, angled), as well as their type. There are many two-dimensional black-and-white as well as colour options available. Black-and-white cameras store pixel values ranging from 0 (black) to 255 (white), with the values in between equating to various shades of grey. They have the advantage of a higher light sensitivity (as there are no colour filters) and can work under lower light conditions and produce images with high sharpness. Colour cameras (i.e. RGB cameras) record the intensity of red, green and blue light in each pixel (Fig. 15.1b, c). Depth-sensing cameras are becoming increasingly available and create a three-dimensional image using a variety of methods including scanning-based triangulation, time of flight, stereoscopic and grid based. Other cameras capture non-visible energy such as those that detect energy emitted in infrared wavelengths (e.g. thermographic cameras). Cameras can be mounted on stable features of the environment and capture images only from one area, or can employ pan, tilt and zoom features. In other cases, cameras could be mounted on tracks or on drones to physically move around environments. Mobile cameras can increase the capture time of individuals in extensive environments and can allow for capture of entire groups of animals across vast landscapes, although they may require additional processing (probably machine learning) to account for changing backgrounds. Cameras have also been mounted on the animals themselves to help validate recording by other sensors, as in the cases of detecting grazing by sheep and horses (Terra-Braga *et al.*, 2018; Nunes *et al.*, 2020). Other camera considerations include their frame rate, resolution, weatherproof rating, field of view, low-light or night-vision capabilities, and whether they offer remote connectivity and control. Faster-moving animals might require higher-quality cameras to accurately track them or capture nuances of their behaviour.

- *Processor.* After images have been acquired, they will need to be processed to extract the relevant information. The processor(s) chosen can greatly affect the performance of a machine vision system. In some cases, a general-purpose computer may be sufficient, while in others a vision-specialized device will be needed. Graphics processing units (GPUs) can perform parallel processing on pixel data, accelerating data-processing times. Most depth-sensing systems include a built-in GPU, and there are a variety of advanced processors commercially available.
- *Software.* Software for machine vision applications range from open-source codes to commercially available products. Before getting started, it is important to consider the availability and quality of customer support, whether there is a graphical user interface and whether coding skills are needed to use the system, the accuracy and precision of the software, and the type of validation that will be required for the system to be used under the required conditions.
- *Output device.* An output device converts data to a human-readable form. In most cases, this will be a computer monitor that allows users to view output from the image analysis.

Examples of machine vision used in research

Machine vision systems for monitoring animals continue to develop. While there are commercially and freely available programs, many users end up dissatisfied with their performance under farm conditions. Research groups may choose instead to develop algorithms that are tailor-made for their specific test conditions, species and objectives, as it proves difficult to obtain high-performance accuracy, even in relatively controlled test environments using previously developed programs (Wurtz *et al.*, 2019b). Developing an algorithm, however, is not easy. Chen *et al.* (2020) developed a deep-learning method to detect aggressive episodes in group-housed pigs. The current iteration of this automated detection method is based on almost a decade of previous work. To apply this method to individual animals, or pigs of different breeds, ages or stocking densities, further training of the program would be necessary. Extensive validation would thus be required before this method could be used in other research settings or on commercial farms.

15.4 Pitfalls

15.4.1 Interdisciplinary communication

Development of precision livestock technology for measuring animal welfare requires communication among individuals with diverse backgrounds. Often, animal welfare scientists rely on engineers or computer scientists to develop sensors and the algorithms required for processing data collected by them. Computer scientists rely on animal welfare scientists to provide the 'gold standard' for references used in algorithm training and validation. The process of manually labelling key variables from collected data is a critical translation step between engineers and animal experts (Tullo *et al.*, 2013). This collaborative process of developing and validating algorithms can present challenges, as individuals with different educational backgrounds, farm experience and ultimate goals try to work together. Furthermore, labelling datasets is labour intensive and time-consuming and should be accounted for in project timelines to avoid unexpected delays in project advancement. Other frustrations may arise as engineers/computer scientists and animal welfare/behaviour scientists are unfamiliar with the terminology used by each group (see Chapter 2, this volume). In addition, groups' expectations may not align with what is practical or possible. On the one hand, animal welfare scientists often lack an understanding of the algorithms and hardware and their limitations. On the other, many engineers are unfamiliar with the constraints that come along with working with live animals in harsh environments – or at the scale of many modern commercial farms.

Finally, there are many stakeholder groups involved in the development of PLF technologies. Researchers, producers, breeders, veterinarians, welfare auditors, feed companies, retailers, consumers and other groups all have an interest in these technologies and varying priorities about what should be developed and implemented on farms. In addition, different funding agencies may have different priorities for how valuable resources should be spent. For example, some research groups are focused on developing cutting-edge technology and may struggle to win grants if their research is not innovative enough. This has led to many research groups starting PLF projects from scratch rather than building on previous work. It can also be difficult to obtain interest and funds to translate existing technology into practical solutions that might have a better chance of being implemented on farms and accepted by farmers.

15.4.2 Challenging environments

Farms have many elements that make integrating precision livestock tools difficult. Dirt, dust, faeces, urine, flies, mice, moisture, ammonia and sunlight can damage sensitive computer equipment, sensors and cameras. Weatherproof housing can provide some protection against the elements, but longevity of equipment may still be an issue. For sensors designed to collect physiological samples such as saliva, breath or urine, contamination by other body fluids or excreta from the sensor-wearing individual or group members interacting with the sensor (Misselbrook *et al.*, 2016) must be taken into account to ensure that only the samples of interest are collected.

The vast majority of sensing technology being brought on to farms was developed for other uses and was not intended for use other than in clean, climate-controlled environments. This requires a lot of creativity in terms of how to protect this valuable equipment and ensure its longevity. Furthermore, there is a demand for low-cost, low-tech tools to improve the usability of PLF systems on smaller farms (lacking economies of scale) or in parts of the world where robust, inexpensive, readily available solutions are needed.

Body-worn sensors can provide a wealth of individualized information relevant to animal welfare. However, it can be difficult to design sensors that will not be destroyed by the animals wearing them. Pigs, for example, are highly inquisitive animals and are equally powerful, and pigs kept in groups can quickly destroy any body-worn sensor placed on a pen mate that is accessible to their snouts.

Infrared technology allows remote temperature monitoring of groups of animals, but this technology struggles in farm environments due to factors such as heat reflected from the walls, floors or feeders. Additional extraneous factors such as faeces, urine and mice can also add to temperature variability (Cook *et al.*, 2018). Programs need to be adaptable, filtering out information that is not relevant from ever-changing and unpredictable conditions.

15.4.3 Limited adaptability

Algorithms that are tailor-made for a specific experimental context may not be adaptable to others. For example, algorithms developed in laboratory settings may not be able to be used in other environments. Many factors encountered on farms may not be anticipated during development. Algorithms also need to be validated on larger groups of animals. Currently, studies assessing the accuracy and precision of algorithms have had small sample sizes and have not demonstrated the ability to scale-up to the group sizes typically found on commercial farms. Animals present another unique challenge in that there can sometimes be wide variations in coat colour and patterns. This is the case with Holstein cattle, while some breeds or species are very homogenous in appearance, such as Black Angus cattle, making it difficult to tell these animals apart using biometric data. Fur appearance (texture and colour) can have large effects on location accuracy when using time-of-flight cameras, especially when animals are moving (Salau *et al.*, 2015). The fact that young animals grow rapidly as they age, changing in both body size and how they move, presents an additional challenge, as algorithms will have to be retrained periodically to remain accurate.

There is a trade-off that accompanies creating models for livestock prediction models. Models that can be adaptable to animals as they age or environments as they change over time are more computationally taxing and may have reduced accuracy. Simple model structures may be more accurate but are limited in their application (Wathes *et al.*, 2008).

Livestock systems are continuing to get larger, as are group sizes of animals. In most cases, it is not economical to equip every animal in the population with sensors. In some cases, PLF may be more expensive than an individual animal. In these cases, a subsample of the group may be selected to serve as sentinels for early detection of health problems or to detect issues within the barn requiring attention (e.g. a clogged feed line or faulty fan). More research is necessary to determine the appropriate subsample sizes, as well as how to select animals to appropriately represent the group. While the ideal implementation of PLF would gather individual information across the whole group, data from a subset of individuals would be an improvement over barn system-level monitoring and more reliable at detecting nuanced early-warning signs of a problem than animal caretakers.

15.4.4 The need to consider the effect of PLF on behaviour

When collecting behavioural information, it is critical that the collection method does not alter the animals' behaviour or contribute to poor welfare (McMahon *et al.*, 2011). This is especially important with body-worn sensors. Laying hens and broilers are much smaller than other livestock species, so the size of the sensor must be small enough not to impact their movement or comfort (Daigle *et al.*, 2012). Chickens are also prone to feather pecking, especially towards individuals within the flock that stand out. It needs to be ensured that sensors added to chickens do not inadvertently make them a victim of aggression or of increased interest from conspecifics, which could alter their behaviour, thus invalidating the data obtained. Chickens present another challenge due to their size relative to the weight and size of many sensors (and their batteries). If sensors are attached to animals using collars, leg bands or harnesses, there is a risk that these devices may cause the animals to become caught on features of the environment. Animals may require a period of acclimation (minutes to weeks) to adjust to wearing the sensor and return to normal behaviour (Daigle *et al.*, 2012; Buijs *et al.*, 2018; Stadig *et al.*, 2018).

Some PLF sensors can be rather invasive (e.g. sensors placed under the skin, or thermometers placed rectally or vaginally). Additionally, the fitting of animals with body-worn sensors often requires animals to be restrained, which could cause stress, such as pedometers placed on legs, in ears or around the necks of cattle. Sensors with short battery lives require recapturing of animals to replace the sensor or the battery (Halachmi *et al.*, 2019). It is critical to push for technologies that are minimally invasive and require minimal human contact, especially when being used to monitor animal welfare.

15.4.5 Interpretation of data

Some argue that the biggest challenge facing PLF is how to translate the massive amount of data generated into meaningful information (Vranken and Berckmans, 2017). Decisions need to be made regarding how the output values relate to welfare and what values fall within an acceptable range in order to trigger alerts or recommend actions. Data quickly accumulate and can be costly to store. In addition, processing of data can be computationally intensive. Programs that process information in real time might be preferred; however, there is often a trade-off in complexity of the algorithm to keep computing power needs manageable. Much more work is necessary to determine the most effective way to find meaningful patterns from the data and how to reveal information at the individual level. Another hurdle is determining what information should be retained. Storing the raw data in its entirety is probably not worthwhile but the question remains as to what information is worth saving and storing.

15.4.6 Ethical and social dilemmas

At its core, PLF is a conglomeration of data as well as real-time data processing and AI decision making, and therefore it is imperative that developers and users of PLF technology carefully consider ethical boundaries for how much power AI has to execute decisions (Russell *et al.*, 2015; Adams-Progar *et al.*, 2017). Many algorithms rely on machine learning to interpret data and lead decision making on farms. This differs from traditional models where users know which variables are at play. The underlying variables used in machine learning remain unknown, making verification of the model nearly impossible.

Issues surrounding ownership rights to the data also exist when producers form partnerships with technology companies. If data are transmitted electronically, there is concern over how to keep this information secure from cyber attacks. There is also the ethical dilemma of replacing the human workforce, which is often drawn from marginalized or economically vulnerable communities, with technology. While many working at the forefront of PLF development advocate using the technology to supplement or complement the efforts of stockpeople (Guarino *et al.*, 2017), there is no guarantee that PLF will always be adopted with this goal in mind. Furthermore, PLF may become better than human stockpeople as AI improves its ability to make predictions and make real-time adjustments more rapidly and to a finer degree (Werkheiser, 2020).

Consumers and other stakeholders may be hesitant to embrace PLF, as there is poor public perception of active management strategies of livestock (i.e. instrumental use of animals). Farmers may be hesitant to invest in PLF technologies. For many technologies, the time period to obtain a return on investment is largely uncertain. On top of this, incomplete development of technology leading to poor equipment reliability has led to rejection by early adopters in some cases (Wathes *et al.*, 2008). Finally, PLF will also change the nature of the interactions between humans and animals, which could have implications for both animal and farmer welfare (Wathes *et al.*, 2008; Werkheiser, 2018).

15.5 Conclusions and Tips

Rapid technological advances offer increasing promise for automated, non-invasive monitoring, measurement, and management of animal behaviour and welfare using PLF technologies. An increasing array of sensors are available that could provide more nuanced insight into animal experiences. Coupled with advances in battery life, decreasing sensor size and cost, and increasing robustness, sensors have the potential to expand how much data we can collect, for how long and from how many animals. However, while some sensors are commercially available specifically for use with livestock and poultry in research or commercial precision livestock applications and can be used off the shelf, many were originally intended for other purposes. Thus, most require effort from animal science researchers to adapt them for use on farm in animal welfare studies. This requires validation for use with a particular species and in a particular system, interpretation of sensor output that has previously been manipulated by algorithms or novel coding to process raw data to transform it into meaningful information. Sensors and identification tags attached to animals can provide animal-level information ranging from location to resource use to details about specific activities, vocalizations or chemical secretions. Body-worn sensors must not interfere with the animal, must be robust enough to survive investigation by other animals, must have sufficient battery life to collect data for the period of interest and should be easily recovered at the end of the animal's life. Remote sensing is typically used to monitor the distribution and movement of entire groups of animals or to detect the presence of specific types of behaviour in a group. Although advances are being made in recognition of individual unmarked animals, it is more difficult to link remote sensing observations back to specific individuals. Additionally, in the case of cameras, limited fields of view or occlusion may make it difficult to monitor the entire animal environment. Beyond the technological considerations, establishing a multidisciplinary team or partnership with colleagues in computer science and electrical engineering is advisable in many cases. It is important to ensure that you have good communication and that colleagues from other disciplines have a clear understanding of the species and the environment in which they live. Visits to the farm by the entire team are recommended to establish a common understanding of the constraints.

In conclusion, while PLF has powerful potential, we are not yet at the stage where such technology can always be used without expertise in or partnership with computer science and engineering. And, in some cases, the technology or the ability to process data does not yet exist that can answer all of our questions.

Further Reading

Suggestions for further reading are given below, as well as those shown in bold in the reference list. These include pieces that provide a general overview of PLF, as well as comprehensive reviews that focus on PLF used with particular species (i.e. swine, poultry and dairy cattle) or approaches (i.e. wearable sensors and machine vision).

Benjamin, M. and Yik, S. (2019) Precision livestock farming in swine welfare: a review for swine practitioners. *Animals* 9, 133.

Li, N., Ren, Z., Li, D. and Zeng, L. (2020) Automated techniques for monitoring the behaviour and welfare of broilers and laying hens: towards the goal of precision livestock farming. *Animal* 14, 617–625.

Maharjan, P. and Liang, Y. (2020) Precision livestock farming: the opportunities in poultry sector. *Journal of Agricultural Science and Technology A* 10, 45–53.

Mottram, T. (2016) Animal board invited review: precision livestock farming for dairy cows with a focus on oestrus detection. *Animal* 10, 1575–1584.

References

Adams-Progar, A., Fink, G., Walker, E. and Llewellyn, D. (2017) Security and privacy issues in the Internet of cows. In: Song, H., Fink, G.A. and Jeschke, S. (eds) *Security and Privacy in Cyber-Physical Systems: Foundations, Principles, and Applications.* Wiley, Chichester, UK, pp. 375–398.

Adrion, F., Kapun, A., Holland, E.-M., Staiger, M., Löb, P. *et al.* (2017) Novel approach to determine the influence of pig and cattle ears on the performance of passive UHF-RFID ear tags. *Computers and Electronics in Agriculture* 140, 168–179.

Andriamasinoro, A., Bindelle, J., Mercatoris, B. and Lebeau, F. (2016) A review on the use of sensors to monitor cattle jaw movements and behavior when grazing. *Biotechnologie, Agronomie, Société et Environnement* 20, 273–286.

Bailey, D.W., Trotter, M.G., Knight, C.W. and Thomas, M.G. (2018) Use of GPS tracking collars and accelerometers for rangeland livestock production research. *Translational Animal Science* 2(1), 81–88. DOI: 10.1093/tas/txx006.

Banerjee, D., Biswas, S., Daigle, C. and Siegford, J.M. (2012) Remote activity classification of hens using wireless body mounted sensors. *Ninth International Conference on Wearable and Implantable Body Sensor Networks*, IEEE Computer Society, Washington, DC, pp. 107–112.

Banerjee, D., Daigle, C.L., Dong, B., Wurtz, K., Newberry, R.C. *et al.* (2014) Detection of jumping and landing force in laying hens using wireless wearable sensors. *Poultry Science* 93(11), 2724–2733. DOI: 10.3382/ps.2014-04006.

Berckmans, D. (2017) General introduction to precision livestock farming. *Animal Frontiers* 7, 6–11.

Boileau, A., Farish, M., Turner, S.P. and Camerlink, I. (2019) Infrared thermography of agonistic behaviour in pigs. *Physiology & Behavior* 210, 112637. DOI: 10.1016/j.physbeh.2019.112637.

Broom, D.M. (2008) Welfare assessment and relevant ethical decisions: key concepts. *Annual Review of Biomedical Sciences* 10, 79–90.

Buijs, S., Booth, F., Richards, G., McGaughey, L., Nicol, C.J. *et al.* (2018) Behavioural and physiological responses of laying hens to automated monitoring equipment. *Applied Animal Behaviour Science* 199, 17–23. DOI: 10.1016/j.applanim.2017.10.017.

Chen, C., Zhu, W., Steibel, J., Siegford, J., Wurtz, K. *et al.* (2020) Recognition of aggressive episodes of pigs based on convolutional neural network and long short-term memory. *Computers and Electronics in Agriculture* 169, 105166.

Chien, Y.-R. and Chen, Y.-X. (2018) An RFID-based smart nest box: an experimental study of laying performance and behavior of individual hens. *Sensors* 18(3), 859. DOI: 10.3390/s18030859.

Christensen, L.S., Brehm, K.E., Skov, J., Harlow, K.W., Christensen, J. *et al.* (2011) Detection of foot-and-mouth disease virus in the breath of infected cattle using a hand-held device to collect aerosols. *Journal of Virological Methods* 177(1), 44–48. DOI: 10.1016/j.jviromet.2011.06.011.

Cook, N.J., Bench, C.J., Liu, T., Chabot, B. and Schaefer, A.L. (2018) The automated analysis of clustering behaviour of piglets from thermal images in response to immune challenge by vaccination. *Animal* 12(1), 122–133. DOI: 10.1017/S1751731117001239.

Daigle, C.L. and Siegford, J.M. (2014) When continuous observations just won't do: developing accurate and efficient sampling strategies for the laying hen. *Behavioural Processes* 103, 58–66. DOI: 10.1016/j.beproc.2013.11.002.

Daigle, C.L., Banerjee, D., Biswas, S. and Siegford, J.M. (2012) Noncaged laying hens remain unflappable while wearing body-mounted sensors: levels of agonistic behaviors remain unchanged and resource use is not reduced after habituation. *Poultry Science* 91(10), 2415–2423. DOI: 10.3382/ps.2012-02300.

Ellen, E.D., van der Sluis, M., Siegford, J., Guzhva, O., Toscano, M.J. *et al.* (2019) Review of sensor technologies in animal breeding: phenotyping behaviors of laying hens to select against feather pecking. *Animals* 9(3), 108. DOI: 10.3390/ani9030108.

Guarino, M., Norton, T., Berckmans, D., Vranken, E. and Berckmans, D. (2017) A blueprint for developing and applying precision livestock farming tools: a key output of the EU-PLF project. *Animal Frontiers* 7, 12–17.

Halachmi, I., Guarino, M., Bewley, J. and Pastell, M. (2019) Smart animal agriculture: application of real-time sensors to improve animal well-being and production. *Annual Review of Animal Biosciences* 7, 403–425. DOI: 10.1146/annurev-animal-020518-114851.

Hampton, J.O., MacKenzie, D.I. and Forsyth, D.M. (2019) How many to sample? Statistical guidelines for monitoring animal welfare outcomes. *PLoS One* 14(1), e0211417. DOI: 10.1371/journal.pone.0211417.

Hansen, M.F., Smith, M.L., Smith, L.N., Salter, M.G., Baxter, E.M. *et al.* (2018) Towards on-farm pig face recognition using convolutional neural networks. *Computers in Industry* 98, 145–152.

Liu, L.S., Ni, J.-Q., Zhao, R.-Q., Shen, M.-X., He, C.-L. *et al.* (2018) Design and test of a low-power acceleration sensor with Bluetooth Low Energy on ear tags for sow behaviour monitoring. *Biosystems Engineering* 176, 162–171.

Lokhorst, C. (1996) Automatic weighing of individual laying hens in aviary housing systems. *British Poultry Science* 37(3), 485–499. DOI: 10.1080/00071669608417880.

Marsot, M., Mei, J., Shan, X., Ye, L. and Feng, P. (2020) An adaptive pig face recognition approach using Convolutional Neural Networks. *Computers and Electronics in Agriculture* 173, 105386.

McMahon, C.R., Collier, N., Northfield, J.K. and Glen, F. (2011) Taking the time to assess the effects of remote sensing and tracking devices on animals. *Animal Welfare* 20, 515–521.

Misselbrook, T., Fleming, H., Camp, V., Umstatter, C., Duthie, C.-A. *et al.* (2016) Automated monitoring of urination events from grazing cattle. *Agriculture, Ecosystems and Environment* 230, 191–198.

Nääs, I.A., Garcia, R.G. and Caldara, F.R. (2014) Infrared thermal image for assessing animal health and welfare. *Journal of Animal Behaviour and Biometeorology* 2, 66–72.

Neethirajan, S., Tuteja, S.K., Huang, S.-T. and Kelton, D. (2017) Recent advances in wearable sensors for animal health management. *Sensing and Bio-Sensing Research* 12, 15–29.

Norton, T. and Berckmans, D. (2018) Precision livestock farming: the future of livestock welfare monitoring and management? In: Butterworth, A. (ed.) *Animal Welfare in a Changing World.* CAB International, Wallingford, UK, pp. 130–144.

Nunes, L., Ampatzidis, Y., Costa, L. and Wallau, M. (2020) Horse foraging behavior detection using recurrent neural networks. *International Symposium on Artificial Intelligence and Mathematics,* Fort Lauderdale, Florida, 6–8 January 2020. Association for the Advancement of Artificial Intelligence, Menlo Park, California. Available at: http://isaim2020.cs.ou.edu/papers/ISAIM2020_Agriculture_Nunes_etal.pdf (accessed 17 June 2021).

Quwaider, M., Daigle, C.L., Biswas, S.K., Siegford, J.M. and Swanson, J.C. (2010) Development of a wireless body-mounted sensor to monitor location and activity of laying hens in a non-cage housing system. *Transactions of the ASABE* 53, 1705–1713.

Richter, S.H. and Hintze, S. (2019) From the individual to the population – and back again? Emphasising the role of the individual in animal welfare science. *Applied Animal Behaviour Science* 212, 1–8.

Rifkin, R.E., Grzeskowiak, R.M., Mulon, P.-Y., Adair, H.S., Biris, A.S. *et al.* (2019) Use of a pressure-sensing walkway system for biometric assessment of gait characteristics in goats. *PLoS One* 14(10), e0223771. DOI: 10.1371/journal.pone.0223771.

Ruiz-Garcia, L. and Lunadei, L. (2011) The role of RFID in agriculture: applications, limitations and challenges. *Computers and Electronics in Agriculture* 79, 42–50.

Russell, S., Dewey, D. and Tegmark, M. (2015) Research priorities for robust and beneficial artificial intelligence. *AI Magazine* 36, 105–114.

Salau, J., Bauer, U., Haas, J.H., Thaller, G., Harms, J. *et al.* (2015) Quantification of the effects of fur, fur color, and velocity on time-of-flight technology in dairy production. *SpringerPlus* 4, 144. DOI: 10.1186/s40064-015-0903-0.

Siegford, J.M., Berezowski, J., Biswas, S.K., Daigle, C.L., Gebhardt-Henrich, S.G. *et al.* (2016) Assessing activity and location of individual laying hens in large groups using modern technology. *Animals* 6(2), 10. DOI: 10.3390/ani6020010.

Song, Y., Min, J., Yu, Y., Wang, H., Yang, Y. *et al.* (2020) Wireless battery-free wearable sweat sensor powered by human motion. *Science Advances* 6(40), eaay9842. DOI: 10.1126/sciadv.aay9842.

Stadig, L.M., Rodenburg, T.B., Ampe, B., Reubens, B. and Tuyttens, F.A.M. (2018) An automated positioning system for monitoring chickens' location: effects of wearing a backpack on behaviour, leg health and production. *Applied Animal Behaviour Science* 198, 83–88.

Terra-Braga, M., Ferraz-de-Oliveira, M.I., Lopes-de-Castro, J.A. and Sales-Baptista, E. (2018) Sheep nocturnal grazing behavior using infrared point-of-view camera – a preliminary study. In: Grant, R.A., Allen, T., Spink, A. and Sullivan, M. (eds) *Measuring Behavior. 11th International Conference on Methods and Techniques in Behavioral Research, Manchester Metropolitan University, 5–8 June 2018*, pp. 338–341. Available at: https://www.measuringbehavior.org/files/2018/MB2018%20Proceedings.pdf (accessed 17 June 2021).

Tullo, E., Fontana, I. and Guarino, M. (2013) Precision livestock farming : an overview of image and sound labelling. *European Conference on Precision Livestock Farming*, KU Leuven, Leuven, Belgium, pp. 30–38.

Turner, M.J.B., Gurney, P. and Belyavin, C.G. (1983) Automatic weighing of layer-replacement pullets housed on litter or in cages. *British Poultry Science* 24, 33–45.

Ungar, E.D. and Rutter, S.M. (2006) Classifying cattle jaw movements: comparing IGER behaviour recorder and acoustic techniques. *Applied Animal Behaviour Science* 98, 11–27.

Vranken, E. and Berckmans, D. (2017) Precision livestock farming for pigs. *Animal Frontiers* 7, 32–37.

Wathes, C.M., Kristensen, H.H., Aerts, J.M. and Berckmans, D. (2008) Is precision livestock farming an engineer's daydream or nightmare, an animal's friend or foe, and a farmer's panacea or pitfall? *Computers and Electronics in Agriculture* 64, 2–10.

Werkheiser, I. (2018) Precision livestock farming and farmers' duties to livestock. *Journal of Agricultural and Environmental Ethics* 31, 181–195.

Werkheiser, I. (2020) Technology and responsibility: a discussion of underexamined risks and concerns in precision livestock farming. *Animal Frontiers* 10(1), 51–57. DOI: 10.1093/af/vfz056.

Wurtz, K., Camerlink, I., D'Eath, R.B., Fernández, A.P., Norton, T. *et al.* (2019a) Recording behaviour of indoor-housed farm animals automatically using machine vision technology: a systematic review. *PLoS One* 14(12), e0226669. DOI: 10.1371/journal.pone.0226669.

Wurtz, K., Norton, T., Siegford, J. and Steibel, J. (2019b) Assessment of open-source programs for automated tracking of individual pigs within a group. *Proceedings of the 53rd Congress of the International Society for Applied Ethology*, Wageningen Academic Publishers, Wageningen, The Netherlands, p. 289.

Xin, H. and Liu, K. (2017) Precision livestock farming in egg production. *Animal Frontiers* 7, 24–31.

Index

Note: The page numbers in bold and italics represents tables and figures respectively.